Sustained-Release Injectable Products

Edited by

Judy Senior

and

Michael Radomsky

CRC Press
Taylor & Francis Group
Boca Raton London New York

CRC Press is an imprint of the
Taylor & Francis Group, an **informa** business

CRC Press
Taylor & Francis Group
6000 Broken Sound Parkway NW, Suite 300
Boca Raton, FL 33487-2742

First issued in paperback 2019

© 2009 by Taylor & Francis Group, LLC
CRC Press is an imprint of Taylor & Francis Group, an Informa business

No claim to original U.S. Government works

ISBN-13: 978-1-57491-101-5 (hbk)
ISBN-13: 978-0-367-39877-4 (pbk)

This book contains information obtained from authentic and highly regarded sources. While all reasonable efforts have been made to publish reliable data and information, neither the author[s] nor the publisher can accept any legal responsibility or liability for any errors or omissions that may be made. The publishers wish to make clear that any views or opinions expressed in this book by individual editors, authors or contributors are personal to them and do not necessarily reflect the views/opinions of the publishers. The information or guidance contained in this book is intended for use by medical, scientific or health-care professionals and is provided strictly as a supplement to the medical or other professional's own judgement, their knowledge of the patient's medical history, relevant manufacturer's instructions and the appropriate best practice guidelines. Because of the rapid advances in medical science, any information or advice on dosages, procedures or diagnoses should be independently verified. The reader is strongly urged to consult the relevant national drug formulary and the drug companies' and device or material manufacturers' printed instructions, and their websites, before administering or utilizing any of the drugs, devices or materials mentioned in this book. This book does not indicate whether a particular treatment is appropriate or suitable for a particular individual. Ultimately it is the sole responsibility of the medical professional to make his or her own professional judgements, so as to advise and treat patients appropriately. The authors and publishers have also attempted to trace the copyright holders of all material reproduced in this publication and apologize to copyright holders if permission to publish in this form has not been obtained. If any copyright material has not been acknowledged please write and let us know so we may rectify in any future reprint.

Library of Congress Cataloging-in-Publication Data

Sustained-release injectable products / edited by Judy H. Senior and Michael Radomsky
      p. ; cm.
   Includes bibliographical references and index.
   ISBN-13: 978-1-5749-1101-5 (alk. paper)
   ISBN-10: 1-5749-1101-5 (alk. paper)
   1. Drugs--Controlled-release. 2. Injections, Intravenous. I. Senior, Judith. II. Radomsky, Michael. [DNLM: 1. Delayed-Action Preparations. 2. Drug Carriers. 3. Drug Delivery Systems. 4. Infusions, Parenteral. QV 785 S9643 2000]

RS201.C64 S88 2000
615'.7--dc21   99-056702

Visit the Taylor & Francis Web site at
http://www.taylorandfrancis.com

and the CRC Press Web site at
http://www.crcpress.com

Contents

Preface

Sustained-release versions of drugs are an essential part of the formulations repertoire of the Pharmaceutical Scientist in the new millennium. Many important products in this field originated as pioneering research in academic institutions. A critical component of their becoming successful products is the science of pharmaceutical formulation development and the subsequent drug development process, fuelled by the efforts of the industry to satisfy clinical needs.

The scientific work of drug development conducted by companies is largely unpublished partly because publishing is not a priority, especially when regulatory submissions are underway. We are therefore especially privileged to collect together in this volume contributions by excellent scientists who day-to-day nurture drugs through the delicate process of drug development as it applies to sustained-release drug delivery.

Why should the development process for a sustained-release version of a drug be different from its unmodified counterpart? While some aspects of the process may appear similar, there are important differences. This volume illustrates answers by using examples of current types of sustained-release systems for local injectable applications. Taking it one step further, the second half of the book brings together common threads of scientific aspects that apply to any sustained-release formulation of a pharmaceutical product, such as the scale-up, safety, biocompatibility, analytical challenges, quality assurance, and specific regulatory factors.

ACKNOWLEDGMENTS

This volume would not have been possible without the generous contributions of the valiant authors, who energetically, patiently, and conscientiously stuck with the process of writing, editing, proofing, and indexing, amid many other pressing priorities.

Judy Senior would especially like to thank her parents, David and Elisabeth Whitehead. They lovingly gave me the tools of discipline, patience, and determination, encouraging me to take my passion for life and turn my dreams into reality. Also my friends and advisors, especially Pramod Gupta, Robert I. Pinsker, Francis Cannon, Paige Lucetti, Ann Thevenin, Pat Drago, András Gruber, Hal Handley, Gregory Gregoriadis, Deborah A. Eppstein, and my colleagues at Syntex, SkyePharma (formerly DepoTech), and Maxim Pharmaceuticals.

Michael Radomsky would like to thank Greg Allen, Andrea Thompson, Bob Spiro, Bach Phan, Junghae Scott, Lynda Sanders, Larry Zeitlin, and Kevin Whaley for their manuscript reviews, technical expertise, and stimulating and encouraging discussions. I am very grateful that I have worked with a number of bright and energetic scientists at Syntex, Orquest, and now Epicyte; these professional and sometimes personal relationships have helped me grow as both a scientist and a human being. Finally, I would also like to thank my parents for their lifelong support and encouragement during both the productive and difficult chapters in my life.

We both would like to thank Dr. Gupta, whose initial sparks ignited this volume. Dr. Radomsky is also grateful that Dr. Senior took the initiative to get him involved in this project. We are also indebted to many small companies that inspire the industry with their innovation, speed to market, and nurturing of employees.

They are where much of the work described in this volume was promulgated.

Finally, we would both like to thank Amy Davis for taking on the initial book idea and Jane Steinmann and Liz Prigge for their expert guidance during the process of realizing the book and bringing it to the attention of readers around the world. We look forward to offering the knowledge and experience collected in this volume for the benefit of a worldwide audience to advance the goal of having effective therapies available sooner through the timely development of effective drugs that reduce suffering in the twenty-first century and beyond.

Judy Senior
Michael Radomsky
January 2000

Contributors

Judy H. Senior, Editor

Dr. Judy Senior is a specialist in drug delivery and drug development cycle management. Over the past 16 years, she has published widely in the fields of liposomes and lipid-based drug delivery systems. Dr. Senior received her BSc in biochemistry from the University of East Anglia and her PhD in drug delivery from the University of London. In 1989, she began to develop liposome-based drug formulations at Syntex Corporation (now Roche Bioscience), where she subsequently gained broad experience in the drug development process with a wide variety of drugs, dosage forms, and delivery systems. In 1993, Dr. Senior joined DepoTech Corporation (now SkyePharma plc) as project leader and later project manager for the company's development programs. Currently at Maxim Pharmaceuticals, Dr. Senior is developing delivery systems for the Maxamine™ and MaxDerm™ family of products. Dr. Senior is active in the AAPS Western Regional planning committee and in a local San Diego pharmaceutical discussion group. She is also a regular scientific contributor in the field of controlled-release and sustained-release technologies.

Michael Radomsky, Editor

Dr. Michael Radomsky is a project leader at Epicyte Pharmaceuticals, Inc. He received his BS in chemical engineering from the Rose Hulman Institute of Technology in 1987 and a PhD in 1991 from The Johns Hopkins University. From 1991 to 1995, he was a staff researcher at Syntex Inc. (now Roche Bioscience) in the Drug Delivery Research and Formulation Development departments. As senior scientist at Orquest from 1995 to 1999, he developed products for the local delivery of proteins in orthopaedic applications. At Epicyte, he is working with transgenic plants to produce antibodies for disease treatment and prevention. His multidisciplinary research interests have included sustained- and controlled-release pharmaceuticals, polymers for controlled drug delivery, mathematical modeling of drug transport and drug delivery systems, design of delivery systems for peptide and protein drugs, local drug delivery, tissue engineering, and antibody technology.

Paul Burke

Dr. Paul Burke received his BS in chemistry from Harvey Mudd College in 1986 and his PhD from MIT in 1992. He is presently laboratory head of sustained-release pharmaceutics at Amgen. From 1993 to 1997, Dr. Burke designed microsphere formulations of several protein therapeutics and antisense oligonucleotides for Alkermes. He coinvented a multiweek formulation of erythropoietin (currently undergoing clinical testing) and developed a patented cryogenic encapsulation process. In addition to protein microencapsulation, his research interests include the characterization of controlled-release systems using magnetic resonance spectroscopy.

Geneva Chen

Dr. Geneva Chen received her BS in chemistry in 1988 from Beijing United University and her PhD in pharmaceutics from the University of Houston in 1995. Dr. Chen was formerly employed by Aronex Pharmaceuticals, Inc., as a research scientist and served as a postdoctoral research associate at the University of Houston. She has also served as pharmacokinetics lecturer at the University of Houston. Dr. Chen is author and coauthor of many technical articles.

Daan J. A. Crommelin

Professor Daan Crommelin received a degree in pharmacy from the University of Groningen in 1975 and a PhD from the University of Leiden in 1979. After serving as a postdoctoral fellow with Prof. W. I. Higuchi at the University of Michigan, he became full professor and head of the Department of Pharmaceutics in 1984. In 1992, he became managing scientific director of the Utrecht Institute for Pharmaceutical Sciences (UIPS) and in 1993 deputy managing director of the Groningen Utrecht Institute for Drug Exploration (GUIDE). Also in 1993, he was appointed adjunct professor in the Department of Pharmaceutics and Pharmaceutical Chemistry at the University of Utah. In 1995, he became scientific director of OctoPlus, BV, a company focused on providing pharmaceutical formulation know-how. He has published over 200 original articles, reviews, and book chapters. He has served as either an editor or on the editorial advisory board of numerous pharmaceutical journals. He is a fellow of the American Association of Pharmaceutical Sciences (AAPS). He was winner of the Maurice-Marie Janot award in 1995 and received the International Award of the Belgian Society of Pharmaceutical Sciences in 1997. He is involved in organizing meetings in the fields of biopharmaceutics, drug targeting, and the development of biotechnological products all over the world.

Richard L. Dunn

Dr. Richard Dunn received his BS in chemistry from the University of North Carolina and his PhD in organic/polymer chemistry from the University of Florida. He then joined the Beaunit Corporation, synthesizing and characterizing nylon, polyester, polypropylene, and viscose polymers into high-performance and specialty fibers. In 1979, Dr. Dunn joined the Applied Sciences

Department of the Southern Research Institute (Birmingham, Ala.), directing multidisciplinary programs in polymer chemistry; polymer engineering; biomaterials; biomedical engineering; membrane development; and controlled release of biologically active agents from fibers, film, and devices, especially those fabricated from biodegradable polymers.

In 1987, Dr. Dunn joined Vipont Research Laboratories, Inc. (now Atrix Laboratories), and directed the development of a local antibiotic delivery product for the treatment of periodontal disease and a biodegradable barrier membrane for periodontal tissue regeneration. Dr. Dunn invented the in situ polymer system that is used in these projects. Currently, as senior vice president, he is responsible for managing projects involving a proprietary biodegradable polymer system (Atrigel®) for use as a medical device or to deliver biologically active agents. Current applications include the prevention of surgical adhesions and the delivery of local anesthetics to relieve postoperative pain, antineoplastic agents for treating solid tumor cancers, LHRH peptides for prostate cancer, and growth factors for tissue regeneration.

Alexander Florence

Dr. Alexander (Sandy) Florence holds a BSc in pharmacy and a PhD from the University of Glasgow and a DSc from Strathclyde. Dr. Florence is currently dean and professor of pharmacy at the School of Pharmacy, University of London. Prior to this appointment in 1989, he was professor of pharmaceutics at the University of Strathclyde. He has published several textbooks as well as authoring over 230 papers, chapters, and reviews in the area of drug delivery and surfactant and polymer systems.

Maninder Hora

Dr. Maninder Hora received his PhD in 1980 in bioengineering (materials) and was a Fulbright Scholar from 1979 to 1981. Dr. Hora's industrial research experience includes positions at Ayerst-Wyeth Laboratories and SmithKline Beecham Laboratories. He joined the Chiron Corporation in 1986 and became senior director of Pharmaceutical R&D in 1997. His research interests and experiences encompass the fields of pharmaceutical drug delivery and formulation development. He has approximately 30 publications, 15 patents, and numerous presentations at national and international meetings to his credit. He is a member of the Controlled Release Society (CRS), the American Association of Pharmaceutical Scientists (AAPS), and the American Chemical Society (ACS).

Kunio Kawamura

Dr. Kunio Kawamura received his PhD in pharmaceuticals from the University of Tokyo in 1959. His career at Takeda Chemical Industries includes managing the Quality Control Department in Osaka, directing corporate quality assurance, and managing (as deputy general manager) the Production Department. As project leader for Lupron Depot®, a sustained-release injectable dosage form, he was responsible for the development and construction of the production facility.

As an advisor to the World Health Organization (WHO), Dr. Kawamura helped draft the WHO GMP. He also served on the board of directors of the Parenteral Drug Association (PDA) from 1994 to 1996. Since 1995, as a special advisor for Otsuka Pharmaceutical Co., Ltd., he has been responsible for worldwide GMP–associated technical and regulatory affairs. Dr. Kawamura is also an accreditation auditor for the Japan Accreditation Board for Conformity Assessment.

Joseph Kost

Dr. Joseph Kost completed all of his educational training at the Israel Institute of Technology, receiving a BS in 1973, an MSc in chemical engineering in 1975, and a DSc in biomedical engineering in 1981. Dr. Kost was the head of the Center for Biomedical Engineering at Ben Gurion University from 1988 to 1993 and head of the Program for Biotechnology from 1993 to 1995. He is currently professor of Chemical Engineering at Ben Gurion University and Chief Scientific Officer of Sontra Medical USA. He serves on the editorial boards of several journals and has published 3 books, 23 book chapters, 57 papers, 120 abstracts, and 25 patents, principally in the area of biomaterials. In

1996, Dr. Kost was awarded the Juludan Prize by the Israel Institute of Technology for outstanding scientific research achievements. Dr. Kost was also awarded the Clemson Award by the Society for Biomaterials in recognition of his outstanding contributions to applied biomaterials research.

Johanna K. Lang

Dr. Johanna Lang received her PhD degree in chemistry from the University of Graz, Austria, in 1982. After completing postdoctoral research in biochemistry at the University of Graz and the University of California, Berkeley, she pursued an industrial career in start-up companies in the United States and Germany. In 1987, she joined Liposome Technology, Inc. As head of the assay development group, she designed and implemented release, stability, and bioanalytical assays for the company's key products, Doxil® (liposomal doxorubicin) and Amphocil® (liposomal amphotericin B), now marketed by Abbott Laboratories. From 1993 to 1996, at Pharmacylics, Inc., Dr. Lang directed formulation development, assay development, and GMP manufacturing for a line of synthetic porphyrin-type lanthanide complexes, now in advanced clinical testing for photodymanic therapy of cancer and heart disease. Subsequently, she was Director of Product Development for IDEA GmbH in Munich, Germany. In 1999, Dr. Lang became an independent consultant based in Fremont, California. She is a member of the San Francisco Bay Area Biomedical Consultant's Network and the American Association of Pharmaceutical Scientists (AAPS). She served on organizing committees for AAPS meetings and chaired the Northern California Pharmaceutical Discussion Group (NCPDG).

Jung-Chung Lee

Dr. Jung-Chung Lee received his MS in 1980 and his PhD in 1983 in pharmaceutical chemistry from the University of Kansas. Starting in 1984, Dr. Lee worked in formulation development at Syntex, Roche, and Oread Labs and became director of Pharmaceutical Development at Cytel in 1997. Dr. Lee has been involved in conventional and controlled-release dosage form design and development in both human and veterinary pharmaceuticals for over 15 years and is currently director of Pharmaceutical Development at Cellegy Pharmaceuticals, Inc. Dr. Lee is a member of the American Association of Pharmaceutical Scientists (AAPS) and the Controlled Release Society (CRS).

LinShu Liu

Dr. LinShu Liu received his BS in 1976 from South China Normal University (Guang Zhou, China), his MSc in polymer chemistry in 1982 from South China Normal University, and his PhD in polymer chemistry in 1990 from Kyoto University (Japan). From 1990 to 1992, Dr. Liu was as a postdoctoral associate with Professor R. Langer at MIT (Cambridge, Mass.). Dr. Liu is a coeditor of 1 book and author or coauthor of 2 book chapters, 21 papers, 45 abstracts, and 6 patents; most of these publications are in the area of biomaterials and the controlled release of drugs.

Natalie McClure

Dr. Natalie McClure received her PhD in organic chemistry from Stanford University and is presently vice president of Regulatory Affairs and Quality Assurance at Matrix Pharmaceutical, Inc. With over 12 years of experience in regulatory affairs, she has experience in all stages of the regulatory process, from pre-IND to postmarketing support. Before branching out into regulatory affairs, Dr. McClure was a process development chemist at Syntex Research.

Sudaxshina Murdan

Dr. Sudaxshina Murdan received a BPharm from the University of Nottingham and was appointed lecturer in pharmaceutics at the School of Pharmacy, University of London, in 1998, following a brief postdoctoral fellowship with Professor A. T. Florence after earning a PhD from the same school. Her current research interests include organogels, amphiphilogels, inverse vesicles and responsive systems for drug delivery, and transdermal immunization. Dr. Murdan is a member of the Royal Pharmaceutical Society of Great Britain, the American Association of

Pharmaceutical Scientists (AAPS), the European Federation of Pharmaceutical Scientists, and the Controlled Release Society (CRS).

Christien Oussoren

Dr. Oussoren studied pharmacy in Utrecht and graduated in 1990. After a short stint as a pharmacist, she began research and obtained her PhD from the University of Utrecht in 1996. In January 1997, she was appointed a faculty member at the University of Utrecht. Her main research interests are in the field of biodistribution and in vivo application of colloidal drug carrier systems.

Leo Pavliv

Mr. Leo Pavliv is a registered pharmacist and received an MBA from Rutgers University. He has 15 years of experience in developing pharmaceutical and biological products, including small molecules, proteins, and peptides in a variety of dosage forms. He has managed the development process of products from preformulation, formulation, and scale-up through manufacturing. He established a clinical supplies group, which was responsible for the manufacturing, packaging, labeling, and distribution of clinical supplies. He has also designed, equipped, and managed a GMP–compliant clinical manufacturing facility.

Michael L. Putnam

Dr. Michael Putnam received his RPh BPhS from the University of Iowa in 1982 and a PhD in pharmaceutics from the same school. For 13 years, he has been involved in veterinary product research and development. Dr. Putnam is currently director of Pharmaceutical Development, Analytical Services, and Technical Services at Fort Dodge Animal Health.

Scott Putney

Dr. Scott Putney received his PhD and postdoctoral training from MIT. He is presently group leader in protein engineering at Lilly Research Laboratories. From 1991 to 1998, he was vice president of protein and molecular biology at Alkermes, Inc., where he led the effort to develop a sustained-release formulation of human growth hormone now awaiting marketing approval from the FDA. From 1983 to 1991, he was vice president of molecular biology at RepliGen Corporation, where he led an effort to develop a vaccine for HIV-1.

Ramachandran Radhakrishnan

Dr. Ramachandran Radhakrishnan received his PhD from Wayne State University in Detroit, Michigan, and was a postdoctoral fellow for 8 years in the laboratory of Dr. H. G. Khorana at MIT, studying lipid-lipid and lipid-protein interactions in biomembranes. As senior scientist at Liposome Technology, Inc., he was responsible for developing liposomal formulations for ocular, inhalational, and parenteral applications. He has been with Chiron Corporation for the past 8 years, dealing with the formulation and delivery of recombinant proteins, vaccines, and genes. He is affiliated with the American Society for Gene Therapy and the American Association of Pharmaceutical Scientists (AAPS). He has authored or coauthored over 40 publications and 12 patents.

Gary Riley

Dr. Gary Riley received his BVS from Sydney University in 1965, an MVS from the University of Melbourne in 1970, and a PhD in comparative pathology from the University of Missouri in 1972. He is a diplomate of the American College of Veterinary Pathologists and the American Board of Toxicology. From 1975 to 1981, Dr. Riley was associate professor of veterinary pathology at Iowa State University. He worked as an experimental pathologist from 1982 to 1993, first in contract research and later as director of Pathobiology at the Medical Research Division of

Lederle Laboratories. Since 1993, Dr. Riley has been at Alkermes Inc., where he has held positions in regulatory toxicology, pharmacokinetics, and the biology of sustained-release systems. He currently is vice-president of toxicology and applied biology at Alkermes.

Mantripragada B. Sankaram

Dr. Mantripragada Sankaram has a BS and an MS in chemistry and a PhD in molecular biophysics from the Indian Institute of Science. From 1985 to 1989, he was an Alexander von Humboldt and Max Planck fellow at the Max Planck Institute for Biophysical Chemistry in Goettingen, Germany, working on membrane structure and lipid-protein interactions. From 1989 to 1994, he was an assistant professor in biochemistry at the University of Virginia School of Medicine, where he further expanded his research to domains in biological membranes and the interaction of peptides, proteins, and cholesterol with membranes. He held several scientific and management positions at DepoTech Corporation from 1992 to 1999 and was involved in the development of sustained-release injectable products. Currently he directs the Research and Analytical Development departments at SkyePharma plc and is engaged in the research and development of novel drug delivery systems for sustained release. He is a member of the American Association for the Advancement of Science (AAAS), the American Association of Pharmaceutical Scientists (AAPS), the American Chemical Society (ACS), the Biophysical Society, and the Controlled Release Society (CRG).

Gert Storm

Dr. Gert Storm studied biology at the University of Utrecht and received his PhD in 1987. His research interest is in the field of drug targeting. From September 1988 until June 1989, he was a visiting scientist at Liposome Technology Inc. (Menlo Park, Calif.) and visiting assistant professor at the School of Pharmacy, Dept of Pharmaceutics, University of California, San Francisco. From February 1990 until September 1991, he was senior research scientist at Pharma Bio-research Consultancy BV in

Zuidlaren, where he contributed to the design, coordination, and evaluation of clinical pharmacological studies. Dr. Storm has published over 120 original articles, reviews, and book chapters and is involved in organizing conferences in the field of advanced drug delivery. He is member of the editorial advisory board of two journals, acts as a consultant to a number of pharmaceutical companies, and teaches "Liposome Technology", a course organized by the Center for Professional Advancement.

Art Tipton

Dr. Art Tipton received his BS in chemistry from Spring Hill College (Mobile, Ala.) in 1980 and then worked for 3 years at the Southern Research Institute. At the University of Massachusetts, he received his PhD in polymer science and engineering in 1988 and subsequently worked at Atrix Laboratories as a senior polymer scientist and manager of the Polymer Science Department prior to joining Southern BioSystems and Birmingham Polymers in 1993. Dr. Tipton's experience includes research and development at the preclinical and clinical stages as well as the scale-up and manufacture of final products. His research interests include human and veterinary delivery from biodegradable systems, the synthesis of novel biodegradable materials, and systems that function as both medical devices and controlled-delivery matrices. Dr. Tipton holds 16 U.S. patents, numerous corresponding foreign equivalents, and has more than 40 publications in the area of controlled release and biomaterials. In addition to overseeing the executive direction of Birmingham Polymers, Dr. Tipton has participated in the development and introduction of three commercial products. He is a member of the American Chemical Society (ACS), the American Association of Pharmaceutical Scientists (AAPS), the Controlled Release Society (CRS), and the Society for Biomaterials.

Stephen F. Tuck

Dr. Steven Tuck received his BSc in chemistry from Imperial College, University of London in 1983 and his PhD in physical–organic chemistry from the same school in 1986, where he studied the endogenous nitrosation of gastric secretions.

Postdoctoral research was conducted at The Johns Hopkins University School of Medicine and University of California, San Francisco, from 1987 to 1992. Then Dr. Tuck became principal scientist in Analytical Development at Chiron Corporation. Since November 1997, Dr. Tuck has been the director of Process Development and Formulation at Dynavax Technologies Corporation, where he is responsible for all CMC activities associated with immunostimulatory DNA development programs.

Praveen Tyle

Dr. Praveen Tyle has more than 15 years of worldwide pharmaceutical industry experience. Dr. Tyle was a product development scientist at American Cyanamid Company and Novartis (formerly Sandoz Pharmaceuticals Corporation) before becoming senior director of Pharmaceutical Development at Agouron Pharmaceuticals, Inc., in 1991. At Agouron, he was responsible for the development of preclinical and clinical products. In 1997, Dr. Tyle became the chief technical officer at Aronex Pharmaceuticals, Inc. (The Woodlands, Tex.) before joining Pharmacia & Upjohn in Kalamazoo, Michigan, as vice president of pharmaceutical development. Dr. Tyle also serves as a member of the Board of Directors of Lipomed, Inc. (Cambridge, Mass.) and the scientific advisory board of Biovector Therapeutics, S.A. He is a scientific advisor to Merck KGaA Biopharmaceuticals in Germany and an adjunct professor of pharmacy at the University of Houston. He holds several U.S. patents in the areas of drug development and delivery systems.

1

Rationale for Sustained-Release Injectable Products

Judy H. Senior
Maxim Pharmaceuticals Inc.
San Diego, California

People do not like injections. As a young child, I distinctly remember the stab of pain as the scary needle attacked my upper arm during a mass vaccination program against smallpox. Previous experiences with vaccinations against polio (which was soon to be available in an oral form) and diphtheria had not prepared me for the derm-abrasion required for the smallpox vaccination, and I was soon unconscious on the floor, much to my solicitous mother's dismay. Fortunately, vaccinations such as these do not require frequent dosing. But in the many cases where multiple injections are unavoidable, it is preferable to give them as infrequently as possible. Thus, a key rationale for developing a sustained-release version of an injectable drug product is to decrease its dosing frequency. Another key rationale is to increase the length of treatment possible resulting from a single dose; many examples are given in this volume.

Avoiding injections altogether by giving medicines orally or in an aerosol form to the lung/nose may be neither possible nor desirable in many cases. Needleless injections are not yet widely available and are unlikely to be a panacea.

This book is about sustained-release injectable products delivered by the parenteral route (i.e., delivered other than via the digestive tract), which are intended to provide a drug depot involving subcutaneous (SC), intramuscular (IM), or local injection. For these applications, the active drug enters the bloodstream gradually, thus prolonging the duration of drug action. The term *local injection* refers to products administered into local compartments where the drug action takes place, without the drug necessarily being absorbed into the circulation. Examples include delivery to the brain/spinal column (intracerebrospinal/intraventricular), joints (intra-articular), eye (intraocular), and regional target sites such as tumors or already infected/inflamed/painful tissues or those that might become diseased. To put the discussions into context, comparisons may be made with routes of administration not specifically covered in this book, such as the intravenous (IV), oral, topical, nasal, buccal, sublingual, and pulmonary routes. Also not covered are devices such as those designed to control the rate of subcutaneous drug administration using a syringe, needle, and/or pump system.

What types of drugs are suitable for sustained-release technology? While many drugs used in sustained release applications are water soluble, less water-soluble and even water-insoluble drugs can be successfully formulated. Indeed, many insoluble drugs lend themselves to this approach, thus making a virtue out of necessity. Whether a drug is suitable for formulation into a sustained-release delivery system may also depend on the following: The half-life of the drug must be long enough to permit a realistic dosage regimen (Tice and Cowser 1984). The drug must be capable of being released from the delivery system at a rate that provides a therapeutic dose of the drug over the desired time of action. Also, effective treatment must not depend on a trough of drug concentration between doses.

Here are some examples of therapeutic areas where sustained/controlled release is being used or considered in clinical trials. Other examples are given throughout this volume.

- Treatment of localized diseases such as many forms of cancer, viral infections such as cytomegalovirus (CMV) retinitis, and other infectious diseases.

- Localized pain relief such as inflammation in joints: Prolonged duration of action is particularly valuable when the injection itself is painful and/or difficult to administer.

- Other types of pain relief: In prolonging the effectiveness of pain relief, comparisons may be made when an opiate in sustained release or free form is administered via different delivery routes such as patient-controlled analgesia (PCA) (Slattery and Boas 1985) or epidurally (Senior 1998).

- Local anesthetic action may keep the area under analgesia longer when given in a sustained release formulation.

- Treatment of local infection in tissue.

- Treatment of local area where bone regrowth is required, e.g., bone regeneration/healing after a significant trauma.

- Slow release of drug into the circulation where prolonged action is needed, e.g., antipsychotic agents, slow-release forms of insulin, and other hormone treatments.

RATIONALE FOR USING SUSTAINED RELEASE FOR SYSTEMIC/LOCAL DELIVERY OF DRUGS

An important reason for designing a sustained-release version of a drug is to optimize its pharmacokinetic (PK) profile hence efficacy while minimizing any toxicity due to the drug itself. Which aspects of the drug's PK behavior in vivo are particularly relevant? One goal is to prolong the drug's effectiveness for as long as possible or for as long as the treatment is desired. Another is to avoid very high and potentially toxic levels of drug in the circulating blood or local injection site immediately after dosing. In the case of oral dosing, after a tablet is swallowed, and first pass metabolism occurs, the drug concentration peaks as the dose reaches the bloodstream; this peak can be very high if an extended duration of action is required. This initial drug peak usually represents an element of overexposure of the patient to the drug that should be minimized, especially for toxic drugs. In many types of sustained-release drug delivery, this initial peak can be avoided because the drug is released slowly from the delivery system and achieves therapeutic doses for relatively long time periods. Dosing by controlled IV or SC infusion can offer many of the same advantages as sustained release, but new problems arise (Madan 1985), e.g., the necessity for an indwelling catheter (line). The overall dose of drug administered by infusion may be greater than that required for a sustained-release version, depending on the PK of the drug in question. Furthermore, some kind of drug-specific sustained-release effect is often obtained with drugs administered intramuscularly or subcutaneously simply due to the nature of the route of injection. In other cases, it is important to formulate the drug so that it is not immediately cleared from the injection site of the local compartment, e.g., local slow release in the eye for treatment of CMV retinitis. This is where drug delivery technology can enhance the performance of the drug and provide maximum benefit to the patient.

Although most people would prefer to take medicine orally— usually in the form of a pill or capsule swallowed as infrequently as possible or formulated into a suspension or solution for pediatric use, there are many instances where oral medication may *not* be desirable, appropriate, or technically feasible. For example, an orally administered drug may not be absorbed without loss of activity (e.g., insulin and other protein/peptide medicines); or it may have poor bioavailability due to poor solubility in water or low permeability through biological membranes. Other examples are drugs where the oral route does not allow the desired dosing regimen, e.g., contraceptive effect of synthetic hormone released from Norplant® or Depo-Provera®, or the anticancer effects of Lupron® depot.

Some examples where sustained-release drug delivery has distinct therapeutic advantages are as follows:

- Minimizing systemic toxicity and maximizing effectiveness by injecting directly into the region where a sustained therapeutic response is desired, e.g., administration of corticosteroids into painful joints, or for treatment directly into the affected region, e.g., intraocular administration (Peyman and Ganiban 1995).

- The drug is only required to be active in a specific region of the body, e.g., in local anesthesia/analgesia. Thus systemic exposure to the drug is minimized or eliminated.

- Dosing frequency can be reduced without compromising the effectiveness of the treatment (Carter et al. 1988).

- The dosing compliance is increased. This is an issue when effective treatment requires multiple doses of a drug. Compliance is especially difficult to obtain when the dosing frequency is high or the patient has difficulty in swallowing or remembering to take the medication. Lack of compliance usually means lack of effective treatment.

- Frequently, pharmacoeconomic arguments can be made in favor of the dosage form. For example, the overall cost of the treatment may be dramatically reduced if the overall costs of patient visits, physicians' and nurses' time, and so on are considered.

- There may be patent and other commercially attractive reasons for considering sustained-release delivery systems.

The disadvantages of sustained-release drug delivery include the following:

- Injectable medicines cannot usually be removed once administered.

- Most drug delivery systems are most advantageous when they use a drug that is very potent/effective.

- It is necessary to study whether any new toxicities will be induced when the drug remains longer in the body.

- It is necessary to study whether there is any "burst effect," i.e., immediate release of a large bolus of drug immediately after injection, and what effect this might have, if any.

- It may be necessary to study whether there is any delay in the dispersal of residual drug from the injection site beyond that required for effective therapy.

- The requirement that any toxicity specific to the carrier itself must be examined (Storm et al. 1993) if previous studies with the carrier system are inadequate.

THE TECHNOLOGIES: SPICING IT UP

If variety is the spice of life, then drug delivery is a tasty business to be in. Sometimes the field is criticized for having too many ingenious inventions chasing too few applications. Yet do cooks complain that there are too many recipes? This book presents a rich menu of technologies, each with its own distinct flavor and purpose. A broad range of technologies is reviewed, from "simple" salt forms to "complex" polymeric delivery systems. Whether there is a particular drug delivery problem to solve or more to be learned about a specific technology, this volume calls forth the experience and expertise of scientists active in industrial and/or academic pharmaceutical settings.

The technology section of this volume begins with a chapter on insoluble drugs, salt forms, and prodrugs. Pharmaceutical companies, whether generic drug companies or "big pharma," may use this approach when formulating a difficult-to-solubilize drug product. Any sustained drug release usually relies on what the system will "naturally" allow. Chapter 4 includes a list of products on the market that use insoluble drug complexes to provide sustained drug release. Insoluble salt forms of drugs are not available for all compounds and are not suitable for all therapeutic situations. It is, however, a good starting place to review practical issues in the sustained release of drugs.

A complementary chapter on oily suspensions quickly gets down to the mechanics of preparing an effective oily formulation of an insoluble drug. The properties of suitable vehicles are probed, and numerous examples are cited. The fate of the oily suspension in vivo is thoroughly reviewed. Examples illustrate the observation that many of the most effective commercial sustained-release products combine the water-insoluble form of a drug compound with an oily suspension to further slow down drug absorption from the (usually) IM injection site (e.g., Depo-Provera®).

Emulsions and liposomes provide a somewhat more sophisticated level of technology and have products on the market worldwide. How are they more effective than the first two approaches? Is it worth the extra effort and complexity? What is known about the in vivo behavior of these systems, and how might this apply to other technologies? The two chapters on emulsions and liposomes address these and many other questions to show that liposomes and emulsions expand the range of drugs that can be formulated for sustained release. Rather than relying on the existence of an insoluble salt complex, almost any drug can be formulated—water-soluble drugs being especially suitable.

Another cluster of interesting biomaterials are the natural polymers. This includes such fascinating drug delivery vehicles as collagen, albumin, and gelatin, which like liposomes and lipid-based particles offer naturally occurring molecules as aids to encapsulation and drug delivery. There are several products close to market using this approach. It is generally thought that systems based on biopolymers, which are naturally present in the body, will have fewer biocompatibility issues than might be expected with synthetic polymers. Of course, it is not as simple as that, as this useful chapter exposes.

Synthetic polymers are a tried and tested technology, with several notable commercial successes so far. With even more effective biodegradable polymers being developed, this chapter and a chapter covering in situ gelling formulations, open up further options for sustained-release applications. Loved by chemists, some synthetic polymers remain objects of suspicion for biologists concerned with biocompatibility and long-term toxicological effects. These issues are addressed along with a discussion of technical obstacles in manufacturing illustrating how, in fact, many water-soluble and water-insoluble drugs can be turned into sustained-release therapies.

THE COMMON HURDLES: LESSONS FROM COMPLEMENTARY AND COMPETING TECHNOLOGIES

The drug delivery systems collected in this volume face many common hurdles from concept to finished product. The introductory chapters describe the principles of product development (Chapter 2) and list many successful products that are on the market or close to it (Chapter 3). After the detailed technology chapters, a series of chapters embraces universal themes applicable to sustained-release systems. Collected side by side, these chapters illustrate that competing and complementary technologies all have common hurdles to overcome on their way to the marketplace.

The first of these universal themes is the challenge of drawing together from diverse experiences some general guidelines for choosing a specific drug delivery system. Familiar issues relating to the characteristics of the drug and the delivery system and some not so familiar ideas are collected and assessed in Chapter 11. Real-life case studies illustrate some important applications of the technologies covered in this volume.

Much can be learned from animals. In a thought-provoking chapter (Chapter 12), sustained-release injectables given to food-producing and companion animals are reviewed. For example, consider the different implications of giving an IM injection into each of these classes of animals. An IM injection is not usually acceptable for food-producing animals, since the meat may be tainted by a residual drug depot. By contrast, an IM injection would be perfectly acceptable for a companion animal, although the release characteristics of a given dose of a product may differ dramatically for a hamster than those for a horse.

Vaccines and gene delivery present their own delivery challenges (Chapter 13). These two large areas of research offer perspectives on two extremes: vaccines are one of the oldest biologicals with regulatory approval for human use, and gene delivery is on the cutting edge of modern biotechnological research. Yet both applications feature a requirement for sustained action rather than sustained release of material. Recent advances in each of these technologies are reviewed by experienced scientists in this lively chapter.

Proteins and peptide delivery technology have their own special requirements, including the need for avoidance of denaturing conditions and the ability to efficiently encapsulate and retain full biological activity of relatively high molecular weight molecules. PK issues are well represented in Chapter 14 with an emphasis on applications using microsphere technology. Are some types of delivery systems more suitable for protein and peptide delivery than others? Which are the most useful applications and routes of administration? Judge for yourself in this informative chapter.

What about toxicity issues? People are increasingly aware that drugs may have unwanted side effects and desire to avoid them. The more toxic or potentially toxic the medicine, the more physicians and patients prefer to minimize potential ill effects and direct the medicine exactly to the place where it is needed (e.g., the knee, eye, or infection site). At the same time, it is clearly important to minimize and quantify any new toxicities or potentially harmful effects that might be attributable to the delivery system itself or due to the change in toxicokinetics of the drug via the delivery system. Biocompatibility and toxicology of degradable and nonbiodegradable polymeric sustained-release drug delivery systems are discussed by authors who have clearly lived through the process (Chapter 15).

The technical challenges of manufacturing and scaling up sustained-release drug products are complex and varied. This is particularly so since the drug product must be sterile for parenteral use. In many cases, sustained-release products cannot be terminally sterilized and must be prepared under aseptic conditions. This adds to the development schedule timeline and requires many activities related to microbiology, system validation, sterile process validation, in vitro drug release, sterility assurance, specifications, construction of facilities, and current Good Manufacturing Practice (cGMP) compliance. A description of scale-up, validation, and manufacturing issues uses microspheres as a living example of the processes required (Chapter 16).

Quality control methods and specifications development may seem unexciting to some people, but what could be more depressing than failing a regulatory inspection, such as a "pre-approval inspection" (PAI), because some aspect of this area was overlooked? Is it always unwise to spend time and money treating a sustained-release version of an old drug like a new chemical entity (NCE)? Are shortcuts desirable or even possible when a previously approved chemical entity is used? These and many other relevant issues are addressed in this authoritative chapter by a veteran of sustained-release drug development (Chapter 17).

Pulling it all together, the last chapter is a concise and well-informed appraisal of the regulatory issues specific to sustained-release drug delivery. In general, the safety and efficacy of a sustained-release or depot version of a drug must be addressed as if it were an NCE, whether or not the drug itself is an NCE. However PK/pharmacodynamic (PD)/toxicity testing and studies to support efficacy claims may be designed, executed, and interpreted better if information about the toxicity, bioavailablity, and PK of the drug compound itself is already known.

THIS BOOK

This book presents technical options that enable the reader to assess developments in the technology, to review the opportunities in this fast-growing field, to evaluate which systems might provide the best tool for the job, and to provoke thought on formulation issues for drugs in development. For example, no one wants to have frequent injections, especially if the injection site is inaccessible or uncomfortable.

Consider the following hypothesis: Unless emergency treatment is required (e.g., in diabetes), there is every reason to use a sustained-release injectable dosage form for as many therapeutic applications as possible. Indeed, is the future of most injectable products sustained release?

Scientists and physicians have come up with many very ingenious ideas for the development of sustained-release drug delivery products. The problems created while turning these ideas into commercial products are often intellectually challenging. Yet the solutions are emotionally satisfying when patients and physicians benefit, and pharmaceutical companies can achieve profits that justify the whole exercise financially. It is a tremendous privilege to be a part of an industry where everybody can win.

REFERENCES

Carter, D. H., M. Luttinger, and D. L. Gardner. 1988. Controlled release parenteral systems for veterinary applications. *J. Controlled Release* 8:15–22.

Madan, P. L. 1985. Sustained release drug delivery systems: Part V, parenteral products. *Pharm. Manuf.* (June): 51–57.

Peyman, G. A., and G. J. Ganiban. 1995. Delivery systems for intraocular routes. *Adv. Drug Delivery Reviews* 16:107–123.

Senior, J. H. 1998. Medical applications of multivesicular lipid particles. In *Medical applications of liposomes,* edited by D. D. Lasic and D. Paphadjopoulos. Amsterdam: Elsevier, pp. 733–750.

Slattery, P. J., and R. A. Boas. 1985. Newer methods of delivery of opiates for relief of pain. *Drugs* 30:539–551.

Storm, G., C. Oussoren, P. A. M. Peeters, and Y. Barenholz. 1993. Tolerability of liposomes in vivo. In *Liposome technology,* 2nd ed., edited by G. Gregoriadis. Boca Raton, Fla.: CRC Press.

Tice, T. R., and D. R. Cowser. 1984. Biodegradable controlled-release patenteral systems. *Pharm. Tech.* (November): 26–36.

2

Product Development Principles of Sustained-Release Injectable Formulations

Michael Radomsky
Epicyte Pharmaceutical Inc.
San Diego, California

The development of a sustained-release formulation is not entirely different from the development of a traditional pharmaceutical dosage form. In fact, there are many more similarities than differences. Although certain "shortcuts" may be possible, unique technical, regulatory, and clinical challenges exist. The successful development of sustained-release injectable products involves thorough planning and coordination of a number of scientific disciplines throughout the process.

Because sustained-release technology can be at the forefront of advances in science and medicine, novel challenges may arise in determining the correct course of action for development. A "commonsense" approach should be embraced when designing the development plan (i.e., determine what makes sense from medical, scientific, manufacturing, and regulatory perspectives). Project teams should utilize the scientific literature, frequent brainstorming sessions, the most relevant regulations and guidelines, feedback from regulators and leaders in the field, and their own data and information for efficient product development; there is no single source for all relevant information. Based on the available information, deviations from the standard/traditional pharmaceutical development plan will often be appropriate. Sponsors should be prepared to defend their position with both data and theoretical arguments. Table 2.1 outlines the features of a fast-track and lengthy product development program.

This chapter briefly outlines the potential development strategies and the technical areas involved in product development. While each development program is unique, broad generalizations and reviews are presented here. Details of these technical areas are covered in depth in the chapters that comprise the bulk of this book and the bibliography given at the end of this chapter.

PRODUCT DEVELOPMENT STRATEGY

It can be an enormous undertaking to begin the development of a sustained-release product. Smaller, less risky steps may be appropriate; it may be most advantageous to be the first to market with a less than optimal formulation rather than gaining approval for a "me too" drug with an optimal formulation. A line extension or proof of concept study with a less elegant formulation may be one appropriate course of action. One strategy may be to utilize a more traditional formulation (tablet, injectable, etc.) to establish proof of concept, feasibility, safety, efficacy, or even drug approval before the more difficult task of developing a sustained-release formulation is undertaken. If establishing efficacy with a traditional formulation is not possible, a formulation with the fewest technical challenges could be utilized. An extruded polymer drug matrix that requires implantation or an implantable pump, although not the most desirable formulation from a patient or physician perspective, can provide sustained release of the drug for the required time and may be a desirable

Table 2.1. *Features of "Fast Track" (fewer unknowns and challenges) Versus Relatively "Lengthy" (more unknowns and challenges) Product Development Programs*

The characteristics of a development process versus factors that will increase costs and development time are tabulated. Even if a development program is undertaken with all the characteristics of the fast-track column, some challenges are inevitable; development programs with absolutely no unpredicted, costly, or time-consuming challenges are unlikely.

Fast Track (fewer unknowns and challenges)	Lengthy (more unknowns and challenges)
A generic drug	A novel drug (new chemical entity)
Drug previously administered to humans in different indication or delivery system	Drug not previously administered to humans
All materials present in approved products and generally regarded as safe	Novel biomaterial
Simple manufacturing process	Complex manufacturing process
Indicated for life-threatening disease, no existing treatments, or more efficacious than existing therapies	Not more efficacious than existing therapies
Well-established, objective clinical endpoints	Subjective or vague clinical endpoints
Rapid patient enrollment	Slow patient enrollment
Short patient follow-up	Long patient follow-up
FDA relationship throughout development process	No relationship established with FDA before filings
Fewer safety concerns than existing treatments	Significant or unknown safety concerns
Well-characterized, stable product	Poorly characterized, unstable product
Well-controlled manufacturing process	Poorly controlled manufacturing process
Terminal sterilization	Aseptic processing
Compliance with GLPs, GMPs, GCPs	Poor compliance with GLPs, GMPs, GCPs

GLP: Good Laboratory Practice
GMP: Good Manufacturing Practice
GCP: Good Clinical Practice

alternative. However, this strategy may not be possible, and moving quickly into development with a sustained-release injectable formulation may be most expedient and appropriate. Each drug and indication must be evaluated independently before the most appropriate strategy can be determined.

PRODUCT DEVELOPMENT TIMELINE

Figure 2.1 indicates a typical timeline for a sustained-release injectable product and categorizes product development into three phases: research, development, and commercialization. For purposes of this illustration, the end of the research phase occurs when a submission for initiating human clinical studies is made. This marks the beginning of the development phase. The end of the development phase occurs upon product approval, which then signals the start of the commercialization phase. Each organization may have its own definitions for when research ends and development begins, but regardless of the name, the same steps must be completed before product approval. For example, some companies consider product development to begin once material is to be made for the preclinical studies required in order to file for approval to initiate clinical studies.

Although Figure 2.1 presents a product development process with similarities shared among all types of products, each case is unique and presents its own technical challenges. This chapter provides a brief overview of the four main technical areas required for the development of these products: (1) formulation, analytical, and quality control functions; (2) preclinical biology, including safety and efficacy data; (3) manufacturing; and (4) regulatory and clinical affairs. This chapter does not discuss the necessary but primarily business-oriented functions needed for the development of successful pharmaceutical products (e.g., sales, marketing, financing, business development, intellectual property, new product planning, etc.). These functional areas of expertise may be unique to each sustained-release technology but can be similar to those in place for the development of any drug, biologic, or medical device.

Figure 2.1. *Product development timeline.*

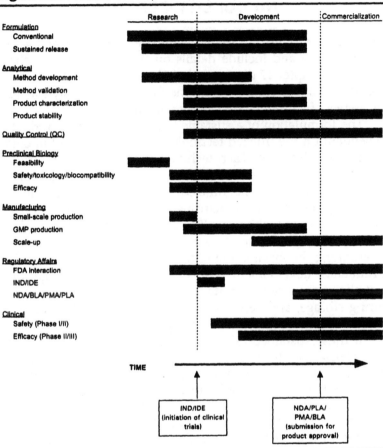

This timeline assumes that a conventional formulation is utilized in the early stages of a program to establish feasibility and safety. The sustained-release injectable formulation is developed for commercialization. No units of time are provided; each phase may take one or more years to complete.

FORMULATION, ANALYTICAL METHODS, AND QUALITY CONTROL

Chapters 4 through 11 discuss technologies appropriate for sustained release and include details on the formulation of these products; Chapter 17 discusses the relevant issues with regard to analytical quality control (QC). The formulation, analytical, and QC issues of a product are closely related.

The formulation of a sustained-release injectable product is dependent on the utilized technology and is discussed in detail throughout this book. Each technology has unique characteristics that require a specific formulation for the indicated application. Excipients may be utilized to improve the stability or bioactivity during manufacturing, storage, or use and vary widely with each technology. The amount and duration of drug delivery are probably the two most important characteristics in determining the formulation of this type of product. The goal of formulation development is to develop a robust dosage form that allows consistent manufacture of multiple batches, is stable during storage, and delivers the correct dose for the indicated duration when administered to the patient.

In these sustained-release dosage forms, it is important to characterize the drug release profile (i.e., determining drug release as a function of time) with appropriate analytical methods. These methods can be used initially to screen formulations and later in QC. The methods should be able to measure the initial burst of drug release and the dose and length of release for the duration of the product. A correlation between in vitro and in vivo release is desirable. For example, this correlation can be extremely valuable for future manufacturing changes. See Chapter 18 for additional details on these correlation studies.

Appropriate analytical methods ensure that the dosage regimen is consistently delivered from the formulation. Analytical functions are typically responsible for method development, stability studies, and method validation. These methods must be stability indicating and capable of detecting and resolving any interference. In general, analytical methods may need to be developed to measure potency, bioactivity, homogeneity, sterility, the in vitro drug release profile, in vitro/in vivo drug release correlation, and excipient characterization. These methods must measure product characteristics that are relevant to the manufacturing and use of the product. Validation of these analytical methods is important to ensure that the data generated are reliable and produce the results that the development team requires.

Stability studies are necessary to determine the shelf life of the formulation under the recommended storage conditions, with product stability defined by physical, chemical, and biological specifications. Compatibility of the packaging components must be determined and encompassed in product stability, and any unique packaging requirements defined.

QC is an extension of analytical development and ensures that the release of product meets all product specifications. The U.S. Pharmacopeia (USP) XXIII, U.S. Food and Drug Administration (FDA) guidelines, and International Conference on Harmonisation (ICH) guidelines describe general expectations for the release testing of drug products. Since these products vary enormously, no single set of release assays will fit all products, and sound scientific judgment must be utilized to determine an acceptable list of product specifications for each application. Chapter 17 should be consulted for more details on QC issues.

PRECLINICAL BIOLOGY
(SAFETY AND EFFICACY)

Preclinical testing is primarily carried out in animal models to evaluate the safety and efficacy of the product. In addition, in vitro (e.g., tissue culture) or other relevant models may be utilized. Chapter 15 discusses the toxicology and biocompatibility of sustained-release injectable products. The objective is to assess the potential use of both the drug and delivery system in a human clinical study by demonstrating in animal models that the product is safe and effective in treating the target disease. Typical studies assess the efficacy and mechanism of action, pharmacokinetics, dose response, and any toxicological response of the drug and delivery system under investigation. The necessary studies for each application must be designed accordingly.

It should be taken into consideration in the drug development planning process, especially in the design of clinical and preclinical studies, that the act of injecting a drug locally may have inherent consequences locally. Frequent subcutaneous (SC) or intramuscular (IM) injection can cause irritation and pain at the injection site from the drug or even from a placebo injection of sterile phosphate-buffered saline. Appropriate studies should be designed to assess the compatibility of the sustained release product with the tissue at the injection site.

A depot or sustained-release version of the parent drug may have modified toxicokinetic behavior and will need to be investigated. The intention of the sustained-release form is obviously to improve and/or prolong efficacy and, if possible, significantly reduce any toxicity concerns. Efficacy of the drug may be altered, favorably or unfavorably, compared to a traditional formulation, and changes in efficacy must be investigated and balanced with the pharmacodynamics and toxicokinetics of the drug in the sustained-release form.

Preclinical and clinical studies should clearly address concerns regarding safety and efficacy and any new or exacerbated toxic effects due to prolonged levels of the drug, local injection site irritation, and other undesirable toxicity. Thus, the benefit of the sustained-release formulation to the patient and the physician always needs to be weighed against any risks.

MANUFACTURING

The manufacturing of a sustained-release product can be quite challenging. Typically, the formulation is first prepared at the bench, often by one scientist with no product specifications, in-process testing, or QC. Moving from this small batch size to large scale is made more difficult because the formulation may have been optimized by those unfamiliar with all of the issues of manufacturing a pharmaceutical product.

Chapter 16 discusses the scale-up, validation, and manufacturing of sustained-release microsphere products. The scale-up, validation, and commercial manufacturing of these products can present unique difficulties, including consistent manufacturing of the novel product on a large scale; establishment of the manufacturing facility and equipment for commercial production; procurement of reliable, high-quality raw materials that may be new to the pharmaceutical industry; establishment of product specifications; development of GMP analytical methods and manufacturing processes; and GMP sterility assurance and validation.

REGULATORY AND CLINICAL AFFAIRS

Chapter 18 discusses the regulatory affairs associated with sustained-release injectable technologies. Phase I studies are typically conducted in healthy, normal human volunteers and are intended to assess the safety of the proposed dosage form. The results from Phase I studies typically generate limited information on the drug and delivery system's safety profile. Phase II studies are usually carried out on patients who require the indicated treatment. The emphasis of a Phase II study is on the safety of the product, although preliminary effectiveness data can also be obtained and the target therapeutic patient population determined. Phase III studies are generally relatively large studies intended to demonstrate the safety and efficacy of the drug in a large and varied patient population at multiple clinical sites. Some of these development studies can be shortened or eliminated if the drug has been developed in other dosage forms.

One of the first considerations in forming a regulatory development plan is to determine if the product will be regulated as a drug, device, or biologic product. These products will be reviewed by the FDA in the Center for Drug Evaluation and Research (CDER), the Center for Devices and Radiological Health (CDRH), or the Center for Biologics Evaluation and Research (CBER); each center has reviewed these types of products. Since the requirements for submission formats are significantly different, the proper designation should be determined early in the development process. The FDA has a series of intercenter agreements to help clarify which center will have primary review responsibility. In addition, a "Request for Designation" can be submitted to the FDA to determine how the product will be regulated.

The first formal regulatory submission in the United States is to initiate human clinical trials (i.e., Phase I), followed by a submission to begin commercialization of the product (typically many years later). Differences in this process exist worldwide, and Chapter 18 outlines the procedures in the United States and Europe. Even though a long stretch of time may exist between formal regulatory submissions, discussions between the sponsor and the regulatory agency should be conducted at every opportunity and can occur prior to Investigational New Drug/Investigational Drug Exemption (IND/IDE) submission, initiation of Phase III studies, before New Drug Application/Biologics Licensing Application/Product License Application/Premarket Approval (NDA/BLA/PLA/PMA) submission, and to discuss any unusual aspects of the drug development program.

CONCLUSION

In developing sustained-release products, excellent scientific knowledge and judgment and a thorough understanding of the regulations from a clinical and regulatory perspective are critical to successful completion of the project. These areas of expertise are required from preclinical testing through postmarketing approval activities. The traditional path for developing pharmaceuticals is a good outline for project planning; however, unique challenges and obstacles will require unique solutions. As we move forward into the next century, fewer unsolvable challenges should be encountered as more and more sustained-release products are designed, developed, and approved in the coming years.

BIBLIOGRAPHY

Akers, M. J. 1994. *Parenteral quality control,* 2nd ed. New York: Marcel Dekker.

Carter, D. H., M. Luttinger, and D. L. Gardner. 1988. Controlled release parenteral systems for veterinary applications. *J. Controlled Release* 8:15–22.

Chess, R. 1998. Economics of drug delivery. *Pharm. Res.* 15:172–174.

Murano, G. 1997. FDA perspectives. In *Specifications for biotechnology products–from IND to PLA*, edited by F. Brown and J. Fernandez. *Dev. Biol. Stand.* 91:3–13.

Wade, M. 1998. Clinical programs in cutting-edge technologies: A guide for beginning the development of novel therapies. *Bio Pharm* 11:52–56.

3

Commercial Sustained-Release Injectable Formulations by Encapsulation

Mantripragada B. Sankaram

Skyepharma plc
San Diego, California

The idea that drug delivery technologies can bring both therapeutic and commercial value to drug products has resulted in great strides in the development of sustained-release injectable products. Some of the best recognized benefits that drug delivery technologies offer include improvement of efficacy and safety of drug substances, expanded labeling, improved pharmacokinetics, extension or renewal of intellectual property rights, and delivery of drug substances that are otherwise undeliverable. While such benefits are the impetus for the development of sustained-release injectable products from a marketing standpoint, the development of drug delivery systems presents a unique set of additional scientific and technological issues to conventional drug development.

This chapter focuses on commercially available, sustained-release drug delivery systems that use encapsulation technologies that are lipid based or polymer based. An overview of the properties of commercially available, sustained-release injectable products is provided, along with a comparison of the corresponding drug substances in a conventional unencapsulated form. It is intended that such an analysis will shed some light on both the advantages and the limitations of sustained-release injectable formulations of drug substances. The information in the tables of this chapter was extracted primarily from two sources (PDR 1998; AHFS 1998).

DRUG PRODUCTS, DRUG SUBSTANCES, AND EXCIPIENTS

A list of commercially available, lipid- and polymer-based drug delivery products for injection and their composition is given in Tables 3.1 and 3.2. The lipid-based drug products are commercially available only for intravenous (IV) injection, because the drug substances in these products, daunorubicin and doxorubicin, elicit toxic responses such as extreme irritation when administered by the intramuscular (IM) or subcutaneous (SC) route. Although the lipid-based drug products of amphotericin B have been approved for IV administration, this drug substance has been given by the intra-articular, intrapleural, and intrathecal routes, as well as by local instillation or irrigation. A discussion of the lipid-based products has been included since it is likely that future formulations of drug substances with similar physicochemical properties will share the basic design elements of these products. Amphotericin B, daunorubicin, doxorubicin, and leuprolide are available as drug products for injection in the unencapsulated form. These products are available as liquid dosage forms (aqueous solutions) for injection or as solid dosage forms (lyophilized powders) that are reconstituted either in sterile Water for Injection (WFI), normal saline, or 5 percent dextrose for injection prior to use for infusions. The dosage forms of the drug products with amphotericin B, doxorubicin, or leuprolide in lipid- or polymer-based delivery systems are

Table 3.1. Drug Substances, Excipients, and Delivery Technologies of Commercially Available Drug Delivery Products for Injection

Drug Product	Drug Substance	Delivery Matrix	Excipients		Delivery Technology
			Other		
Abelcet®	Amphotericin B	DMPC, DMPG	Normal saline		Lipid Complex
AmBisome®	Amphotericin B	HSPC, cholesterol, DSPG, α-tocopherol	Sucrose, disodium succinate Hexahydrate		Liposomes
Amphotec®	Amphotericin B	Cholesteryl sulfate	Tromethamine, disodium edetate Dihydrate, lactose monohydrate, HCl		Lipid Complex
DaunoXome®	Daunorubicin	DSPC, cholesterol	Sucrose, glycine, calcium chloride dihydrate		Liposomes
Doxil®	Doxorubicin	MPEG-DSPE, HSPC, cholesterol	Ammonium sulfate, histidine, sucrose, HCl or NaOH		Stealth Liposomes
Lupron Depot®	Leuprolide	PLGA	Carboxymethylcellulose sodium, Mannitol, acetic acid, polysorbate 80		Microspheres
Lupron Depot®-Ped	Leuprolide	PLGA	Carboxymethylcellulose sodium, mannitol, acetic acid, polysorbate 80		Microspheres

Table 3.1 continued on next page.

Table 3.1 continued from previous page.

Drug Product	Drug Substance	Delivery Matrix	Excipients Other	Delivery Technology
Lupron Depot®-3 Month	Leuprolide	PLA	Carboxymethylcellulose sodium, mannitol, acetic acid, polysorbate 80	Microspheres
Lupron Depot®-4 Month	Leuprolide	PLA	Carboxymethylcellulose sodium, mannitol, acetic acid, polysorbate 80	Microspheres
Zoladex Depot®-1 Month and 3 Month	Goserelin	PLGA	Acetic acid	Microspheres

DMPC: L-α-dimyristoylphosphatidylcholine; DMPG: L-α-dimyristoylphosphatidylglycerol; DSPG: L-α-distearoylphosphatidylglycerol; DSPC: L-α-distearoylphosphatidylcholine; HSPC: hydrogenated soy phosphatidylcholine; PLGA: poly(lactide-glycolide); PLA: poly(lactic acid); MPEG-DSPE: methoxypolyethylene glycol combined with distearoyl-sn-glycerophosphoethanolamine.

Table 3.2. Dosage Form and Potency of Commercially Available, Lipid- and Polymer-Based Drug Delivery Products for Injection

Drug Product	Dosage Form	Potency	Manufacturer	Drug Substance
Abelcet®	Semisolid	5 mg/mL	The Liposome Co.	Amphotericin B
AmBisome®	Solid	50 mg	NeXstar	Amphotericin B
Amphotec®	Solid	50, 100 mg	Sequus	Amphotericin B
DaunoXome®	Semisolid	2 mg/mL	NeXstar	Daunorubicin
Doxil®	Semisolid	2 mg/mL	Sequus	Doxorubicin
Lupron Depot®	Solid	3.75, 7.5 mg	TAP Pharmaceuticals	Leuprolide
Lupron Depot®-Ped	Solid	7.5, 11.25, 15 mg	TAP Pharmaceuticals	Leuprolide
Lupron Depot®-3 Month	Solid	22.5 mg	TAP Pharmaceuticals	Leuprolide
Lupron Depot®-4 Month	Solid	22.5 mg	TAP Pharmaceuticals	Leuprolide
Zoladex® Depot-1 Month and 3 Month	Solid	3.6, 10.8 mg	Zeneca Pharmaceuticals	Goserelin

also available either as solid (lyophilized powder) or semisolid (aqueous suspension) dosage forms. However, all products need to be reconstituted or diluted with an appropriate medium.

Lipid-Based Drug Products

Abelcet® is an amphotericin B lipid complex consisting of amphotericin B and the phospholipids L-α-dimyristoylphosphatidylcholine (DMPC) and L-α-dimyristoylphosphatidylglycerol (DMPG) in a 1:0.7:0.3 mole ratio. The complex has a ribbonlike structure with a diameter of about 2–11 μm. Abelcet® is an aqueous suspension at pH 5–7 containing 5 mg/mL amphotericin B, 3.4 mg/mL DMPC, 1.5 mg/mL DMPG, and 9 mg/mL sodium chloride. The product is available in a single-use 1 mL vial with a potency of 50 mg of amphotericin B per mL of suspension.

AmBisome® contains amphotericin B as the drug substance intercalated into a unilamellar bilayer liposomal membrane. The average diameter of the liposome is less than 100 nm. This drug product consists of hydrogenated soy phosphatidylcholine (HSPC), cholesterol, distearoylphosphatidylglycerol (DSPG), and α-tocopherol as the liposomal excipients; sucrose and disodium succinate hexahydrate are other excipients for maintaining isotonicity and pH. AmBisome® is a lyophilized powder. Upon reconstitution with sterile WFI a yellow, translucent suspension at a pH of 5–6 is obtained. The product is available in a single-use vial with 50 mg of amphotericin B per vial.

Amphotec® is also a lipid-drug complex with amphotericin B but complexed in an equimolar stoichiometry to the delivery vehicle excipient, sodium cholesteryl sulfate; tromethamine, disodium edetate dihydrate, lactose monohydrate, and hydrochloric acid are other excipients for maintaining isotonicity and pH. The lyophilized powder forms a colloidal dispersion upon reconstitution in sterile WFI. The excipients and the drug substance form a bilayer in microscopic, disk-shaped particles with an average diameter of about 115 nm and a thickness of 4 nm. The product is available in a single-use vial with 50 mg or 100 mg amphotericin B per vial.

DaunoXome® is a liposomal formulation of daunorubicin. A citrate salt of the drug substance is encapsulated in liposomes

with an average diameter of 35–65 nm composed of daunorubicin, distearoylphosphatidylcholine (DSPC) and cholesterol in
the molar ratio 1:10:5. Sucrose, glycine, and calcium chloride dihydrate are used as other excipients for maintaining isotonicity,
stability, and pH. DaunoXome® is a sterile, pyrogen free, translucent red liposomal dispersion at pH 4.9–6. It is available in a
single-use vial with 2 mg daunorubicin per mL.

A PEGylated lipid, methoxypolyethylene glycol, combined
with distearoyl-*sn*-glycerophosphoethanolamine (MPEG-DSPE),
hydrogenated soy phosphatidylcholine (HSPC), and cholesterol,
is used to formulate Stealth liposomes into which doxorubicin is
encapsulated to produce Doxil®. This product further contains
sucrose, histidine, ammonium sulfate, and hydrochloric acid or
sodium hydroxide for maintaining isotonicity and pH. Doxil® is a
sterile, translucent red liposomal dispersion at pH 6.5. The product is available in a single-use vial with 2 mg doxorubicin per mL.

Polymer-Based Drug Products

Leuprolide acetate, a synthetic nonapeptide analog of naturally
occurring gonadotropin releasing hormone (GnRH), luteinizing
hormone releasing hormone (LHRH), and gonadorelin is one of
two drug substances that is currently available as a sustained-
release polymer-based injectable product. The delivery vehicle
is formed from the biodegradable copolymers of lactic and glycolic acids [poly (lactide-co-glycolide), (PLGA)] or poly (lactic
acid) (PLA). Other excipients include carboxymethylcellulose
sodium, mannitol, acetic acid, and polysorbate 80. A series of
products namely, Lupron Depot®, Lupron Depot®-Ped, Lupron
Depot®-3 month, and Lupron Depot®-4 month, are available.
The Lupron Depot series of products are available as single-use,
two-vial products. One vial contains a lyophilized powder of
leuprolide acetate in microspheres formed from PLGA or PLA
that may also contain gelatin and mannitol. The other vial contains an appropriate vehicle for reconstituting of the powder
into a suspension. The reconstitution buffer is made from the
other excipients listed in Table 3.1, dissolved in sterile WFI. The
four Lupron Depot® products differ both in potency and in the
type of polymer excipient.

Lupron Depot® is available at a potency of 3.75 mg leuprolide and contains 0.65 mg gelatin, 33.1 mg PLGA, and 6.6 mg mannitol. It is also available at a potency of 7.5 mg leuprolide and twice the above stated amounts of gelatin, PLGA, and mannitol. Lupron Depot®-Ped is available at a potency of 7.5 mg, 11.25 mg, or 15 mg and 2, 3, or 4 times the above stated amounts of gelatin, PLGA, and mannitol. Lupron Depot®-3 month and Lupron Depot®-4 month are available at a potency of 22.5 mg or 30 mg leuprolide, respectively. PLA is used to form microspheres for these two products, and mannitol is present as the other excipient. The reconstitution medium for the Lupron Depot® products contains 7.5 mg carboxymethylcellulose sodium, 75 mg mannitol, 1.5 mg polysorbate 80, and WFI.

A similar polymer-based product, Zoladex Depot®, is a formulation of goserelin acetate, a synthetic decapeptide analog of LHRH, in PLGA. Other excipients include acetic acid. This product is available either for 1-month or 3-month depot delivery. Zoladex Depot® is supplied as a sterile, biodegradable product containing goserelin acetate equivalent to 10.8 mg goserelin. It is designed for SC implantation with continuous release over a 12-week period. Goserelin acetate is dispersed in a matrix of PLGA (12.82-14.76 mg/dose) containing less than 2 percent acetic acid and up to 10 percent goserelin-related substances. It is presented as a sterile, white to cream colored, 1.5 mm diameter cylinder, preloaded in a special single-use syringe with a 14-gauge needle and overwrapped in a sealed, light- and moisture-proof, aluminum foil laminate pouch containing a desiccant capsule. Zoladex Depot® is also supplied as a sterile, biodegradable product containing goserelin acetate equivalent to 3.6 mg of goserelin designed for administration every 28 days.

STORAGE AND HANDLING REQUIREMENTS

The presence of the excipients required to form the drug delivery vehicle establishes storage and handling requirements unique to these compounds. The stability of lipids against hydrolysis, the integrity of the liposomal structure, retention of the encapsulated drug substance in the aqueous core, and the choice of appropriate buffer salts for maintaining isotonicity and pH are critical in the development of a lipid-based sustained-release injectable product. Lipid hydrolysis can be prevented if a solid dosage form is chosen. However, formation of a solid dosage form by lyophilization of a liposome frequently results in a loss of the structural integrity of the liposomal structure and a release of the encapsulated hydrophilic drug substance. As a result, solid dosage forms for lipid-based delivery systems are available only for the hydrophobic drug substance, amphotericin B. In this case, the drug products are stoichiometric complexes of lipids with the drug substance rather than liposomes. For the hydrophilic drug substances daunorubicin and doxorubicin, the drug product is a liposomal suspension. For these products, the lipid composition, pH, and osmolarity are chosen in order to minimize chemical degradation and physical changes relating to liposomal structure, integrity, and drug retention.

Use of PLGA and PLA requires that a solid dosage form be developed. This is because of the susceptibility of the ester linkages in the polymers to hydrolysis. Additionally, readily available diluents, such as water, saline, or dextrose solutions, are not suitable for reconstitution of the lyophilized powder. These diluents frequently produce a massive aggregation of the lyophilized powder resulting in a material that cannot be injected through normal size needles. A suitable reconstitution medium is supplied by the manufacturer for the Lupron Depot® series of products. With the exception of Zoladex®, all lipid- or polymer-based drug products should be reconstituted in order to obtain the correct potency for injection. Table 3.3 provides a summary of the appropriate solutions for reconstitution, stability after reconstitution, and any special precautions to be taken.

DOSAGE AND ADMINISTRATION

IM and SC injections are two major routes of administration for sustained-release injectable products (Kadir et al. 1992). The buttock (gluteus maximus), deltoideus (upper arm), and lateral

Table 3.3. *Storage and Handling Requirements After Reconstitution of Commercially Available Lipid- and Polymer-Based Drug Delivery Products for Injection*

Drug Product	Solutions for Reconstitution or Dilution	Recommended Storage Temperature (°C)	Maximum Duration	Special Precautions
Abelcet®	5% dextrose	2–8°C room temperature	48 h 6 h	Should not be frozen; should not be diluted with saline or other electrolytes
AmBisome®	WFI	2–8°C	24 h	Should not be frozen
Amphotec®	WFI	2–8°C	24 h	Should not be frozen; should not be diluted with saline, dextrose, or other electrolytes
DaunoXome®	5% dextrose	2–8°C	24 h	
	5% dextrose	2–8°C	6 h	Should not be frozen; should not be diluted with any solution other than D5W
Doxil®	5% dextrose	2–8°C	24 h	Prolonged freezing to be avoided; should not use any diluent other than D5W
Lupron Depot®	Diluent supplied by manuf.	RT	24 h	Should not use any diluent other than that supplied by the manuf.
Lupron Depot®-Ped	Diluent supplied by manuf.	RT	24 h	Should not use any diluent other than that supplied by the manuf.
Lupron Depot®-3 Month	Diluent supplied by manuf.	RT	0 h	Should not use any diluent other than that supplied by the manuf.
Lupron Depot®-4 Month	Diluent supplied by manuf.	RT	0 h	Should not use any diluent other than that supplied by the manuf.

thigh (vastus lateralis) are the major sites of IM injection, with the gluteus being the most common site for adults due to its greater muscle mass. Blood flow is highest in the deltoid muscle, resulting in greater systemic absorption. SC injections are administered into the adipose and connective tissue beneath the skin. Relative to muscular tissue, subcutaneous tissue is more loosely organized and poorly perfused with blood. Preferred sites include the arms, legs, and the abdomen. Although SC injections are relatively easier to administer, they are generally limited to nonirritating, hydrophilic drugs. Further, the volume of injection is limited to 0.5 to 2 mL for SC injections, while volumes as high as 6 mL, perhaps even 10 mL, are possible with the IM route.

Table 3.4 is a summary of the recommended usual adult dosage of unencapsulated drug substances and of drug substances in a lipid- or polymer-based drug product. The dosage for Amphotec®, AmBisome®, and Abelcet is greater than the maximum dosage for unencapsulated amphotericin B. Since all three products are recommended to be administered on a daily basis at the indicated doses, it appears that the increase in dosage by lipid complexation is obtained by mechanisms other than sustained release.

For the drug products based on liposomal delivery, Doxil® and DaunoXome®, the dosage is lower than that for their unencapsulated counterparts. Both the dose and the frequency of dosing are lower for DaunoXome® than for daunorubicin, while only the dose is lower for Doxil® compared to doxorubicin. In this case, it appears that the effect of liposomal encapsulation is one of prolonging the circulation time of the drug rather than one of sustained release.

As expected of a sustained-release injectable product, it has been possible to significantly reduce the frequency of dosing required for the unencapsulated drug substance by encapsulation into a polymer-based delivery system. A daily dose of 1 mg leuprolide acetate subcutaneously administered for 1 to 4 months can be replaced by a single IM injection of Lupron Depot®. Similarly, daily treatment by goserelin acetate can be replaced by implantation once a month or once in three months of Zoladex Depot®.

Table 3.4. *Comparison of Recommended Dosage of Amphotericin B, Daunorubicin, Doxoubicin, and Leuprolide as Unencapsulated Drug Products and in Lipid- or Polymer-Based Drug Products (The recommended dosage is the usual adult dosage except for Lupron Depot®-Ped.)*

Drug Substance	Drug Product	Recommended Dosage
Amphotericin B	Unencapsulated	Not more than 1.5 mg/kg/day IV
Amphotericin B	Abelcet®	5 mg/kg single daily dose IV
Amphotericin B	Amphotec®	3-4 mg/kg single daily dose IV
Amphotericin B	AmBisome®	3-5 mg/kg daily dose IV
Daunorubicin	Unencapsulated	60 mg/m^2 daily for 3 days IV; repeat every 3–4 weeks
Daunorubicin	DaunoXome®	40 mg/m^2 as a 60-minute IV infusion once every 2 weeks
Doxorubicin	Unencapsulated	60–75 mg/m^2 single dose every 21 days IV
Doxorubicin	Doxil®	20 mg/m^2 as a 30-minute IV infusion once every 3 weeks
Leuprolide	Unencapsulated	1 mg/daily SC
Leuprolide	Lupron Depot®	7.5 mg once per month IM
Leuprolide	Lupron Depot®-Ped	7.5–15 mg once per month IM or SC
Leuprolide	Lupron Depot®-3 month	22.5 mg once every 3 months IM
Leuprolide	Lupron Depot®-4 month	30 mg once every 4 months IM
Goserelin	Unencapsulated	N/A
Goserelin	Zoladex® Depot-1 Month	3.6 mg every 4 weeks SC
Goserelin	Zoladex® Depot-3 Month	10.8 mg every 12 weeks SC

PRODUCTS IN DEVELOPMENT

A select list of sustained-release injectable products currently in development and advanced clinical trials is given in Table 3.5. Numerous technologies and drug products in earlier stages of development are not included. It should be noted that given the rapid changes that occur in product development programs at any given company and the continual formation of strategic alliances between companies, some of the product development programs may be discontinued, modified, or new programs initiated.

One of the emerging technologies is a novel form of a lipid-based drug carrier called DepoFoam™ (Oussoren et al., Chapter 7). Also referred to as multivesicular liposomes or multivesicular lipid particles, DepoFoam™ is a suspension dosage form of particles with diameters in the range of 10 to 50 μm suspended in an aqueous phase for injection. The particles are composed of nonconcentric lipid bilayer vesicles. The anticancer therapeutic drug substance cytosine arabinoside, when encapsulated into DepoFoam™ particles to form DepoCyt®, significantly reduces the frequency of injections. As of this writing, this drug product has been recommended for approval by the Oncologic Drugs Advisory Committee of the U. S. Food and Drug Administration (FDA) for treating lymphomatous meningitis by intrathecal injection. Another product, DepoMorphine™, is based on the DepoFoam™ technology for delivering morphine by the epidural route. A novel process of microsphere production, ProLease™, is currently under development for delivering human growth hormone (Burke and Putney, chapter 14).

CONCLUSIONS

This short compilation of commercial products is intended to encourage readers interested in the development of sustained-release injectables to investigate further the reasons for the success of these products and incorporate appropriate features of these products into their own development programs. While the list of technologies for developing sustained-release injectable

Table 3.5. *Lipid- and Polymer-Based Drug Delivery Products in Development*

Drug Substance	Drug Product	Technology	Therapeutic Indication	Route of Administration	Manufacturer
Cytarabine	DepoCyt®	DepoFoam™	Leptomeningeal metastases	Intrathecal	DepoTech
Morphine	DepoMorphine™	DepoFoam™	Analgesia	Epidural	DepoTech
hGH	Nutropin Depot®	Microspheres (ProLease™)	Growth deficiency	Subcutaneous	Alkermes

products is large and the drug substances for which sustained-release formulations were developed is broadly encompassing, the list of products that have been commercialized is relatively small. However, both enteral and parenteral drug delivery comprise one of the fastest growing segments of the pharmaceutical industry, as emerging technologies drive growth (MedAdNews 1998; GEN 1998; *Drug Delivery Systems* 1998). According to one projection, all protein and peptide drugs, for example–currently worth more than $10 billion of the world pharmaceutical market and a rapidly growing segment–are candidates for alternative delivery methods. It is anticipated that the total drug delivery market will be worth in excess of $100 billion by 2007 (DR Reports 1998).

REFERENCES

AHFS. 1998. *AHFS drug information.* Bethesda, Md., USA: American Society of Health-System Pharmacists, Inc., p. 3158.

Drug delivery systems: New developments, new technologies. 1998. Norwalk, Conn., USA: Business Communications Co.

DR Reports. 1998. *Drug delivery systems: technologies and commercial opportunities.* Waltham, Mass., USA: Decision Resources, Inc.

GEN. 1998. *Genetic Engineering News* (June 15).

Kadir, F., J. Zuidema, and D. J. A. Crommelin. 1992. Liposomes as drug delivery systems for intramuscular and subcutaneous injections. In *Pharmaceutical particulate carriers in medical applications* edited by A. Rolland. New York: Marcel Dekker, Inc.

MedAdNews. 1998. *The Magazine of Pharmaceutical Business and Marketing* (August). Engel Publishing Partners, West Trenton, New Jersey.

PDR. 1998. *Physicians desk reference,* 52nd ed. Montvale, N.J., USA: Medical Economics Company.

4

Insoluble Salt Forms and Drug Complexes

Judy H. Senior
Maxim Pharmaceuticals
San Diego, California

One of the most straightforward ways to create a depot sustained-release formulation of a water-soluble drug is to use a water-insoluble salt form to make an insoluble drug complex. This concept is not new. Pharmaceutical suspensions have been used for many decades in the controlled release of medicines, most notably for optimizing the duration of insulin activity and more recently for local relief of inflammation using corticosteroids. It is presumed that the insoluble drug depot gradually disintegrates/dissolves at the injection site. Thus, active drug slowly solubilizes out of the complex to enter the systemic circulation and/or provides a reservoir of locally acting drug, e.g., intra-articularly. There are now 30–40 different drug compounds formulated as sterile suspensions with the goal of prolonging drug action and over 100 drug products. The advantages and disadvantages of this approach will be reviewed in this chapter and illustrated by currently marketed products. Some products use excipients to further attenuate the slow release properties of an insoluble drug compound. These excipients include vegetable oil (see Chapter 5) and phospholipids (see Chapter 7). The resulting pharmaceutical suspensions are intended for parenteral administration. However, their relatively large particle size usually precludes their use via the intravenous (IV) route.

GENERAL PHYSICAL AND PHARMACEUTICAL CHARACTERISTICS OF INSOLUBLE DRUG COMPLEXES

Typical physical and pharmaceutical considerations arising for drug products manufactured with the active drug in the form of an insoluble drug complex are described in this section.

Formulation Requirements

The formulation and pharmaceutical development of sterile suspensions and solutions of insoluble drug complexes must meet standard requirements for sterile injectables. In particular, sterile suspensions must be free of particles other than the drug crystals, sterilizable without losing activity, and of controlled potency. Marketed products that successfully use this approach are shown in Table 4.1. Some specific issues with formulating crystals of insoluble drugs into pharmaceutically acceptable products are the ease of resuspension of the particulate matter, the viscosity of the suspension, and content uniformity during testing. An acceptable suspension remains homogeneous for at least long enough to remove and administer a desired dose after agitating the container. Small and uniform particle size generally means that the parenteral suspension will give slow, uniform rates of sedimentation and predictable rates of dissolution and drug release. Uniform particle size reduces the tendency for larger crystal growth during storage. Such a change can cause caking of a suspension, difficult syringeability because of the formation of large particles, and changes in the

Table 4.1. *Water-Insoluble Drug Complexes for Local Sustained-Release in U.S. Pharmaceutical Products[1]*

CORTICOSTEROIDS

Drug Product[2]	Route	Dosage & Administration	Formulation[3]
Betamethasone sodium phosphate & beta-methasone acetate e.g., Celestone® Soluspan® (Schering)	IM, IA, IS, IL (intradermal, not SC), soft tissue	IM[1] 0.5–9 mg daily; IA, IS: large joint 6–12 mg, small joint 1.5–6 mg, tendon sheath 1.5–3 mg. Used as an anti-inflammatory or immunosuppressant agent.	Betamethasone sodium phosphate: 3 mg betamethasone with 3 mg betamethasone acetate per mL, pH 6.8–7.2, with disodium edetate & benzalkonium chloride
Cortisone acetate, e.g., Cortone® acetate (Merck)	IM	25–300 mg daily. Typically 12 h intervals between dosing. The drug is slowly absorbed from the injection site over 24–48 h. Used in replacement therapy in patients with adrenocortical insufficiency.	50 mg/mL: pH 5–7, with benzyl alcohol (9 mg/mL)
Dexamethasone acetate, e.g., Dalalone® (Forest); Decadron® (Merck)	IM, IA, IL, soft tissue	8–16 mg IM; 0.8–1.6 mg IL; 4–16 mg IA or soft tissue. Repeat doses 1–3 weeks as required. Used primarily as an anti-inflammatory or immuno-suppressant agent.	8 mg/mL, 16 mg/mL: pH 5–7.5, with benzyl alcohol and/or sodium bisulfite

Table 4.1 continued on next page.

Table 4.1 continued from previous page.

CORTICOSTEROIDS

Drug Product[2]	Route	Dosage & Administration	Formulation[3]
Hydrocortisone acetate e.g., Hydrocortone® acetate (Merck)	IA, IS, IB, IL, soft tissue	Large joint (e.g., knee) 25–50 mg; small joint 10–25 mg; bursae 25–50 mg; ganglia 10–25 mg; injection into soft tissue 5–12.5 mg; tendon sheath inflammation 25–75 mg. For soft tissue infiltration, injection may be repeated once every 3–5 days (bursae) & every 1–4 weeks (joints). Used as an anti-inflammatory agent.	25 mg/mL, 50 mg/mL: pH 5–7 with benzyl alcohol
Methylprednisolone acetate, e.g., Depo-Medrol® (Pharmacia & Upjohn); Depoject® (Merz)	IM, IA, IL or soft tissue; NOT intrathecal	IM 10–80 mg; IA, IL, soft tissue: large joint 20–80 mg; small joint 4–40 mg: bursa, ganglia, soft tissue, 4–30 mg. Repeat every 1–5 weeks. Used primarily as an anti-inflammatory agent.	20 mg/mL, 40 mg/mL, 80 mg/mL: pH 3.5–7, with benzyl alcohol
Prednisolone acetate, e.g., Key-Pred® (Hyrex)	IM, IA, or soft tissue	IM, 4–60 mg daily at 12 h intervals. Used primarily as an anti-inflammatory or immuno-suppressant agent.	25 mg/mL, 50 mg/mL: pH 5–7.5, with benzyl alcohol

Table 4.1 continued on next page.

Table 4.1 continued from previous page.

CORTICOSTEROIDS

Drug Product[2]	Route	Dosage & Administration	Formulation[3]
Triamcinolone acetonide, e.g., Tac® (Herbert); Kenalog® (Westwood)	IM, IA, IS, IL, (intradermal) or SL, soft tissue, or oral inhalation	IM: 60 mg, additional doses 20–100 mg (usually 40–80 mg) usually at 6-week intervals; SL: 10 mg/mL or 1 mg per injection site 1 or more times per week; IA, IS, soft tissue: 2.5–40 mg; large joint 15–40 mg; small joint 2.5–10 mg; tendon sheath inflammation, 2.5–10 mg. Used primarily as an anti-inflammatory agent.	3 mg/mL, 10 mg/mL, 40 mg/mL: pH 5–7.5, with benzyl alcohol
Triamcinolone diacetate, e.g., Aristocort® (Fujisawa)	IM, IA, IS, IL (intradermal) SL, or soft tissue injection	IM: 40 mg once per week using 40 mg/mL dosage form; IL & SL: 25 mg/mL total dose 5–48 mg depending on size, location, and type of lesion. 2 or 3 injections at 1- to 2-week intervals. Used primarily as an anti-inflammatory agent.	25 mg/mL, 40 mg/mL: pH 4.5–7.5, with benzyl alcohol
Triamcinolone hexacetonide, e.g., Aristospan® (Fujisawa)	IA, IL, SL	IL or SL: 0.5 mg per square inch of affected skin; IA: 2–20 mg depending on size of joint. Repeat dosing every 3–4 weeks. Used primarily as an anti-inflammatory agent.	5 mg/mL, 20 mg/mL: microcrystalline, with benzyl alcohol

Table 4.1 continued on next page.

Table 4.1 continued from previous page.

STEROID HORMONES/CONTRACEPTIVES

Drug Product	Route	Dosage & Administration	Formulation
Estradiol cypionate, e.g., Depo®-Estradiol (Pharmacia & Upjohn)	IM	1–5 mg every 3–4 weeks for management of estrogen deficiency. 1.5–2 mg per month for replacement therapy in female hypogonadism.	1 mg/mL, 5 mg/mL: in oil (e.g., cottonseed) with chlorobutanol
Estradiol cypionate combinations, e.g., depAndrogyn® (Forest)	IM	1–5 mg every 3–4 weeks, for management of estrogen deficiency. 1.5–2 mg per month, for replacement therapy in female hypogonadism.	2 mg/mL with testosterone cypionate; 50 mg/mL in oil with chlorobutanol
Estradiol valerate, e.g., Delestrogen® (Bristol-Myers Squibb)	IM	10–20 mg every 4 weeks as necessary, for management of moderate to severe vasomotor symptoms of the menopause. 30 mg or more every 1–2 weeks, for palliative treatment of advanced prostate cancer.	10 mg/mL, 20 mg/mL, 40 mg/mL: in oil with chlorobutanol
Estradiol valerate combinations with testosterone, e.g., testosterone enanthate and estradiol valerate injection (Steris, Taylor)	IM	10–20 mg every 4 weeks as necessary, for management of moderate to severe vasomotor symptoms of menopause. 30 mg or more every 1–2 weeks, for palliative treatment of advanced prostate cancer.	4 mg/mL with testosterone enanthate 90 mg/mL in oil with chlorobutanol

Table 4.1 continued on next page.

Table 4.1 continued from previous page.

STEROID HORMONES/CONTRACEPTIVES

Drug Product	Route	Dosage & Administration	Formulation
Estrone e.g., Estrone® (Keene)	IM	0.1–0.5 mg 2 or 3 times weekly, for management of moderate to severe vasomotor symptoms of menopause. Up to 2 mg weekly, for management of female hypogonadism, etc. 2–4 mg 2 or 3 times a week, for palliative treatment of advanced prostate cancer.	2 mg/mL, 5 mg/mL: may include benzyl alcohol, benzethonium chloride, parabens, povidone, or combinations thereof
Hydroxyprogesterone caproate, e.g., Hy-Gestrone® (Taylor)	deep IM	Usual dosage: single 375 mg dose may be repeated at 4-week intervals.	Sterile solution in vegetable oil (castor or sesame); 125 mg/mL, 250 mg/mL: with benzyl alcohol and benzyl benzoate
Progesterone	IM	Sterile suspensions no longer commercially available (cause pain at the injection site).	Sterile solution. 50 mg/mL in oil
Medroxyprogesterone acetate, e.g., Depo-Provera® (Pharmacia & Upjohn)	deep IM, into gluteal or deltoid muscle	The dosage is usually 150 mg every 3 months for female contraception. Other uses may require IM administration of up to 500 mg weekly.	Sterile suspension: 150 mg/mL, 400 mg/mL; may include polysorbate 80 and parabens

Table 4.1 continued on next page.

Table 4.1 continued from previous page.

INSULINS

Drug Product	Route	Dosage & Administration	Formulation
Pork insulin combinations "purified," e.g., Mixtard® (Novo Nordisk)	SC (may be given IM in diabetic ketoacidosis)	Variable. A typical adult dose may be 5–10 units of insulin SC before meals and at bedtime. Dose and frequency must be carefully individualized.	30 units/mL with isophane insulin; 70 units/mL with phenol; ≤ 10 ppm proinsulin
Insulin human combinations (recombinant DNA origin), e.g., Humulin® 70/30 (Lilly)	SC	Variable. A typical adult dose may be 5–10 units of insulin SC before meals and at bedtime. Dose and frequency must be carefully individualized.	Human insulin (regular) 30 units/mL with isophane insulin Hu 70 units/mL or Human insulin (regular) 50 units/mL with isophane insulin Hu 50 units/mL
Insulin human (regular) (recombinant DNA origin), e.g., Novolin® R Penfill (Novo Nordisk)	SC	The NovolinPen® insulin delivery device is designed to deliver 2–36 units of insulin in 2-unit increments.	Sterile suspension: delivers 100 units/mL via NovolinPen® or other compatible device
Extended insulin human zinc (recombinant DNA origin), e.g., Humulin® U Ultralente (Lilly)	SC (not IV)	May be used in the morning or twice daily in conjunction with doses of rapid-acting insulin before each meal.	100 units/mL

Table 4.1 continued on next page.

(Table 4.1 continued from previous page)

INSULINS

Drug Product	Route	Dosage & Administration	Formulation
Insulin human zinc (re-combinant DNA origin), e.g., Humulin® L (Lilly); Novolin® L (Novo Nordisk)	SC	The usual adult dose of insulin is 7–26 units given as a single dose 30–60 min before breakfast. A second, smaller dose may be given 30 min before supper or at bedtime.	100 units/mL; may have methylparaben
Isophane insulin human (recombinant DNA origin), e.g., Humulin® N (Lilly); Novolin® N (Novo Nordisk)	SC	See under insulin human zinc.	100 units/mL; may have cresol and phenol
Isophane insulin human (semisynthetic), e.g., Isulatard Human N (Novo Nordisk)	SC	See under insulin human zinc.	100 units/mL; with cresol and phenol
Insulin, isophane (pork/beef or pork origin) "single peak" or "purified"	SC	See under insulin human zinc.	100 units/mL for single or purified; combination has 70 units/mL regular insulin plus 30 units/mL isophane
Insulin zinc "single peak" or "purified," e.g., Insulin, Lente® Purified Pork (Novo Nordisk)	SC	See under insulin human zinc.	100 units/mL; with methylparaben

Table 4.1 continued on next page.

Table 4.1 continued from previous page.

INSULINS

Drug Product	Route	Dosage & Administration	Formulation
Insulin Lispro (recombinant DNA origin), e.g., Humalog® (Lilly)	SC	Rapid or short-acting dosing. May be used with a pump for continuous SC infusion.	Sterile solution: 100 mg/mL; with cresol, glycerin, and zinc oxide

MENTAL DISORDERS

Tranquilizers/Antipsychotics

Drug Product	Route	Dosage & Administration	Formulation
Fluphenazine decanoate, e.g., Fluphenazine decanoate injection (Fujisawa, Pasadena)	IM, SC	The usual initial adult IM or SC dose in management of patients requiring prolonged parenteral antipsychotic therapy is 12.5–25 mg.	Oily solution: 25 mg/mL, with benzyl alcohol 1.2% in sesame oil
Fluphenazine enanthate, e.g., Prolixin Enanthate® (Apothecon)	IM, SC	The usual initial adult IM or SC dose in management of patients requiring prolonged parenteral antipsychotic therapy is 25 mg every 2 weeks.	Oily solution: 25 mg/mL, with benzyl alcohol 1.5% in sesame oil
Haloperidol decanoate, e.g., Haldol® (Ortho-McNeil/Scios Nova)	IM only	Dosage based on 10–15 times the oral dosage not exceeding 100 mg. Usually administered at monthly intervals.	Oily solution: 50 mg/mL, 100 mg/mL; with benzyl alcohol 1.2% in sesame oil

Table 4.1 continued on next page.

Table 4.1 continued from previous page.

MENTAL DISORDERS

Drug Product	Route	Dosage & Administration	Formulation
Thiothixene HCl, e.g., Navane® intramuscular (Roerig)	IM only	For management of mild to moderate psychotic conditions in patients unable or unwilling to take oral medication. Typical dosing is 4 mg 2–4 times per day.	5 mg/mL; contains mannitol
Anticonvulsants			
Magnesium sulfate, e.g., Magnesium sulfate injection (Abbott)	IM	Dose is administered in conjunction with the administration of the IV dosage form (e.g., for management of severe preeclampsia or eclampsia). Typical IM dosing of 4–5 mg uses undiluted 50% dosage form.	10%, 12.5%, 50%: microcrystals, pH 3.5–7

1. Information in this table is compiled from the *AHFS Drug Information* (1998) and *Mosby's Complete Drug Reference* (7th ed., 1997).

2. Many drug substances shown in this table are available by nonproprietary name (i.e., the compounds may also be formulated by the physician or pharmacist).

3. Unless shown otherwise, products are formulated as injectable sterile suspensions.

Abbreviations: IA: intraarticular; IB: intrabursal; ID: intradermal; IL: intralesional; IM: intramuscular; IS: intrasynovial; IV: intravenous; SC: subcutaneous; SL: sublesional drug release rate from the depot after injection. Stabilization of a suspension for the period between manufacture and use may overcome problems with settling and caking. For example, surface active agents such as polysorbate, added early in the process, can stabilize the product by reducing interfacial surface tension between the particles and the vehicle. The most common additives in sterile suspensions are preservatives such as benzyl alcohol, the parabens, or sodium sulfite.

Other key pharmaceutical considerations concern the importance of wetting particles, particle interaction and behavior, and the nature of the crystals formed as a result of the precipitation method used to prepare them (Lachman et al. 1976; Smith et al. 1970).

Considerations in Sterile Manufacture

The manufacture of sterile suspensions offers a range of challenges in managing the flow of suspended particles through the production process to ensure the ease of product syringeability, i.e., the ease with which the product is passed in and out of the syringe, in the clinic (Lachman et al. 1976). This is mostly determined by the viscosity of the product, which is affected by the amount of solids present and the nature of the vehicle. The solids content of parenteral suspensions usually ranges between 0.5 and 5 percent, but may be as high as 30 percent. The particle size and shape of insoluble drug complexes may also affect viscosity. An example of particle size and shape in a commercial formulation of an insoluble drug complex is the approximately 1–5 μm platelike crystals shown in scanning electon micrographs of the corticosteroid triamcinolone hexacetonide (Gordon and Schumacher 1979). Since the shape of crystals of insoluble drugs may vary from a box shape to long thin rods, each form will have its own particular manufacturing problems, especially those shapes that are not conducive to smooth flow in manufacturing processes, which contributes to high viscosity. Finally, the particle size and/or the heat sensitivity of the crystal form controls whether aseptic manufacture will be selected for a sterile drug suspension rather than terminal sterilization by heat.

Pharmacokinetics, Dosage, and Therapeutic Applications

The pharmacokinetics (PK) of water-insoluble salt forms can differ dramatically from the water-soluble parent compound and may also differ in terms of duration of biological activity. For example, Table 4.2 illustrates the differing clinical response (duration of suppression of hypothalamic-pituitary-adrenal [HPA] axis) of two different salt forms of the corticosteroid triamcinolone. Administered intramuscularly in the same dose range, the acetonide salt form of triamcinolone shows a two-to four-week duration of suppression versus a one-week duration for the diacetate. The acetate form of another corticosteroid, methylprednisolone, when administered in a similar dose range, has a duration of suppression of only four to eight days. This is still a much longer duration of response than either parent drug given orally at similar doses (Table 4.2). These differences in duration and type of biological response allow the

Table 4.2. *Comparison of the Duration of Action of Salt Forms of Corticosteroids*[1]

Drug	Dose	Duration of Suppression[2]
Salt form administered as single dose, IM		
Triamcinolone acetonide	40–80 mg	2–4 weeks
Triamcinolone diacetate	50 mg	1 week
Betamethasone mixture	9 mg	1 week
Methylprednisolone acetate	40–80 mg	4–8 days
"Parent" drug administered as single oral dose		
Triamcinolone	40 mg	2 days
Betamethasone	6 mg	3 days
Methylprednisolone	40 mg	1–1.5 days

[1]Compiled from *AHFS Drug Information*, 1998.

[2]Duration of suppression of the HPA axis is measured by response to corticotropin or cosyntropin tests.

physician flexibility in dosing, which is especially useful in inflammation and hormone replacement therapy. The existence of multiple products with different clinical utility but using the same active ingredient may be confusing, however, where product names are similar. Many companies clarify which products are intended for sustained-release applications by using descriptive names such as short-acting, "lente," and so on for insulins.

Toxicity and Safety Considerations

Insoluble drug complexes or salts are likely to have modified toxicokinetic behavior in comparison with the parent drug. Thus, the PK of drug release into the blood and both local and systemic toxicology must be studied for the sustained-release salt formulation. There is a possibility of local irritation caused by the presence of microcrystals, although in practice this is not usually a concern. Similarly, it is unusual for there to be any specific issues with allergic responses to the drug in insoluble form.

A SURVEY OF COMMERCIAL AND EXPERIMENTAL INSOLUBLE DRUG PRODUCTS

A review of commercial products that are intended for sustained release after parenteral administration shows that an insoluble salt form of the drug is typically formulated as a sterile suspension. The following therapeutic areas predominate: anti-inflammatories such as corticosteroids, contraceptive hormones, hormone supplementation, and therapies for mental disorders (Table 4.1). This chapter uses examples of marketed drugs to illustrate how insoluble drug complexes can be formulated, variations in the PK of different salts, the range of therapeutic uses and doses according to the route of administration, the in vivo fate of certain complexes given intramuscularly or subcutaneously, and, finally, some examples of applications and products in development.

Formulation Characteristics

Examples of products that are formulated as sterile suspensions of the microcrystalline salt form of the drug include corticosteroids, insulins, steroid hormones, and drugs to treat mental disorders.

Corticosteroids

Sustained-release versions of corticosteroids are particularly well represented in currently marketed products formulated as sterile suspensions of the water-insoluble salt forms of the parent drug. To illustrate the popularity of sustained-release salt forms of corticosteroids, the 1998 edition of the "Red Book" (AHFS 1998) lists some 40 versions of 9 drug products offered by 16 companies. The corticosteroid drug compounds listed are formulated as sterile suspensions of the acetate salts except for triamcinolone, which is available as acetonide, diacetate, or hexacetonide salts. Formulations may also combine a faster-acting or immediate-acting soluble form of the drug with an insoluble salt designed to sustain the effect of the drug. For example, Celestone® Soluspan® consists of synthetic glucocorticoid betamethasone as a mixture of the water-insoluble acetate salt and the water-soluble sodium phosphate salt (Table 4.1).

Insulins

The formulation of insulin, a peptide hormone, uses a different approach to forming the insoluble drug complex. Insulins for injection are manufactured by precipitating the peptide in the presence of zinc chloride to form zinc insulin crystals. The resulting drug products fall into four categories: rapid-acting, short-acting (regular), intermediate-acting, and long-acting forms. The long-acting and intermediate-acting formulations of the peptide hormone comprise some 10 to 11 types, of which 8 are termed "lente," "semilente," or "extended" (Table 4.1). How are the intermediate-acting and long-acting versions formulated to provide greater sustained release of active peptide after subcutaneous (SC) injection? The intermediate-acting isophane insulin, for example, combines protamine sulfate with a sterile suspension of zinc insulin crystals in buffered Water for Injection (WFI) so that the solid phase of the suspension consists of crystals composed of insulin, protamine, and zinc. The product appearance is a cloudy or milky suspension of rod-shaped crystals free from large aggregates of crystals following moderate agitation. When examined microscopically, the insoluble material in the isophane insulin suspension is crystalline, with not more than a trace amount of amorphous material. It contains an amount of zinc similar to that of regular insulin (10–40 μg of zinc) and with a similar shelf life when extremes of temperature and direct sunlight are avoided. Extended insulin zinc such as "Ultralente®" versions are also prepared by the addition of zinc in the form of zinc chloride so that the solid phase of the suspension is predominantly crystalline. However, the milky or cloudy suspension has a mixture of characteristic crystals of predominantly 10–40 μm in maximum dimension. Each 100 USP units of extended insulin zinc contains 200–250 μg of zinc. These preparations may also contain 0.15–0.17 percent sodium acetate, 0.65–0.75 percent sodium chloride, and 0.09–0.11 percent methylparaben. Typical insulin-based drug products are formulated at pH 7.0–7.8 and have a shelf life of 24 months from date of manufacture when stored at 2–8°C.

Steroid Hormones

A third class of compounds where insoluble salt forms of the drug are commonly used in sustained release are the steroid hormones such as estradiols, medroxyprogesterone, and so on. In most cases, these drug crystals are combined with an oil to maximize the sustained release of the drug. Examples include estradiol cypionate in oil for depot parenteral injection (see Chapter 5), estradiol cypionate combinations (with testosterone) and estradiol valerate, and estradiol valerate combinations (with testosterone). Typical formulations use cottonseed oil as the suspending medium, although at least one product, with polyestradiol phosphate as the active ingredient, is formulated with propylene glycol as a viscosity enhancer. The use of a surface active agent in these products is illustrated by Depo-Provera®, which contains Tween 80®. Antioxidants used in these products include chlorobutanol and the parabens.

Drugs for the Treatment of Mental Disorders

Finally, certain drugs designed to facilitate the treatment of mental disorders, such as the tranquilizer fluphenazine and the anticonvulsant magnesium sulfate have been formulated as insoluble salts. Fluphenazine can be formulated as fluphenazine decanoate or fluphenazine enanthate in oily suspension, and magnesium sulfate is formulated as crystals of the salt suspended in dextrose solution for intramuscular (IM) administration (Table 4.1). Recently, the antipsychotics haloperidol decanoate and thiothixene hydrochloride have also been formulated for deep IM injection.

Comparative Pharmacokinetics, Typical Dosage, and Administration

The corticosteroids, insulins, and steroid hormones are used to address the following questions:

- For the same indication and route of administration in a given dose range, which salt forms act the longest?

- For which therapeutic applications does sustained release make a big difference in PK versus the soluble or oral form?

- Is there any reduction in the overall dose of drug required for treatment, i.e., a drug sparing effect, when an insoluble salt form is administered locally rather than a soluble form administered systemically?

- What are typical toxicity issues with the sustained release salt form of a drug?

- Can the dosing and dosage form be tailored to the therapeutic indication for an insoluble drug complex?

Salt Forms That Are Long Acting for a Given Indication, Route of Administration, and Dose Range

The longest-acting corticosteroid is triamcinolone acetate, which is typically administered intramuscularly at 6-week intervals as an anti-inflammatory agent in the dose range of 40–80 mg. Other corticosteroids can provide anti-inflammatory effects from 1 to 5 weeks after IM injection. For example, dexamethasone acetate at doses of 8–16 mg is administered every 1–3 weeks, and 10–80 mg of methylprednisolone acetate is administered every 1–5 weeks. Other routes of administration, such as local injection into joints, bursae, ganglia, soft tissue, and so on, may be subject to dosing limitations imposed by injection site volume. For example, up to 80 mg of methylprednisolone acetate may be administered every 1–5 weeks if the local injection site or compartment is big enough (e.g., a large joint), a small joint may only allow up to 40 mg to be dosed, with bursa, ganglia, and soft tissue up to 30 mg. Another anti-inflammatory agent, triamcinolone hexacetonide, may be administered intra-articularly or intralesionally every 3–4 weeks at a dose of 0.5–20 mg depending on the size of the compartment.

However, the insoluble drug complex that provides the longest biological activity is the steroid hormone medroxyprogesterone acetate for use in contraception. Depo-Provera®, formulated as an oily suspension, is typically given at a dose of 150 mg every 3 months. Other hormones formulated as microcrystals in the presence of oil are not so long-acting, e.g., estrone, progesterone, but are intended for other therapeutic uses (Table 4.1) and do not necessarily benefit the patient through such prolonged duration of action.

Therapeutic Consequences of Sustained-Release Salt Forms Versus the Soluble or Oral Form

The active ingredient of Depo-Provera®, medroxyprogesterone acetate, is also available in an oral tablet form for once-a-day dosing for contraceptive use. The 3-month formulation provoked much debate when it first became available, prompted by concerns over its safety and the irreversibility of administering a depot form of the drug. These issues still raise concerns for some users, although for many others the formulation remains a useful option for reasons of compliance and economics. Another therapeutic indication where local sustained release can reduce the dosing frequency and systemic exposure to the drug is the treatment of localized painful inflammation of joints, bursae, ganglia, and so on. For example, prednisolone is administered intravenously or intramuscularly *daily* at 4–60 mg as the sodium phosphate salt. When the much less soluble tebutate salt is used, a dosage of 4–40 mg is only required at *2–3 week intervals* for a large joint. By contrast, oral corticosteroids have a much shorter duration of action (Table 4.2).

Is There a Drug Sparing Effect When an Insoluble Salt Form Is Administered Locally?

If a drug can be administered in a sustained-release form, there is a good chance that the overall systemic exposure to the drug will be reduced since the peak and trough drug levels associated with daily or more frequent dosing are avoided. For insoluble drug complexes, such advantages are dependent on the drug, salt form, and formulation selected. An example of an apparent dose sparing effect is provided by the fluphenazine salts. These salts are used for the symptomatic management of psychotic disorders, e.g., schizophrenia, where patients cannot be relied upon to take oral medication. Fluphenazine hydrochloride, the water-soluble form of the drug, is given at one-third to one-half of the oral dose when administered intramuscularly. For sustained release, the longer-acting enanthate and decanoate salts of fluphenazine can be used, which are highly insoluble in water. These salts are formulated in oil for IM administration and typically dosed at 25 mg every *week* rather than 20 mg of the hydrochloride salt *daily*. There is clearly a dose-sparing effect here.

In contrast, methylprednisolone acetate is given at a total weekly dose of 10–80 mg regardless of whether the drug is given orally or intramuscularly. IM dosing itself promotes a sustained response to the medication after a single injection consisting of seven times the *daily* dose of methylprednisolone acetate. Thus, no dose-sparing occurs in this example. When the same drug is given locally at the same dose range, the anti-inflammatory response may be much greater than if the drug had been administered intramuscularly or orally. A typical dose range for methylprednisolone acetate is 20–80 mg once per week into a large joint (small joint, 4–40 mg).

Typical Toxicity Issues Specific to the Sustained-Release Salt Form of a Drug

Toxic effects were observed for a crystalline depot of hydrocortisone acetate (McCarty and Hogan 1964), where the crystalline depot of corticosteroid preparations was shown to occasionally cause crystal-induced arthritis after intra-articular (IA) administration. However, corticosteroid therapy administered intra-articularly has been used successfully in the clinic since the 1950s to ameliorate nonseptic local joint inflammation (Hollander et al. 1951). The anticipated side reactions include the rare introduction of infection into the joint and a steroid crystal-induced postinjection flare (Hollander 1970). More recently, local toxicity was studied for lecithin-coated bupivacaine (Kuzma et al. 1994) and tetracaine (Kline et al. 1992) crystals. No toxic effects were observed in a rat model where the crystal suspensions were injected intradermally (Kuzma et al. 1997).

The main problems with the long-lasting dosage forms result if the patient responds poorly to the medication, whether the drug is formulated as an insoluble complex or if a delivery system is being used. An example of a sustained-release form with which there is now wide experience is Depo-Provera® (medroxyprogesterone acetate). The medication typically persists at the local administration site for 3 months. However, the drug is available in oral form, so it is theoretically possible that the patient could be given the longer-acting dosage once response to the medication is established. (Issues of toxicity and biocompatibility are discussed further in Chapter 15.)

Tailoring the Dosing and Dosage Form to the Therapeutic Indication for an Insoluble Drug Complex

Systemic absorption occurs slowly and is more sustained following IA, intralesional, or soft tissue injection of glucocorticoids and many other insoluble drug complexes. This allows versatility in responding to the clinical requirements for corticosteroids (Table 4.2) and other drugs simply by varying the route of administration and the salt form used. Dexamethasone acetate is absorbed relatively slowly from IM injection sites, with peak plasma concentrations about 8 h after administration. Therefore, IM injection of dexamethasone acetate suspension is not indicated when an immediate effect of short duration is required. In that case, a water-soluble glucocorticoid ester can be administered intravenously in the range of 0.5–24 mg daily. The usual initial adult IM dose of dexamethasone acetate is 8–16 mg of dexamethasone, and additional doses are given at intervals of 1–3 weeks if necessary. For intralesional injection, the usual dose of dexamethasone acetate is 0.8–1.6 mg per injection site. For IA or soft tissue injection, the usual dose is 4–16 mg depending on the location of the affected area and the degree of inflammation, with doses repeated at intervals of 1–3 weeks as required.

Another approach to tailoring the dosage to the therapeutic indication is to combine longer-acting and fast-acting forms into one product. In the case of Celestone® Soluspan®, the sustained-release salt form, betamethasone acetate, is formulated with the soluble salt form, betamethasone sodium phosphate, to provide a product with combined therapeutic activity after IM injection. The combination of a water-soluble and water-insoluble salt form of the drug allows both rapid uptake and sustained release such that anti-inflammatory effects may appear within 1–3 h at a dose of 9 mg, which persists for up to 7 days after IM injection versus 3.25 days for a single, orally administered dose of 6 mg betamethasone (Table 4.2). The same combination of salts can result in anti-inflammatory effects that persist for 1–2 weeks after administration into knee joints, for example (Table 4.2).

The insulins provide excellent examples of how insoluble complexes of the peptide hormone can be tailored to provide different release rates of drug into the bloodstream after SC injection. The different types within each category (regular vs. extended forms) are shown in Table 4.3. The short-acting or regular form is typically administered to a diabetic patient in response to emergency treatment of diabetic ketoacidosis or coma, to initiate therapy in patients with diabetes mellitus when rapid control is required, and in combination with intermediate-acting or long-acting insulin preparations to give better control of blood glucose concentrations in diabetic patients. The intermediate-acting isophane insulin has a typical onset of action of 1–2 h, a peak of 4–12 h, and a duration of 18–24 h. By contrast, a long-acting insulin such as extended insulin "Ultra-lente®" versions have a typical onset of action of 4–8 h, with a peak at 10–30 h and duration of 36 h.

Typical In Vivo Fate of Insoluble Drug Complexes

What happens to the depot of insoluble drug at the local injection site after administration? Does the drug gradually dissolve out of the complex, or is the complex removed from the injection site by cellular activity or migrate into the lymphatics? Is the insoluble complex likely to remain at the injection site for

Table 4.3. *A Matrix of Commercial Insulin Types*[1]

This matrix illustrates the wide selection of insulin types that can be used to tailor to the needs of patients. Factors such as the requirement for rapid action is met by the short or rapid-acting and regular insulins, whereas combination, intermediate, and longer-acting forms can be used in well-controlled patients. Other variations (see Table 4.1) allow changes to be made to accommodate any hypersensitivity reactions that might arise over periods of extended use of insulins from a particular source.

Sources	Type	Duration of Action (h)
Human, pork, beef/pork,	Rapid-acting	3–6.5
	lispro	
Recombinant DNA origin	Regular	6–10
Semisynthetic	beef & pork (single peak)	
	pork (purified)	
	Intermediate-acting	16–24
	zinc	
	isophane	
	Extended-acting	24–28

[1]Compiled from *AHFS Drug Information*, 1998, pp. 2567–258.

an extended period? Some of these questions are addressed by following the fate of depot corticosteroid crystals in vivo after IA injection (Gordon and Schumacher 1979). The insoluble drug complex in question was composed of crystals formed from the branched chain ester of prednisolone and triamcinolone. Either 20 mg of prednisolone tebuate (Hydeltra®, MSD) or 20 mg of triamcinolone hexacetonide (Aristospan®, Lederle) was injected into the knee joints of seven patients with osteoarthritis and painful effusions. Scanning and transmission using electron microscopy revealed intra- and extracellular crystals 24 h and 1 week after the initial injection, although a depot of IA corticosteroid crystals has been seen for up to one month (Kahn et al. 1970). Physical changes in the crystals can be detected after the drug has been in vivo using scanning electron microscopy. No change in the structure of extracellular or intracellular (phagocytosed) crystals was observed for crystals taken from 6 patients 24 h after injection (data was not shown for longer times).

Is there any experimental basis for choosing a particular formulation or salt form to attempt to tailor the drug's sustained-release properties? The general principle seems to be that if a drug is less water soluble, the longer it will persist at the injection site. This is illustrated by a study examining the differences in the long-acting properties of two salts to determine the rate limiting step for their release into and persistence in the blood. Long-acting esters of fluphenazine base, the decanoate and the enanthate were used in combination with sesame oil to compare the relative release rates of ^{14}C-fluphenazine decanoate, ^{14}C-fluphenazine enanthate, and ^{14}C-fluphenazine base when each compound was separately administered intramuscularly in a dog model (Dreyfuss et al. 1976). The overall conclusions were that ^{14}C-fluphenazine decanoate was released from the injection site more slowly than the enanthate ester whereas the ^{14}C-fluphenazine base did not produce any slow-release characteristics. A large fraction of the dose was released from the injection site in the first 12 h after dosing. The slow-release characteristics produced by the enanthate and decanoate esters of fluphenazine base appear to be a consequence of their relative rates of diffusion into the circulation from an injection site. Since both esters are rapidly cleaved once in the circulation in the dog model, metabolism in the plasma does not appear to be as important as the relative rates of diffusion away from the injection site and into the circulation.

Experimental Approaches and Future Products

Sustained-Release Formulations of Long-Acting Anesthetics

Parenterally administered local anesthetics are used for infiltration and nerve block anesthesia (Boedeker et al. 1994; Korsten et al. 1991; Korsten and Grouls 1996), which is achieved by injecting the local anesthetic epidurally, intradermally, subcutaneously, or submucosally across the paths of nerves supplying the area to be anesthetized. Compounds that have been formulated into long-acting anesthetic versions include tetracaine, bupivacaine, n-butyl-p-aminobenzoate (BAB), and others (Kuzma et al. 1997). As an example, tetracaine microcrystals were encapsulated in 10 percent lecithin to look for prolonged therapeutic effects versus tetracaine in solution and phosphatidylcholine (PC; lecithin) without the drug (Boedeker et al. 1994). The microcrystals were made using tetracaine-HI crystals formed from tetracaine HCl using potassium iodide, and the encapsulation in PC was achieved by controlled sonication. The study found that the anesthesia produced by the PC–coated tetracaine microcrystals was prolonged versus the soluble drug form, and no toxic effects were observed. In other approaches, microdroplets were prepared by encapsulating volatile anesthetics such as methoxyflurane in PC. These particles were administered intradermally to provide local anesthesia in rats and human skin for 24 h (Haynes 1992). Also, a suspension of an insoluble local anesthetic, 10 percent BAB in a Tween 80® suspension in saline, was successfully given by epidural administration to patients with intractable cancer pain (Korsten et al. 1991).

Sustained-Release Contraceptive Utility for Steroid Hormones

Long-acting injectable contraceptive formulations of norethindrone (NET) and mestranol (ME) were evaluated in a clinical study (Diaz-Sanchez et al. 1987) conducted for one month after a single IM injection of synthetic steroid crystals (Garza-Flores et al. 1984). The subjects received a combination of 10 mg NET and 1 mg of ME by intragluteal (deep IM) injection. Two particle ranges were evaluated (< 50 μm and 125–177 μm), which showed different dissolution rates in vitro, i.e., the smaller particles dissolved faster in the dissolution medium of ethanol water (35:65 v/v). The data demonstrate that the combined administration of 5.09 ± 0.84 mg of NET plus 0.54 ± 0.06 mg ME inhibited ovulation and corpus luteum formation for 1 month in all subjects. The estimated 24 h release of drug from the injection site was 250 μg of NET and 25 μg of ME, which is similar to the daily intake of low-dose oral combined contraceptives.

Improving the Effectiveness of a Diagnostic Peptide

The compound pentagastrin (Peptavalon®) is commercially available as a parenteral injection dosage form in normal saline, adjusted to pH 7–8 with ammonium hydroxide and/or hydrochloric acid. The peptide (which has 4 amino acids) is used diagnostically to evaluate gastric acid secretion. Peptavalon stimulates gastric acid secretion approximately 10 min after SC injection, with a peak response occurring 20–30 min later and a duration of activity of 60–80 min (Wai et al. 1970).

To evaluate whether a sustained-release form of the peptide could alter the gastric acid secretion profile in a dog model, 4 formulations, including 2 different types of crystals, were designed to release pentagastrin at different rates after SC injection. The formulations were (a) standard formulation in ammonia, (b) standard formulation in ammonia plus propylene glycol, and (c) crystalline pentagastrin formed by adding propylene glycol and normal saline to pentagastrin dissolved in 0.1N ammonia. An equivalent quantity of 0.1N acetic acid was added to neutralize the ammonia, and the solution was stored at 4°C for 24 h to allow the crystals to form. Finally, crystalline pentagastrin was formed as in (c), but 0.1M aluminum potassium sulfate (0.32 mL) was added to the preparation, which was adjusted to pH 5.0 with 0.1N sodium hydroxide, and the volume adjusted to 10 mL with normal saline. The resulting crystals had a particle size distribution of 3–20 μm, measured using a Coulter counter.

Biliary excretion in rats suggests a slower initial release of active drug when administered in any of the crystalline forms. The secretion of gastric acid in dogs after SC injection showed the most pronounced sustained-release effect for crystals prepared as in formulation (d). The secretion of gastric acid in the dog model after SC injection was prolonged to at least 4–5 h whereas the soluble formulations showed little or no activity 2–2.5 h after administration. Thus, the biological response was more than doubled by rendering the peptide into an insoluble drug complex; the particular salt form made a less significant contribution to the duration of response.

ADVANTAGES AND LIMITATIONS OF INSOLUBLE SALTS IN SUSTAINED RELEASE

The conceptually simple approach described in this chapter is relatively widely used in commercial products. The success of formulating a sustained-release dosage form of a parenteral drug by utilizing its water-insoluble salt form or by forming an insoluble complex of the drug, relies primarily on the availability of a suitable salt form of the drug. The second challenge is the pharmaceutical manufacturability of the potential salt form into a drug product. Either of these hurdles will be moot unless the clinical toxicokinetics are suitable to allow product testing for its intended therapeutic purpose.

The advantages of insoluble salt forms in sustained drug release are as follows:

- The formulation approach is simple.

- There may be fewer toxicological concerns due to excipients, such as materials used to create the drug delivery system.

- Pharmaceutical manufacturing concerns are known/predictable.

- PK data are easily interpretable.

The disadvantages of insoluble salt forms in sustained drug release are as follows:

- Applicability is limited to drugs from which suitable insoluble drug complexes and sterile suspensions can be made.

- Some suspension formulations must not be diluted or mixed with other drugs due to adverse effects on the physical nature of the suspension, which could in turn alter the clinical performance.

- Pharmaceutical manufacturing requirements may be unconventional/unpredictable.

- Terminal heat sterilization is not a universal option. Other methods, such as gamma irradiation, must be evaluated on a case-by case basis.

- Sterility assurance needs to be built in via aseptic processing for heat-labile entities.

- PK behavior of the drug must be measured from a practical standpoint. Little can be done to predetermine or tailor the PK of a particular salt form or crystal suspension. Thus, "you get what you get" for each salt form or complex.

FUTURE TRENDS

It is likely that future drug products will continue to be manufactured to take advantage of this relatively simple approach to sustained-release dosing wherever appropriate insoluble salt forms are available and the pharmacological response is adequate. Other approaches that use drug solubility to tailor drug PK involve formulation of the drug in a cyclodextrin or as a prodrug. Although these approaches to date are usually concerned with trying to increase drug solubility, they could theoretically be applied to local sustained delivery of drugs. In this case, the cyclodextrin or prodrug must facilitate the local retention of a drug or be able to target the drug to cellular sites from where the drug can be released.

Cyclodextrins provide a molecular structure in which the drug can be housed and may indirectly have a role in sustained drug delivery, especially when combined with liposomal drug delivery (McCormack and Gregoriadis 1996). Prodrugs have the drug chemically bound to another moiety, usually via a biodegradable linkage that is metabolized in vivo to generate the active drug. The chemical moiety of the prodrug can also be attached to create a pharmacologically inert and nontoxic compound. Once in the body, the prodrug is transformed into the pharmacologically and therapeutically active compound (Giudicelli 1988).

In the future, it is likely that injectable products will combine the benefits of insoluble drug complexes with those of other delivery systems, such as those described elsewhere in this volume.

REFERENCES

AHFS, 1998. *AHFS Drug Information*. Bethesada, Md., USA: American Society of Health System Pharmacists.

Boedeker, B. H., E. W. Lojeski, M. D. Kline, and D. H. Haynes. 1994. Ultralong duration local anesthesia produced by injection of lecithin-coated tetracaine microcrystals. *J. Clin. Pharmacol.* 34:699–702.

Diaz-Sanchez, V., J. Garza-Flores, S. Jimenez-Thomas, and H. W. Rudel. 1987. Development of a low-dose monthly injectable contraceptive system: II. Pharmacokinetic and pharmacodynamic studies. *Contraception* 35:57–68.

Dreyfuss, J., J. M. Shaw, and J. J. Ross, Jr. 1976. Fluphenazine enanthate and fluphenazine decanoate: Intramuscular injection and esterification as requirements for slow-release characteristics in dogs. *J. Pharm. Sci.* 65:1310–1315.

Garza-Flores, J., V. Diaz-Sanchez, S. Jimenez-Thomas, and H. W. Rudel. 1984. Development of a low-dose monthly injectable contraceptive system: I. Choice of compounds, dose and administration route. *Contraception* 30:371–379.

Giudicelli, J. F. 1988. Les promedicaments. *La Presse Medicale* 17:1000–1009.

Gordon, G. V., and H. R. Schumacher. 1979. Electron microscopic study of depot corticosteroid crystals with clinical studies after intra-articular injection. *J. Rheumatology* 6:7–14.

Haynes, D. H. 1992. Phospholipid-coated microcrystals: Injectable formulations of water-insoluble drugs. U.S. Patent 5,091,187.

Hollander, J. L. 1970. Intrasynovial corticosteroid therapy in arthritis. *Md State Med. J.* 19:62–66.

Hollander, J. L., E. J. Brown, R. A. Jessar, and C. Y. Brown. 1951. Hydrocortisone and cortisone injected into arthritic joints. *JAMA* 147:1629–1635.

Kahn, D. B., J. L. Hollander, and H. R. Schumacher. 1970. Corticosteroid crystals in synovial fluid. *JAMA* 211:807–809.

Kline, M. D., B. H. Boedeker, and M. E. Mattix. 1992. Intradermal toxicity of lecithin-coated microcrystalline tetracaine. *Anesthesiology* 77:A800.

Korsten, H. H., and R. J. Grouls. 1996. Long-lasting sensory blockade by an epidurally injected material-based drug delivery system. *Amer. Pain Soc. Bull.* 6:1–4.

Korsten, H. H., E. W. Ackerman, R. J. Grouls, A. A. van Zundert, W. F. Boon, F. Bal, M. A. Crommelin, J. G. Ribot, F. Hoefsloot, and J. L. Sloof. 1991. Long-lasting epidural sensory blockade by N-butyl-p-aminobenzoate in the terminally ill intractable cancer pain patient. *Anesthesiology* 75:950–960.

Kuzma, P. J., M. D. Kline, B. H. Boedeker, and D. H. Haynes. 1994. Tissue toxicity of lecithin-coated bupivacaine microcrystals after intradermal infiltration. *Anesthesiology* 81:A1034.

Kuzma, P. J., M. D. Kline, M. D. Calkins, P. S. Staats. 1997. Progress in the development of ultra-long-acting local anesthetics. *Reg. Anes.* 22 (6):543–551.

Lachman, L., H. A. Lieberman, and J. L. Kanig. 1976. *The theory and practice of industrial pharmacy*, 2nd ed. Philadelphia: Lea & Febiger.

McCarty, D. J., and J. M. Hogan. 1964. Inflammatory reaction after intrasynovial injection of microcrystalline adrenocorticosteroid esters. *Arthritis Rheum.* 7:359–367.

McCormack, B., and G. Gregoriadis. 1996. Comparative studies of the fate of free and liposome-entrapped hydroxypropyl-β-cyclodextrin drug complexes after intravenous injection into rats: Implications in drug delivery. *Biochim. Biophys. Acta* 1291:237–244.

Smith, W. E., J. D. Buehler, and M. J. Robinson. 1970. Preparation and in vitro evaluation of a sustained-action suspension of dextromethorphan. *Pharm. Sci.* 59:776–779.

Wai, K. N., E. E. Gerring, and R. D. Broad. 1970. Formulation and pharmacological studies of a controlled release pentagastrin injection. *J. Pharm. Pharmacol.* 22:923–929.

5

Non-aqueous Solutions and Suspensions as Sustained-Release Injectable Formulations

Sudaxshina Murdan
Alexander T. Florence
Centre for Drug Delivery Research,
School of Pharmacy, University of London
London, United Kingdom

Drugs are formulated in oily vehicles for several reasons: the insolubility of the drug in aqueous media, the irritancy of aqueous formulations of the drug, drug degradation and instability in aqueous media or to secure sustained therapeutic action of the drug following injection. In this chapter, oily solutions and suspensions as sustained-release formulations are discussed. Examples of commercially available, long-acting oil formulations include solutions of the oil-soluble neuroleptics, slow-release hormone preparations for contraception and hormone replacement and depot procaine penicillin G (Table 5.1).

Table 5.1. *Examples of Some Commercially Available Depot Formulations*

Depot Preparation	Oil Used	Recommended Injection Interval	Use
Zuclopenthixol decanoate (Clopixol®, Clopixol Conc.®)	Thin vegetable oil	2–4 weeks	Neuroleptic
Haloperidol decanoate (Haldol Decanoate®)	Sesame oil	4 weeks	Neuroleptic
Pipothiazine palmitate (Piportil Depot®)	Sesame oil	4 weeks	Neuroleptic
Fluphenazine decanoate (Modecate®, Modecate Concentrate®)	Sesame oil	2–5 weeks	Neuroleptic
Flupenthixol decanoate (Depixol®, Depixol Conc.®, Depixol Low Volume®)	Thin vegetable oil	1–4 weeks	Neuroleptic
Norethisterone enanthate (Noristerat®)	Castor oil	8 weeks	Female contraceptive
Testosterone propionate, phenylpropionate and isocaproate (Sustanon®)	Arachis oil	2 weeks	Treatment of androgen deficiency
Testosterone enanthate (Primoteston Depot®)	Castor oil	3–6 weeks for hypogonadism; 2–3 weeks for breast cancer	Treatment of hypogonadism, and breast cancer

Table 5.1 continued on next page.

Table 5.1 continued from previous page.

Depot Preparation	Oil Used	Recommended Injection Interval	Use
Nandrolone decanoate (Deca-Durabolin®, Deca-Durabolin 100®)	Arachis oil	1 week for anaemia; 3 weeks for osteoporosis	Treatment of anaemia and osteoporosis
Hydroxyprogesterone hexanoate (Proluton Depot®)	Castor oil	Weekly during first half of pregnancy	Prevention of abortion in women with a history of habitual abortion
Testosterone propionate (Virormone®)	Ethyl oleate	2–3 times weekly	Replacement therapy and breast carcinoma in post-menopausal women
Testosterone enanthate		1 week	Male contraceptive

Most of these slow-release preparations are administered intramuscularly, the favoured sites being the gluteal, deltoid and vastus lateralis muscles. Subcutaneous (SC) injection is not normally used for non-aqueous preparations due to the pain and irritation they can cause at the injection site. Only a few references to SC administration of oily injectables exist in the literature, generally when the route was studied experimentally. These will be included in this review, where appropriate. To avoid tissue damage with non-aqueous injections, injection volumes must be limited, and, indeed, should not exceed 2 mL for intramuscular (IM) and 1.5 mL for SC administrations. In addition, the injection site should be rotated to prevent scarring, which itself can cause a reduction in tissue vascularity and, consequently, inefficient drug absorption.

Table 5.1 shows the range of durations of the depot effect of some long-acting injectables. Despite the desirability of sustained therapeutic activity, the longest lasting depot may not always be the optimal choice with respect to treatment management and patient acceptability. For example, a long-acting neuroleptic formulation is appropriate for patients on maintenance therapy, while a shorter-acting one may be preferable for those patients with rapidly changing symptoms (Dencker and Axelsson 1996). There is also a high acceptability of monthly injectable contraceptives, even though longer-acting, bimonthly and quarterly agents are available (Toppozada 1977; Abrego de Aguilar et al. 1997). Treatment with the monthly agents can be discontinued more easily if desired, as in the event of side effects. Side-effects generally subside more quickly when shorter-acting contraceptive agents are used (Toppozada 1994).

Depot oily injectables theoretically show the advantages of long-acting formulations, embodied in reduced dosing frequency achieving optimal plasma level profiles, a reduction in the total dose of drug and fewer side-effects, thus providing a more efficient utilisation of drug and improved patient compliance. Such advantages have been especially beneficial in the treatment of schizophrenic patients who often fail to maintain an often complex, daily oral medication regimen once discharged from hospital. The oil formulations in which a lipophilic drug ester is dissolved in a non-aqueous medium such as sesame oil enable a much reduced dosing regimen; in the case of hospitalised patients, staff time is reduced and daily medication administration struggles can be avoided. Hospital stays are also shortened, and following discharge, patients can receive continuous medication and can be followed up in out-patient clinics. Weiden et al. (1995) have suggested that depot neuroleptics may have a role in encouraging patients to attend out-patient clinic programs during the critical in-patient to out-patient transition period. Schizophrenic relapses and readmissions may thus be significantly reduced (Kane 1995; Citrome et al. 1996). Another significant advantage of depot neuroleptic formulations is the lower cost of treatment. Hale and Wood (1996) calculated that an annual savings of £10.5 million could be achieved by shifting only 10 percent of the community-based schizophrenic population of the United Kingdom from oral to depot medication.

Oil formulations do not automatically provide sustained-release characteristics. For example, the antimalarial artemisinin has been found to be absorbed at a faster rate from an oily suspension compared to an aqueous suspension (Titulaer et al. 1990). Another example was reported by Slevin et al. (1984), who showed similar absorption profiles of bleomycin from a saline solution and a suspension in sesame oil. A depot effect is achieved with the careful choice of drug form and vehicle. In this chapter, we will discuss the basic formulation issues revolving around the properties of the vehicle and the drug, physicochemical and physiological factors affecting absorption and the therapeutic activity of oil-based injections.

THE OIL VEHICLE

Non-aqueous solvents available for use as oily parenteral vehicles include fixed oils, e.g., olive oil, corn oil, sesame oil, arachis oil, almond oil, peanut oil, poppyseed oil, soya oil, cottonseed oil and castor oil. Of these, sesame oil is the preferred oil as it is the most stable due to its content of natural antioxidants, except in light. The fixed oils (esters of glycerol and fatty acids) are normally well tolerated; however, some patients may have allergic reactions to vegetable oils, and specific oils should always be listed on the product label. Isopropyl myristate, ethyl oleate, benzyl benzoate, polyoxyethylene oleic triglycerides ("Labrafils"®), thin vegetable oil (fractionated coconut oil, Viscoleo®) and the liquid polyethylene glycols (PEGs) are some of the synthetic alternative vehicles available. Ethyl oleate is sometimes preferred over fixed oils due to its lower viscosity, hence better syringeability and injectability, which allows easier administration. The properties of some of these oily vehicles have been compiled in Table 5.2. Spiegel and Noseworthy (1963) reviewed the use of non-aqueous solvents in parenteral products and list some lesser-used vehicles, e.g., ethyl lactate and glycerin.

In the preparation of a sustained-release, non-aqueous formulation, the vehicle has to be chosen with care, considering the drug to be dissolved or suspended, the desired release profile, the condition to be treated and the possible side-effects of the drug/oil formulation. Back in 1943, Brown et al. listed the requirements of an ideal oil for the formulation of IM injections:

- Chemically, the oil should be stable and should not react with the medication to form toxic products. Fixed oils must be free from rancidity and must not contain mineral oils or solid paraffins, as these are not metabolised by the body and might eventually cause tissue reaction and even tumours.

- Biologically, the oil should be inert, non-toxic, non-antigenic, non-irritant, biocompatible, pyrogen free, and it should be absorbed from tissues after administration, leaving no residues. Any breakdown products should also be non-toxic and be absorbed from the injection site. The vehicle should have no pharmacological action of its own nor potentiate the activity of the medicament.

- Physically, the oil should be a good solvent or dispersing medium for the drug. A high loading capacity is desired such that sufficient drug can be administered without the injection of excessive volumes of oil. The vehicle should also remain fluid over a fairly wide range of temperatures and should not have a high viscosity for good syringeability and injectability.

Rubber plunger tips of plastic syringes may absorb oil and swell, with a consequent increase in the force needed to expel

Table 5.2. *Properties of Some Oily Vehicles Used in Parenteral Formulations*

Oil	Description	Typical Properties	Viscosity	Stability and Storage	Safety
Sesame oil	Clear, pale yellow liquid with a slight, pleasant odour and a bland taste	Solidification pt. ~ -5°C; insoluble in water, slightly soluble in ethanol.	43.37 cps	Stable and does not readily turn rancid. Store in a well-filled, airtight container at temperatures not exceeding 40°C. Protect from light.	Non-toxic orally.
Isopropyl myristate	Transparent, colourless, almost odourless, mobile liquid with a bland taste	Solidification pt. ~ 3°C; boiling pt. 140.2°C; immiscible with water, miscible with fixed oils.	7 cps at 25°C	Resistant to oxidation and hydrolysis. Does not become rancid. Store in a closed container at room temperature.	Very low toxicity, irritability. No sensitising properties in rabbits and guinea pigs.
Ethyl oleate	Pale yellow to almost colourless, mobile, oily liquid with a taste resembling that of olive oil. Slight but not rancid odour.	Solidification pt. ~ -32°C; boiling pt. 205–208°C; immiscible with water, miscible with fixed oils and fats.	≥ 5.15 cps; less viscous than fixed oils.	Oxidises on exposure to air. Store in a cool place in a small, well-filled, airtight container. Protect from light.	No reports of intramuscular irritation during use have been recorded.

Table 5.2 continued on next page.

Table 5.2 continued from previous page.

Oil	Description	Typical Properties	Viscosity	Stability and Storage	Safety
Cottonseed oil	Pale yellow or yellow, oily liquid, odourless with a bland, nutty taste.	Solidification pt. ~ 0 to -5°C; below 10°C, particles of solid fat may separate from the oil.	39.19 cps	Stable. Store in a well-filled, airtight, light-resistant container at temperatures not exceeding 40°C.	
Corn oil	Clear, light yellow, oily liquid having a faint characteristic odour and taste.		37.36 cps[b] 38.83 cps[a]	Stable. Store in an airtight, light-resistant container at temperatures below 40°C.	Non-toxic after oral administration.
Castor oil	Nearly colourless or slightly yellow transparent viscid oil with a slight odour and taste which is bland at first, but afterwards slightly acrid.			Store at temperatures not exceeding 15°C in well-filled, airtight containers. Protect from light. Oil intended for use in the manufacture of parenterals must be kept in glass containers.	

Table 5.2 continued on next page.

Table 5.2 continued from previous page.

Oil	Description	Typical Properties	Viscosity	Stability and Storage	Safety
Arachis oil	Colourless or pale yellow liquid with a faint nutty odour and a bland, nutty taste.	At about 3°C, the oil becomes cloudy and partially solidifies at lower temperatures; miscible with oils.	39.44 cps[b] 42.96 cps[a]	Stable. Store in an airtight, light-resistant container. Avoid exposure to excessive heat. On exposure to air, it thickens very slowly and may become rancid. Oil intended for use in the manufacture of parenterals must be kept in glass containers.	
Thin vegetable oil (synonyms: fractionated coconut oil, medium chain triglycerides)	Colourless or pale yellow odourless or almost odourless oily liquid.	Practically insoluble in water; miscible with vegetable oils.		Protect from light.	

Table 5.2 continued on next page.

Table 5.2 continued from previous page.

Oil	Description	Typical Properties	Viscosity	Stability and Storage	Safety
Polyethylene glycols (PEGs)	Liquid PEGs are clear, colourless or slightly yellowish viscous liquids, with a slight, characteristic odour. They taste bitter and slightly burning.	PEG 200 super-cools; solidification pts: PEG 300: −15 to −8°C, PEG 400: 4 to 8°C, PEG 600: 20 to 25°C. Soluble in water in all proportions to form clear solutions; miscible in all proportions with other PEGs.	Viscosity (measured by capillary viscometer) PEG 200: 39.9 PEG 300: 68.8 PEG 400: 90.0 PEG 600: 131 mm^2/sec at 25°C	Chemically stable in air. PEGs do not support microbial growth or become rancid. Stainless steel, aluminium, glass or lined steel are preferred storage containers. Store in well-closed containers. Oxidation may occur if PEGs are exposed for long periods to temperatures above 50°C.	Toxicity is low. SC dosages of PEG 400 up to 10 mL in rats caused no permanent damage. The activity of drugs dissolved in PEGs may be potentiated.
Benzyl benzoate	Colourless, oily liquid with a pleasant aromatic odour	Boiling pt. 323°C; insoluble in water, miscible with fixed oils		Store at temperatures below 40°C in well-filled, airtight containers. Protect from light.	

a: Supplier—Capital

b: Supplier—Welch, Home & Clark Co.

the syringe contents (Halsall 1985; Dexter and Shott 1979). Such oil absorption by the rubber plunger tip may alter the formulation through loss of oil. The use of glass syringes for the injection of oil-based formulations is therefore strongly recommended. Further considerations include water and body fluid miscibility, the degree of flammability, availability, source of supply and constant purity (Spiegel and Noseworthy 1963).

Fate of the Oil Vehicle Following Administration

After IM administration, the oil formulation forms a localised depot at the injection site (Shaffer 1929), whose spread depends on the formulation, its viscosity and surface tension, the needle size and the force used during injection. Dissolved drug molecules then partition from the oil formulation into the aqueous interstitial fluid and are subsequently absorbed into the blood (Figure 5.1). Obviously, the persistence of the oil depot is important in determining the period over which the drug is released from the formulation.

The oil itself is very slowly cleared from the injection site (Deanesly and Parkes 1933; Howard and Hadgraft 1983), and this enables it to act as a drug reservoir. Clearance of the oil is thought to occur via absorption through capillary blood vessels (Howard and Hadgraft 1983); by lymphatic absorption (Svendsen et al. 1980); by phagocytosis, possibly related to the inflammatory response of the tissues to the vehicle (Bisgard and Baker 1940; Rees et al. 1967; Ballard 1968), and by metabolism *in situ* followed by absorption (Svendsen and Aaes-Jørgensen 1979). It is most probable that *in vivo* absorption of the oil vehicle occurs via all of these pathways, the contribution of each depending on the nature of the oil itself and the formulation, which might contain surfactants, and could well depend on the volume and frequency of dosing and the administration site due to the anatomical difference in the nature and distribution of capillary and lymph vessels.

The rate of clearance of the oil vehicle has been found to depend on the nature of the oil; for example, arachis oil was found to be cleared significantly slower than ethyl oleate following IM and SC administration (Howard and Hadgraft 1983). The higher

Figure 5.1. *Route of absorption of lipophilic drugs into the bloodstream from an oil solution and oil suspension.*

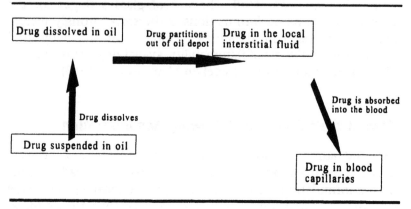

viscosity of arachis oil, which would limit its spreading, was thought to be responsible for this effect. Absorption of oil also depends on the presence of additives; for instance, the clearance of hexadecane from water-in-oil-in-water (w/o/w) emulsions was found to be significantly enhanced upon increasing the amount of the lipophilic surfactant, sorbitan monooleate, from 2.5 to 10 percent in the formulation (Omotosho et al. 1989). This was thought to be due to increased oil absorption in the blood and the lymphatic system as a solubilised surfactant system and possible enhanced dispersion of hexadecane droplets in the muscle by the higher concentration of lipophilic surfactant.

Although a slow clearance rate of the oil is desirable to achieve sustained release, the oil must preferably be totally absorbed from the site of injection, without leaving residues. Too slow of clearance of the oil following chronic dosing may lead to oil accumulation in multiple-depot sites and give rise to side-effects, e.g., formation of cysts (Emery et al. 1941), pulmonary microembolisation after translocation of oil droplets (Svendsen et al. 1980) and localised lymphadenopathy (in this case, enlargement of the inguinal lymph nodes) (Ahmed and Greenwood 1973) (see "Adverse Effects of Sustained-Release Oil Solutions and Suspensions").

DRUG ABSORPTION FROM OIL SOLUTIONS AND SUSPENSIONS

Drugs administered in a non-aqueous formulation must be released from the formulation, absorbed into the blood circulation and transported to the target site(s) before it becomes bioavailable, as shown in Figure 5.1. Dissolved drug from oil solutions partitions from the oil medium into the aqueous interstitial fluid before diffusing into the local blood capillaries. Drug partitioning into the local interstitial fluid can be thought of as a dynamic equilibrium between the drug in the oil phase and that in the aqueous phase, with a characteristic equilibrium constant, the apparent partition coefficient. Drug partitioning from the oil phase to the aqueous medium is believed to be the rate-limiting step controlling the drug release from oil *solutions*. It follows that the absorption rate and, hence, the depot characteristics can be controlled by manipulating the factors affecting the partition coefficient, for example, the nature of the oil or the lipophilicity of the active through the use of prodrugs.

Hirano et al. (1981, 1982) have reported that drug absorption from an oil solution follows first-order kinetics after IM and SC administration, such that

$$dC/dt = -k_a C \tag{1}$$

where k_a is the first-order absorption rate constant and C is the drug concentration in the oil.

Absorption of poorly water-soluble drugs from oil suspensions occurs in the same manner, except that the drug particles, in theory, dissolve in the oil medium before the dissolved molecule partitions into the local interstitial fluid, as shown in Figure 5.1. The suspended particles act as a drug reservoir, continuously dissolving to replenish what is being lost. In this case, the dissolution of the drug particles in the oil phase is the rate-limiting step, and the mean dissolution rate is defined by the Noyes-Whitney equation,

$$\text{Dissolution rate} = DA(Cs - C)/\delta \tag{2}$$

where D is the diffusion coefficient of the dissolved drug molecules in the medium, A is the surface area of drug particles exposed to the solvent, Cs is the saturation solubility of the drug in the medium, C is the concentration of dissolved drug in the medium and δ is the thickness of the hydrodynamic diffusion layer surrounding the drug solids.

The absorption rate of drugs from oil suspensions can thus be influenced by manipulating drug, particle and oil medium properties. Oily suspensions potentially offer a slower rate of drug absorption compared to oily solutions, and absorption is thought to have a zero-order profile (Chien 1982). However, when the drug particles suspended in the oil phase have a relatively high aqueous solubility, the release of drug from the non-aqueous depot can involve the direct apposition of the drug crystals to the aqueous phase and their dissolution in the latter without prior dissolution in the oil phase, as shown in Figure 5.2 (Crommelin and de Blaey 1980a, b; Mendelow et al. 1989). With relatively high concentrations of water-soluble drug, a percola-tion effect will be evident from Figure 5.2. This effect may be en-hanced if surfactants are employed to stabilise the suspension, as there may be inverse micelle formation in the oil phase which will aid the transport of water through the oil phase.

Factors Influencing Drug Absorption from Oil Solutions and Suspensions

Formulation Factors

Lipophilicity of the Vehicle. The vehicle exerts a profound effect on the absorption kinetics of lipophilic drugs from oil solutions, particularly where partitioning of drug from the formulation is the rate-limiting step. Usually, the more lipophilic the vehicle, the slower the release rates. A high affinity of the drug for the oil means a slower rate of partitioning of the drug from the formu-lation into the aqueous interstitial fluid at the injection site and, hence, a slower absorption rate and a longer depot effect. Hirano et al. (1981) found a direct correlation between the experimental partition coefficient of drug measured *in vitro* and the *in vivo* ab-sorption rate constant (Table 5.3). A similar relationship between the partition coefficient and the elimination half-life of testos-terone propionate from the muscle was reported by Al-Hindawi et al. (1987) (Table 5.4).

Figure 5.2. *Release of water-soluble drug particles.*

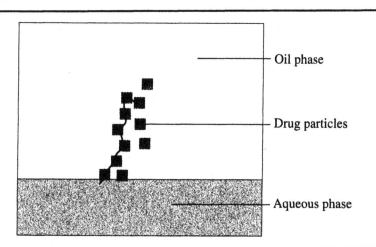

Oil phase

Drug particles

Aqueous phase

Water-soluble drug particles can be released from the non-aqueous medium by direct apposition of the drug crystals to the aqueous phase. At high concentrations of drug particles, a percolation effect is evident; this enables water penetration into the oil formulation.

The simple correlation between oil lipophilicity and the drug absorption rate may, however, sometimes be overcome by other factors. For example, androsterone was found to be twice as effective at stimulating the growth of prostate glands and seminal vesicles when administered in the water-miscible solvent propylene glycol compared to the water-immiscible olive oil vehicle (Ballard and Nelson 1975). This is thought to be due to the precipitation of the steroid from the propylene glycol vehicle caused by the dilution of the vehicle with tissue fluids following injection. A crystalline deposit may thus be produced, whose slow dissolution rate enables sustained absorption and biological effect of the drug (Chien 1982).

Lipophilicity of the Drug. Increasing the lipophilicity of drugs by chemical modification is another means of reducing the rate of partitioning of drugs from oil solutions into the local interstitial media and enabling sustained biological activity (Sinkula 1978).

Table 5.3. *The Direct Correlation Between the Distribution Coefficient and First-Order Absorption Rate Constant of p-Hydroxyazobenzene After the Subcutaneous Administration of Various Oil Solutions in Rats*

Oil	Distribution Coefficient (K)	First-Order Absorption Rate Constant (k, hr^{-1})
Simethicone	17	3.00 (0.41)
Isopropyl myristate-simethicone (1:9, v/v)	54	1.66 (0.12)
Isopropyl myristate-simethicone (2:8, v/v)	160	0.98 (0.08)
Isopropyl myristate-simethicone (4:6, v/v)	510	0.30 (0.01)
Sesame oil	1,300	0.21 (0.01)
Isopropyl myristate	2,900	0.13 (0.01)
Medium chain triglycerides	3,500	0.084 (0.007)
Diethyl sebacate	15,000	0.029 (0.003)

Modified from Hirano et al. (1982b). The distribution coefficient is the ratio of the drug concentration in oil to that in 0.9% saline at equilibrium (37°C); standard error in parentheses.

Table 5.4. *The Correlation Between the Partition Coefficient of Testosterone Propionate and Its Elimination Half-Life in the Muscle After Intramuscular Injection in Rats*

Solvent	Partition Coefficient Kp ($\times$ 10^{-3})	Half-life of Testosterone Propionate in Muscle (h)
Ethyl oleate	6.3 (0.4)	10.3 (0.8)
Octanol	5.3 (0.5)	9.7 (0.1)
Isopropyl myristate	4.3 (0.2)	7.8 (0.3)
Light liquid paraffin	1.5 (0.3)	3.2 (0.3)

Modified from: Al-Hindawi et al. (1987). The partition coefficient is ratio of drug concentration in oil to that in water at 37°C. Standard deviations are given in parentheses.

For example, the esterification of a number of molecules, e.g., testosterone, nandrolone, estradiol and various neuroleptics, has been used to enhance their lipophilicity. Oily solutions of the more lipophilic esters can then act as slow-release formulations. Following administration, the more lipophilic prodrug ester slowly diffuses from the oil depot; esterases in the local interstitial fluid rapidly hydrolyse the esters back to the parent drug before the latter is transported away from the injection site.

When testosterone is injected as a solution in oil, only a small, transient androgenic effect is observed due to rapid metabolism and excretion of the steroid. Enhanced drug plasma levels with testosterone esters were achieved, the duration of action of the esters in rats increasing with increasing fatty acid chain length (acetate and formate << propionate < butyrate < valerate esters). In higher chain esters, especially the decanoate and benzoate, the rise in effect was greatly retarded and the peak was diminished (Miescher et al. 1936). The increase in fatty acid chain length is accompanied by a decrease in the aqueous solubilities of the esters, a corresponding increase in the oil/water partition coefficient and, thus, a longer depot effect. Other examples include the longer duration of action of testosterone undecanoate compared to the enanthate ester (Partsch et al. 1995), a prolonged and retarded release profile of nandrolone hexoxyphenylpropionate ester (aromatic ring with 18 carbons) compared to the decanoate ester (Belkien et al. 1985) and an elimination half-life of 7 days for perphenazine enanthate compared to 27 days for perphenazine decanoate (Dencker and Axelsson 1996).

To the formulator, a rapid *in vitro* measurement of the partition coefficient of drugs may give an indication of the *in vivo* depot potential of these drugs. A high experimental partition coefficient value is equivalent to the slow partition of drug molecules from the oily vehicle into the interstitial fluid, a slow rate of drug absorption and, hence, a longer *in vivo* depot action. Indeed, the times of maximum effect of the lower fatty acid testosterone esters were found to be logarithmically related to the logarithms of their distribution coefficients (James 1972). Further Hansch analysis of the anabolic activities of some nandrolone esters showed a binomial relationship between log anabolic activity and log ethyl oleate–water distribution coefficients (Chaudry and James 1974).

Particle Size of Drugs in Suspensions. In oily suspensions, where the dissolution rate of water-insoluble drug particles is the rate-limiting step in drug absorption, particle size, which determines the total surface area available for drug dissolution, has a profound impact on drug absorption and biological activity. The surface area increases when particle size decreases according to

$$S_v = 6/d \qquad (3)$$

where S_v is the specific surface area and d is the average particle diameter. Larger particles of water-insoluble drugs thus offer a relatively smaller surface for drug dissolution, with consequent lower dissolution rate, as can be seen from equation 2. Larger particles may therefore be advantageous when formulating sustained-release suspensions. For example, therapeutic blood levels could be achieved for longer duration (48 h instead of 24 h) when large procaine penicillin G particles (size range of 150–175 μm) in sesame oil were intramuscularly administered in rabbits compared to micronized particles (particle size < 5 μm) (Buckwalter and Dickison 1958). An increase in particle size is, however, accompanied by an increase in the sedimentation rate, and suspension stability may be reduced. A compromise must therefore be reached between the particle size and the sedimentation rate.

The macrocrystal principle (i.e., large particles dissolve more slowly than small ones) can be overcome by other excipients in the formulation. For example, when a sesame oil vehicle was gelled with aluminium stearate, micronized procaine penicillin G particles (< 5 μm in diameter) were absorbed at a slower rate, and drug levels in the blood could be detected for a substantially longer time period compared to larger particles (Buckwalter and Dickison 1958). In the gelled vehicle, the procaine penicillin G particles are thought to be coated with the water-repellent aluminium stearate and are, therefore, less easily reached by aqueous tissue fluids. Dissolution and drug absorption are thus delayed. The micronized drug particles offer a larger surface area which can be protectively coated with the aluminium stearate, and a correspondingly longer delay in drug absorption occurs (Lippold 1980).

In the case of water-soluble drug particles suspended in an oily formulation, the release rates also show marked dependence on particle size, with higher release rates being achieved with larger crystals (Crommelin and de Blaey 1980a, b; Mendelow et al. 1989). Water-soluble drug particles are thought to be released from oily formulations by transport to the water-oil interface, passage through the interface and dissolution of the drug crystals in the aqueous phase. Sedimentation of the particles through the oil medium to the oil-water interface was found to be the rate-limiting step in drug release from these formulations (Crommelin and de Blaey 1980a). A larger particle size would thus result in a higher sedimentation rate according to Stokes' law and, subsequently, faster drug release. A percolation hypothesis, whereby water penetrates into the oily base through channels provided by the dissolution of water-soluble crystals, has been suggested (Mendelow et al. 1989). Water penetration into the formulation is important for further drug dissolution and release. The dissolution of larger crystals enables greater water percolation into the oily base compared to the dissolution of smaller crystals, thus enhanced release rates of the water-soluble drug occur.

Viscosity of the Solvent. Reports on the influence of solvent viscosity on drug release seem to be contradictory. Buckwalter and Dickison (1958) reported that viscosity per se does not appear to influence absorption since some rather viscous products are absorbed faster than less viscous ones. Hirano et al. (1981, 1982) reported that the viscosity of solvents did not have any influence on the absorption of p-hydroxyazobenzene, p-aminoazobenzene and o-aminoazotoluene from oil solutions after IM and SC injection in rats (Table 5.5). The rate-limiting step in drug absorption from oil solutions is the rate of drug partitioning from the oil solvent into the aqueous interstitial fluid; this movement is not affected by the viscosity of the oil medium. Theoretically, therefore, viscosity should not affect the absorption rate of drugs except by affecting the spread of the depot following administration and, hence, its overall surface area. For example, when ethyl oleate was substituted for sesame oil as a parenteral hormone vehicle, increased hormone activity was observed (Dekanski and Chapman 1953). In this case, the lower viscosity of the ethyl oleate vehicle was thought to be responsible. After administration, ethyl oleate is likely to spread to a greater extent

Table 5.5. *The Lack of Correlation Between Viscosity and First-Order Absorption Rate Constant of p-Hydroxyazobenzene After the Subcutaneous Administration of Various Oil Solutions in Rats*

Oil	Viscosity (cps)	First-Order Absorption Rate Constant (k, hr^{-1})
Simethicone	16.0	3.00 (0.4^1)
Isopropyl myristate-simethicone (1:9, v/v)	13.2	1.66 (0.12)
Isopropyl myristate-simethicone (2:8, v/v)	10.7	0.98 (0.08)
Isopropyl myristate-simethicone (4:6, v/v)	7.6	0.30 (0.01)
Sesame oil	35	0.21 (0.01)
Isopropyl myristate	3.6	0.13 (0.01)
Medium chain triglycerides	15	0.084 (0.007)
Diethyl sebacate	3.9	0.029 (0.003)

Modified from Hirano et al. (1982), with standard error in parentheses.

compared to sesame oil, with a corresponding increase in the surface area available for drug release from the formulation and consequent drug absorption.

In oil suspensions containing water-insoluble drug particles, where the absorption rate is dissolution controlled, the viscosity of the oil phase in which the drug diffuses influences the diffusion coefficient of drug molecules according to the Stokes-Einstein equation,

$$D = kT/6\pi\eta r \qquad (4)$$

where D is the diffusion coefficient of the drug, k is the Boltzmann constant, T is the absolute temperature, η is the viscosity of the vehicle and r is the molecular radius. If the viscosity of the oily medium is increased by gelling agents, then the bulk viscosity does not reflect the influence on diffusion. Rather, the important parameter is the actual viscosity of the medium in which the drug diffuses, the microviscosity, which is difficult to measure. A high microviscosity results in a lower diffusion coefficient of the dissolved molecule which, in turn, hinders further dissolution of the drug particles.

Viscosity will affect diffusion of the drug only if the microviscosity of the medium changes. However, the bulk viscosity of oil formulations has been increased by gelation of the oil phase in an attempt to achieve sustained release. For example, beeswax–peanut oil systems have shown higher blood levels of penicillin for up to 7 h after administration, compared to 4 h with the ungelled peanut oil formulation (Romansky and Rittman 1944). Oil formulations gelled with aluminium monostearate have also shown sustained plasma levels of penicillin compared to ungelled oils (Buckwalter and Dickison 1958). *In vitro* studies on injectable gels (consisting of Labrafil® and glyceryl ester of fatty acids) showed that increasing the wax concentrations from 5 percent w/w to 20 percent w/w significantly decreased the drug release from the gels, as shown in Figure 5.3 (Gao et al. 1995a). The reduced diffusivity of the drug was attributed to both an increase in tortuosity and an increase in the bulk viscosity of the gels. After SC administration of the gels in rats, prolonged release rates of contraceptive steroids, resulting in completely blocked rat estrous cycles, were achieved (Gao et al. 1995b). Drug release was thought to follow the slow biodegradation of the gels at the injection site. In addition, rats treated with the gel formulations did not show signs of systemic toxicity, such as alopecia shown by rats treated with oil formulations containing the same drug (ethinyl estradiol) loadings (Gao et al. 1995b).

Increasing the viscosity of formulations in an attempt to enhance the depot effect is, however, limited by practical considerations such as the syringeability and injectability of the formulation. Too high a viscosity will result in higher forces being required to inject the gels, which should be avoided to prevent pain on injection. Injectability can be assessed by a test for syringeability, where a controlled pressure ranging from 10 to 15 kPa is applied directly to the formulation. The time taken to deliver a standard volume of the formulation is then measured. Flow rates can thus be calculated (Chien et al. 1981).

Figure 5.3. *The effect of increasing wax (Precirol) concentrations on the cumulative amount of levonorgestrel released from formulations.*

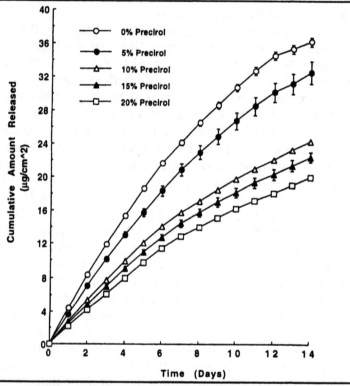

Each value represents the mean and SEM of 3 samples. Source: Gao et al. (1995).

The gelation of oils and the subsequent increase in bulk viscosity does not always result in a slower release of incorporated entities, as we found following the IM administration of sorbitan monostearate/polysorbate 20/isopropyl myristate gels (unpublished results). In this case, similar release rates were found from the gel and the oil formulations. This was thought to be caused by a rapid disintegration of the gel formulations due to interactions with the local interstitial fluid following administration. The latter explanation is based on *in vitro* studies when an aqueous phase was made to contact with the gel (Murdan et al. *1999*). Upon contact, the aqueous phase was seen to rapidly penetrate into the gel via the surfactant (sorbitan monostearate/polysorbate 20) network, which, as well as being the gel skeleton, provides convenient conduits for the movement of the aqueous phase into the organic gel medium (Figure 5.4a). Such an aqueous invasion causes the gel to break into smaller fragments. Emulsification was also observed at the gel surface, and oil droplets were seen to bud off (Figure 5.4b). Thus the original gel formulation is probably disintegrated via interactions with body fluids soon after administration, with consequent rapid release of incorporated solute.

Injection Volume. The volume injected affects the size of the depot formed. A small volume offers a relatively larger surface area across which dissolved drug can be released into the interstitial fluid; therefore, absorption rates are higher. Figure 5.5 shows the effect of injection volume (V_0) on the IM absorption of p-hydroxyazobenzene in sesame oil. Hirano et al. (1981) have also shown that the first-order absorption rate constant in oil solutions is inversely proportional to the cube root of the injection volume (Figure 5.6). Such a correlation is thought to arise as a result of the spheroidal shape assumed by an oily formulation after injection.

For the sustained release of drugs, therefore, large volumes are preferable. This is illustrated by the higher biological activity

Figure 5.4a. *The penetration of aqueous toluidine blue solution into the organogel through the tubular network.*

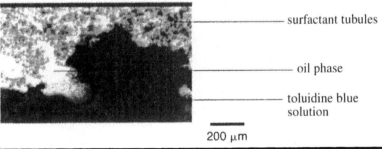

surfactant tubules

oil phase

toluidine blue solution

200 μm

Figure 5.4b. *Emulsification occurs between the oil phase and water at the gel surface. The gel consequently erodes.*

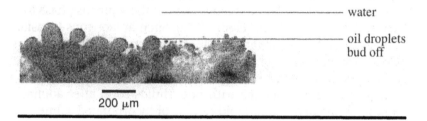

water

oil droplets
bud off

200 μm

of testosterone and a larger increase in the weight of the prostate glands and seminal vesicles in rats when larger volumes were used (Table 5.6) (Ballard and Nelson 1975). Other authors have reported contradictory results. Honrath et al. (1963) reported a larger increase in the weight of the seminal vesicles and of the prostate glands when 0.2 mL of testosterone in sesame oil was administered subcutaneously compared to 0.8 mL. When testosterone propionate was used in the same vehicle, however, the larger volume (0.8 mL) gave the highest biological response (Table 5.7).

Figure 5.5. *Effect of injection volume (V_0) on the intramuscular absorption of p-hydroxyazobenzene in sesame oil. Source: Hirano et al. (1982).*

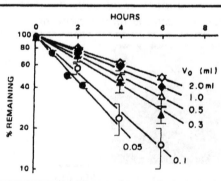

Figure 5.6. *Relationship between the first-order absorption rate constant (k) and the injection volume (V₀). Standard error bars are plotted. Source: Hirano et al. (1982).*

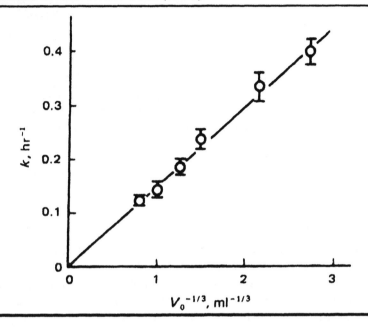

Table 5.6. *Effect of the Volume of Arachis Oil on the Biological Activity of Testosterone in Rats*

	Biological Activity[b]	
Volume[a] (mL)	Prostate Glands (mg)	Seminal Vesicles (mg)
2	61	27
5	183	118
10	219	163

[a]Subcutaneous administration of 2 mg of testosterone in either 2, 5 or 10 mL of oil to rats in 10 equal daily injections.

[b]The weight of organs one day after the last injection.

Source: Chien (1982).

Table 5.7. *Effect of the Formulation Volume on the Androgenic Activities of Testosterone and Its Propionate Ester*

Androgen[a]	Seminal Vesicle Weight[b]		Ventral Prostate Weight[b]	
	0.2 mL	0.8 mL	0.2 mL	0.8 mL
Testosterone	28	10	44	26
Testosterone propionate	58	140	124	160

[a]A total of 5 mg of androgen was dissolved in either 0.2 or 0.8 mL of sesame oil and injected as a single dose subcutaneously into each castrated male rat.

[b]The weight changes (in mg/100 g) of the seminal vesicles and of the ventral prostate at the end of 10 days.

Source: Honrath et al. (1963).

Physiological Factors

Injection Site. Plasma levels of a number of drugs, e.g., lidocaine (Cohen et al. 1972; Schwartz et al. 1974), have been found to differ significantly depending on the muscles used for injection. Such differences in drug absorption have been linked to different blood perfusion levels in the muscles. From Figure 5.1, we can appreciate that the rate of drug absorption depends on the blood perfusion at the injection site. A high perfusion rate (i.e., fast clearance of the drug from the injection site) creates sink conditions, and diffusion of the drug from the interstitial fluid into the blood vasculature is encouraged. In turn, drug partition from the formulation into the interstitial fluid is also encouraged. The opposite holds true for poorly perfused sites.

Resting blood flow in different muscles is significantly different (Evans et al. 1974); deltoid muscle blood flow (11.6 mL/100 g/min) is significantly greater than the *gluteus maximus* blood flow (9.6 mL/100 g/min), with the *vastus lateralis* having intermediate flow (10.8 mL/100 g/min). It follows that for sustained release, the formulation must preferably be administered into the *gluteus maximus*. The gluteal site is indeed the recommended injection site for depot neuroleptics and hormone formulations. The thick muscle mass at the gluteal site also permits the administration of large volumes.

As far as absorption from different SC sites is concerned, Nora et al. (1964) found significant differences in the absorption half-life of an aqueous insulin formulation in the human arm and thigh. However, Hirano et al. (1981) reported that absorption profiles of p-hydroxyazobenzene in sesame oil from the dorsal and abdominal sites in rats were virtually identical. It is difficult to explain these contradictory results, but taking into account the different vehicles, species, drug and injection sites, they are perhaps not surprising. Different sites may be used for SC injection in experimental animals (e.g., dorsum, abdomen, scruff and side) by different researchers, and the injection site must therefore be taken into account when the results are being compared. To our knowledge, there have not been many investigations on the dependency of drug absorption on the SC injection site, and this gap remains to be filled.

Injection Depth. It is recommended that depot oil formulations be injected deep into the muscle, away from major nerves and arteries. Experience with aqueous formulations has shown that intended IM injections may end up being intralipomatous ones when short needles are used at the gluteal sites (Cockshott et al. 1982). In a study by Dundee et al. (1974), plasma diazepam levels were found to be higher after injection by a doctor who used a 4 cm needle compared to injection by a group of nurses who usually employ 3 cm needles (Figure 5.7). The nurses may have injected into the gluteal fat, with consequent poor absorption of the drug. Care must be exercised when administering sustained-release oil formulations intramuscularly such that the gluteal fat is avoided and to ensure that patients receive the appropriate dose.

Other Factors. Other factors that influence drug absorption after IM and SC injection include physiological factors. Exercise increases the muscle blood flow; as more drug is swept away from the localised absorption site, the drug may be absorbed and dispersed faster. This can be an important factor in causing adverse high levels of drugs such as neuroleptics in ambulant patients. For instance, after the administration of a procaine penicillin G oil suspension, the drug concentrations in blood were maintained above the effective 0.039 units/mL for 33 h in bedridden

Figure 5.7. *Plasma diazepam levels 90 min after intramuscular injection by one doctor and several nurses showing the influence of injection depth on drug absorption. Source: Dundee et al. (1974).*

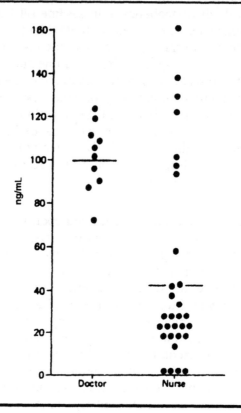

pneumonia patients compared to only 12 h in ambulatory patients (Chien 1982). Rubbing the injection site following drug administration also disperses the injected oil formulation, with a concomitant increase in surface area available for drug release. Pain on injection can also affect absorption levels. Trauma at the injection site causes the release of histamine which constricts blood vessels; this is followed by a reduced rate of absorption, particularly immediately following injection (Schou 1958a, b). Other factors include age; the tissue condition, e.g., scarring with associated reduced vascularity may result in drug accumulation at injection site with poor absorption; disease states such as circulatory shock, hypotension, congestive heart failure and myxoedema where blood flow to skeletal muscle is reduced; body temperature, which affects vasodilation and vasoconstriction; the presence of mediator enzymes, e.g., hyaluronidase; and so on. The reader is referred to the excellent reviews by Wagner (1961), Ballard (1968), and Zuidema et. al. (1988, 1994) for extensive general discussion on absorption aspects after parenteral administration.

ADVERSE EFFECTS OF SUSTAINED–RELEASE OIL SOLUTIONS AND SUSPENSIONS

Like all long-acting preparations, sustained-release injectables suffer from one major drawback. In cases where subjects may wish to discontinue treatment, for example, after side-effects, treatment cannot be stopped immediately, as would be the case with a daily regimen. In addition, side-effects of long-acting formulations may be more severe and last longer. For example, bimonthly and quarterly depot contraceptive agents suffer from an increased incidence of cycle irregularities, and the adverse effects subside at a slower rate compared to those of monthly injectables (Toppozada 1977). Depot neuroleptics have also demonstrated such long-term unwanted reactions. It is generally believed that extrapyramidal side-effects of antipsychotic drugs are related to a blockade of dopamine D_2 receptors in the basal ganglia (Nordstrom et al. 1992; Farde et al. 1992). Der Meer et al. (1993) reported 2 cases where a large percentage of D_2 receptors (83 percent and 50 percent) were still blocked 1.5 months after the last injection of fluphenazine decanoate. The patients displayed prominent extrapyramidal side-effects (akinesia with rigidity and tremor in one patient, continuous choreoathetoid movements in the other), though they did not show any psychotic signs on admission. In contrast, the return to normal receptor availability after withdrawal of oral therapy has been reported to occur within 5–15 days (Baron et. al.

1989), or even within 3 days (Cambon et al. 1987). Similar long-term antidopaminergic effects were reported by Wistedt et al. (1981), who noted that plasma prolactin levels fell progressively for about 1 year after the withdrawal of flupenthixol decanoate injections in a female patient. It seems that in some individuals, the elimination half-life of neuroleptics is much longer than generally believed, and side-effects may occur long after the cessation of therapy.

The importance of depot preparations is well recognised in the maintenance therapy of schizophrenic patients, as they save staff time and shorten hospital stays because patients may continue to receive their medication in out-patient clinics and be treated in the community (Weiden et al. 1995; Kane 1995; Citrome 1996). It is vital in such cases that out-patient services are well organised so that patients can be regularly assessed, and any changes in their condition may be acted upon immediately. Alarcon and Carney (1969) reported 16 cases where the patients became severely depressed, 5 of whom committed suicide shortly after the administration of a slow-release fluphenazine injection. This report illustrates how reduced contact between a patient and the physician can be fatal if the patient's condition is not monitored closely. Surveillance and follow-up of patients on depot antipsychotic drugs are crucial to enable the occurrence of any depressive episodes to be promptly detected and treated. The conversion of schizophrenia into acute depression following the administration of neuroleptics has been reported (Tewfik 1965); this may even be advantageous if the depressive episode responds to electroconvulsive therapy (ECT) and can thus be resolved (Hamilton 1965).

Other side-effects of sustained-release oil formulations include reactions to the oil vehicle. Fixed oils extracted from nuts may cause reactions in patients who are allergic to nuts. Therefore, the nature of the oil vehicle must always be labelled on parenteral oil formulation packs. Side-effects may also occur following the administration of large volumes or repeated injections. Ahmed and Greenwood (1973) reported the occurrence of lymphadenopathy in a 17-year-old patient after repeated oil-based injections. Histological examination of the excised lymph node revealed the presence of numerous oil cysts which were surrounded by foreign body giant cells. The occurrence of encapsulated oil cysts in experimental animals after the IM and SC injection of oils has also been reported previously (Emery et al. 1941).

Svendsen and Aaes-Jørgensen (1980) found that after IM injection in experimental animals, the oily vehicle is absorbed to some extent in regional lymph nodes. After repeated oil injections, the oil-retaining capacity of the nodes is exceeded, and pulmonary microemboli may arise. Macrophages, leucocytes, and small mononuclear interstitial cells are present in the vicinity of these oil microemboli. Oil cysts surrounded by a thin reactive granulomatous zone were also found at the injection site, and haemosiderosis appeared in some lymph nodes. Pulmonary microemboli in humans receiving large doses of neuroleptic oily depot formulations have also been observed (Svendsen et al. 1980).

In a systematic study on the tolerability of IM injections of testosterone enanthate ester in a castor oil vehicle (Mackey et al. 1995), episodes of sudden-onset, non-productive cough were found to be the only systemic side-effect of a small percentage of these injections. The cough was thought to be due to a pulmonary microembolism. Most of the injections did not cause any complaints, only minor local side effects, mostly pain and bleeding, were common. The gluteal site, having the lowest resting blood flow (Evans et. al. 1974), was less prone to bleeding compared to deltoid or thigh injections but was more painful, with pain being associated with sitting or lying on the injection site. When all of the side-effects were considered, the gluteal site had fewer complaints than the deltoid and thigh sites. The authors concluded that, when administered by an experienced nurse, deep IM injection of testosterone enanthate in a castor oil vehicle is generally safe and well tolerated but causes relatively frequent minor side-effects, including pain and bleeding. This conclusion probably applies to the majority of oil formulations administered intramuscularly. Some local pain and discomfort after IM administration is unavoidable following the introduction of a needle and deposition of fluid, which may contribute to pain by causing tissue distension, disruption, or chemical irritation (Travell 1955; Svendsen 1983).

CONCLUSIONS

There is a need for new and probably synthetic non-aqueous media for pharmaceutical use, especially non-toxic solvents for drugs which can be readily formulated as solutions. Because of the limited range of materials available, most non-aqueous formulations will be suspensions, where behaviour *in vivo* cannot always be predicted with simple *in vitro* tests because of the range of factors discussed in this chapter that affect drug release. The greater understanding of inverse micelle formation and inverse vesicles may lead to a resurgence of interest in non-aqueous depot systems. It is clear that the modern literature on non-aqueous formulations is relatively sparse; but with new materials and new systems, a renaissance is due.

REFERENCES

Abrego de Aguilar, M., L. Altamirano, D. A. Leon, R. Caridad de Fung, A. E. Grillo, J. D. J. Gonzalez, J. R. L. Canales, J. Del Carmen Mojica Sanchez, J. L. Pozuelos, L. Ramirez, R. Rigionni, J. S. Salgado, L. Torres, G. Vallecillos, E. J. Zambrano, and C. Zea. 1997. Current status of injectable hormonal contraception, with special reference to the monthly method. *Adv. Contraception* 13:405–417.

Ahmed, A., and N. Greenwood. 1973. Lymphadenopathy following repeated oil-based injections. *J. Pathol.* 111:207–209.

Al-Hindawi, M. K., K. C. James, and P. J. Nicholls. 1987. Influence of solvent on the availability of testosterone propionate from oily, intramuscular injections in the rat. *J. Pharm. Pharmacol.* 39:90–95.

Alarcon, R., and M. W. P. Carney. 1969. Severe depressive mood changes following slow-release intramuscular fluphenazine injection. *Br. Med. J.* 3:564–567.

Ballard, B. E. 1968. Biopharmaceutical considerations in subcutaneous and intramuscular drug administration. *J. Pharm. Sci.* 57:357–378.

Ballard, B. E., and E. Nelson. 1975. Prolonged-action pharmaceuticals. In *Remington's pharmaceutical sciences*, 15th ed., edited by A. Osol, A. R. Gennaro, H. S. Hutchinson, E. A. Swinyard, M. R. Gibson, W. Kowalick, L. F. Tice, S. C. Harvey, A. N. Martin, and C. T. Van Meter. Easton, Penn, USA: Mack Publishing Co., pp. 1618–1643.

Baron, J. C., J. L. Martinot, H. Cambon, J. P. Boulenger, M. F. Poirier, V. Caillard, J. Blin, J. D. Huret, C. Loc'h, B. Maziere. 1989. Striatal dopamine receptor occupancy during and following withdrawal from neuroleptic treatment: Correlative evaluation by positron emission tomography and plasma prolactin levels. *Psychopharmacology* 99:463–472.

Belkien, L., T. Schurmeyer, R. Hano, P. O. Gunnarson, and E. Nieschlag. 1985. Pharmacokinetics of 19-nortestosterone esters in normal men. *J. Steroid Biochem.* 5:623–629.

Bisgard, J. D., and C. Baker. 1940. Experimental fat metabolism. *Am. J. Surg.* 47:466–478.

Brown, W. E., M. D. Violet, M. Wilder, and P. Schwartz. 1943. A study of oils used for intramuscular injections. *J. Lab. Clin. Med.* 29:259–264.

Buckwalter, F. H., and H. L. Dickison. 1958. The effect of vehicle and particle size on the absorption, by the intramuscular route, of procaine penicillin G suspensions. *J. Am. Pharm. Ass.* XLVII:661–676.

Cambon, H., J. C. Baron, J. P. Boulanger, C. Loc'h, E. Zarifian, and B. Maziere. 1987. In vivo assay for neuroleptic receptor binding in the striatum: Positron tomography in humans. *Br. J. Psychiatry* 151:824–830.

Chaudry, M. A. Q., and K. C. James. 1974. A Hansch analysis of the anabolic activities of some nandrolone esters. *J. Med. Chem.* 17:157–161.

Chien, Y. W. 1982. Parenteral controlled-release drug administration. In *Novel drug delivery systems*, edited by Y. W. Chien. New York: Marcel Dekker, Inc., pp. 219–310.

Chien, Y. W., P. Przybyszewski, and E. G. Shami. 1981. Syringeability of nonaqueous parenteral formulations–development and evaluation of a testing apparatus. *J. Parent. Drugs Ass.* 35:281–284.

Citrome, L., J. Levine, and B. Allingham. 1996. Utilization of depot neuroleptic medication in psychiatric inpatients. *Psychopharm. Bull.* 32:321–326.

Cockshott, W. P., G. T. Thompson, L. J. Howlett, and E. T. Seeley. 1982. Intramuscular or intra-lipomatous injections. *N. Engl. J. Med.* 307:356–358.

Cohen, L. S., J. E. Rosenthal, D. W. Horner, Jr., J. M. Atkins, O. A. Matthews, and S. J. Sarnoff. 1972. Plasma levels of lidocaine after intramuscular administration. *Am. J. Cardiol.* 29:520–523.

Crommelin, D. J. A., and C. J. de Blaey. 1980a. In vitro release studies on drugs suspended in non-polar media. I. Release of sodium chloride from suspensions in liquid paraffin. *Int. J. Pharm.* 5:305–316.

Crommelin, D. J. A., and C. J. de Blaey. 1980b. In vitro release studies on drugs suspended in non-polar media. II. The release of paracetamol and chloramphenicol from suspensions in liquid paraffin. *Int. J. Pharm.* 6:29–42.

Deanesly, R. and A. S. Parkes. 1933. Note on the subcutaneous absorption of oils by rats and mice, with special reference to the assay of oestrin. *J. Physiol.* 78:155–160.

Dekanski, J., and R. N. Chapman. 1953. Testosterone phenyl propionate (TPP): Biological trials with a new androgen. *Br. J. Pharmacol.* 8:271–277.

Dencker, S. J., and R. Axelsson. 1996. Optimising the use of depot antipsychotics. *CNS Drugs* 6:367–381.

Der Meer, C. H., T. Brucke, S. Wenger, P. Fischer, L. Deecke, and I. Podreka. 1993. Two cases of long term dopamine D2 receptor blockade after depot neuroleptics. *J. Neural Trans. [GenSect]* 94:217–221.

Dexter, M. B., and M. J. Shott. 1979. The evaluation of the force needed to expel oily injection vehicles from syringes. *J. Pharm. Pharmacol.* 31:497–500.

Dundee, J. W., J. A. S. Gamble, and R. A. E. Assaf. 1974. Plasma diazepam levels following intramuscular injection by nurses and doctors. *Lancet* 2:1461.

Emery, F., C. S. Matthews, and E. L. Schwabe. 1942. The absorption of stilbestrol and theelin from cysts of sesame and peanut oils. *J. Lab. Clin. Med.* 27:623–627.

Evans, E. F., J. D. Proctor, M. J. Fratkin, J. Velandia, and A. J. Wasserman. 1974. Blood flow in muscle groups and drug absorption. *Clin. Pharm. Ther.* 17:44–47.

Farde, L., A-L. Nordström, F-A. Wiesel, S. Pauli, C. Halldin, G. Sedvall. 1992. Positron emission tomographic analysis of central D_1 and D_2 dopamine receptor occupancy in patients treated with classical neuroleptics and clozapine. *Arch. Gen. Psychiatry.* 49:538–544.

Gao, Z-H., A. J. Shukla, J. R. Johnson, and W. R. Crowley. 1995a. Controlled release of a contraceptive steroid from biodegradable and injectable gel formulations: In vitro evaluation. *Pharm. Res.* 12:857–863.

Gao, Z.-H., W. R. Crowley, A. J. Shukla, J. R. Johnson, and J. F. Reger. 1995b. Controlled release of contraceptive steroids from biodegradable and injectable gel formulations: In vivo evaluation. *Pharm. Res.* 12:864–868.

Hale, A. S., and C. Wood. 1996. Comparison of direct treatment costs for schizophrenia using oral or depot neuroleptics: A pharmacoeconomic analysis. *Br. J. Med. Econ.* 10:37–45.

Halsall, K. G. 1985. Calciferol injection and plastic syringes. *Pharm. J.* 235:99.

Hamilton, M. 1965. Discussion on "Background to a drug trial" by G. I. Tewfik. *Clin. Trials J.* 2:158.

Hirano, K., T. Ichihashi, and H. Yamada. 1981. Studies on the absorption of practically water-insoluble drugs following injection. I: Intramuscular absorption from water immiscible oil solutions in rats. *Chem. Pharm. Bull.* 2:519–531.

Hirano, K., T. Ichihashi, and H. Yamada. 1982. Studies on the absorption of practically water-insoluble drugs following injection. V: Subcutaneous absorption in rats from solutions in water immiscible oils. *J. Pharm. Sci.* 71:495–500.

Honrath, W. L., A. Wolff, and A. Meli. 1963. The influence of the amount of solvent (sesame oil) on the degree and duration of action of subcutaneously administered testosterone and its propionate. *Steroids* 2:425–428.

Howard, J. R., and J. Hadgraft. 1983. The clearance of oily vehicles following intramuscular and subcutaneous injections in rabbits. *Int. J. Pharm.* 16:31–39.

James, K. C. 1972. R_m values and biological action of testosterone esters. *Experientia* 28:479–480.

Kane, J. M. 1995. Dosing issues and depot medication in the maintenance treatment of schizophrenia. *Int. Clin. Psychopharm.* 10 (Suppl. 3):65–71.

Lippold, B. C. 1980. Depot preparations. *Pharm. Int.* 1:60–63.

Mackey, M.-A., A. J. Conway, and D. J. Handelsman. 1995. Tolerability of intramuscular injections of testosterone ester in oil vehicle. *Human Reprod.* 10:862–865.

Mendelow, A. Y., A. Forsyth, A. J. Baillie, and A. T. Florence. 1989. In vitro release of nickel sulphate of varying particle size from paraffin bases. *Int. J. Pharm.* 49:29–37.

Miescher, K., A. Wettstein, and E. Tschopp. 1936. The activation of the male sex hormones. *Biochem. J.* 30:1977–1990.

Murdan, S., G. Gregoriadis, and A. T. Florence. 1999. Interaction of a nonionic surfactant-based organogel with aqueous media. *Int. J. Pharm.* 180:211–214.

Nora, J. J., W. D. Smith, and J. R. Cameron. 1964. The route of insulin administration in the management of diabetes mellitus. *J. Pediatr.* 64:547–551.

Nordström, A.-L., L. Farde, and C. Halldin. 1992. Time course of D_2 dopamine receptor occupancy examined by PET after single oral doses of haloperidol. *Psychopharmacology* 106:433–438.

Omotosho, J. A., T. L. Whateley, and A. T. Florence. 1989. Release of 5-fluorouracil from intramuscular w/o/w multiple emulsions. *Biopharm. Drug Disp.* 10:257–268.

Partsch, C-J., G. F. Weinbauer, R. Fang, and E. Nieschlag. 1995. Injectable testosterone undecanoate has more favourable pharmacokinetics than testosterone enanthate. *Eur. J. Endocrin.* 132:514–519.

Romansky, M., and G. Rittman. 1944. A method of prolonging the action of penicillin. *Science* 100:196–198.

Schou, J. 1958a. Self-depression of the subcutaneous absorption of drugs due to release of histamine. *Nature* 182:324.

Schou, J. 1958b. The influence of histamine and histamine liberation on subcutaneous absorption of drugs. *Acta Pharm. Tox., Kbh.* 15:43–54.

Schwartz, M. L., M. B. Meyer, B. G. Covino, R. M. Narang, V. Sethi, A. J. Schwartz, and P. Kamp. 1974. Antiarrhythmic effectiveness of intramuscular lidocaine; influence of different injection sites. *J. Clin. Pharmacol.* 14:77–83.

Shaffer, L. W. 1929. The fate of intragluteal injections. *Arch. Dermatol. Syphilol.* 19:347–364.

Sinkula, A. A. 1978. Methods to achieve sustained drug delivery—the chemical approach. In *Sustained and controlled release drug delivery systems*, edited by J. R. Robinson. New York: Marcel Dekker, Inc., pp. 412–555.

Slevin, M. L., V. J. Harvey, G. W. Aherne, N. K. Burton, A. Johnston, P. F. M. Wrigley. 1984. Delayed-release bleomycin—comparative pharmacology of bleomycin oil suspension and bleomycin in saline. *Cancer Chemother. Pharmacol.* 13:19–21.

Spiegel, A. J., and M. M. Noseworthy. 1963. Use of nonaqueous solvents in parenteral products. *J. Pharm. Sci.* 52:917–927.

Svendsen, O. 1983. Local muscle damage and oily vehicles: A study on local reactions in rabbits after intramuscular injection of neuroleptic drugs in aqueous or oily vehicles. *Acta Pharmacol. Toxicol.* 52:298–304.

Svendsen, O., and T. Aaes-Jørgensen. 1979. Studies on the fate of vegetable oil after intramuscular injection into experimental animals. *Acta Pharmacol. Toxicol.* 45:352–378.

Svendsen, O., S. J. Dencker, R. Fog, A. O. Gravem, and P. Kristjansen. 1980. Microscopic evidence of lymphogenic absorption of oil in humans receiving neuroleptic oily depot preparations intramuscularly. *Acta Pharmacol. Toxicol.* 47:157–158.

Tewfik, G. I. 1965. Background to a drug trial. *Clin. Trials J.* 2:150–152.

Titulaer, H. A. C., J. Zuidema, P. A. Kager, J. C. F. M. Wetsteyn, C. B. Lugt, and F. W. H. M. Merkus. 1990. The pharmacokinetics of artemisinin after oral, intramuscular and rectal administration to volunteers. *J. Pharm. Pharmacol.* 42:810–813.

Toppozada, M. 1977. The clinical use of monthly injectable contraceptive preparations. *Obstetr. Gynecol. Survey* 32:335–347.

Toppozada, M. K. 1994. Existing once-a-month combined injectable contraceptives. *Contraception* 49:293–301.

Travell, J. 1955. Factors affecting pain of injection. *JAMA* 158:368–371.

Wagner, J. G. 1961. Biopharmaceutics: Absorption aspects. *J. Pharm. Sci.* 50:359–387.

Weiden, P., B. Rapkin, A. Zygmunt, T. Mott, D. Goldman, and A. Frances. 1995. Postdischarge medication compliance of inpatients converted from an oral to a depot neuroleptic regimen. *Psychiatric Services* 46:1049–1054.

Wistedt, B., D. Wiles, and T. Kolakowska. 1981. Slow decline of plasma drug and prolactin levels after discontinuation of chronic treatment with depot neuroleptics. *Lancet* 1:1163.

Zuidema, J., F. A. J. M. Pieters, and G. S. M. J. E. Duchateau. 1988. Release and absorption rate aspects of intramuscularly injected pharmaceuticals. *Int. J. Pharm.* 47:1–12.

Zuidema, J., F. Kadir, H. A. C. Titulaer, and C. Oussoren. 1994. Release and absorption rates of intramuscularly and subcutaneously injected pharmaceuticals (II). *Int. J. Pharm.* 105:189–207.

6

Emulsions for Sustained Drug Delivery

Maninder Hora
Stephen Tuck
Chiron Corporation
Emeryville, California

Emulsions are ubiquitous in our everyday lives. Obvious examples of emulsions include foods such as milk and ice cream, paints, and petroleum products. The healthcare industry also exploits emulsions, examples being creams for topological application of active substances and total nutrient admixtures (TNAs) for intravenous (IV) use. The *Oxford American Dictionary* defines an emulsion as a creamy liquid in which particles of oil or fat are evenly distributed. In more scientific terms, an emulsion consists of a dispersion of small particles (or droplets) of one liquid into another mutually inert and immiscible liquid. The dispersed and bulk liquids are respectively known as the internal and external phases of an emulsion. The two liquids are usually a hydrophobic oil phase and an hydrophilic aqueous phase. Emulsions are usually characterized as oil-in-water (o/w) and water-in-oil (w/o) emulsions, depending on the nature of the dispersed phase. Multiple emulsions are more complex, as they contain more than one dispersed phase in the system. Both w/o/w and o/w/o have been described (Florence and Whitehill 1982).

Emulsions are thermodynamically unstable systems. The finely dispersed particles of the internal phase express an enormous interfacial surface area, creating a large surface energy. In order to reduce the interfacial surface area of the system, the particles tend to come together to create fewer but larger particles. Emulsion stability can be increased by the use of present-day surfactants. The current repertoire of synthetic and naturally occurring surfactants, in conjunction with improved emulsion manufacturing techniques, has resulted in the development of emulsion-based products with a commercially acceptable stability profile (~ 2 years stability).

HISTORICAL BACKGROUND

Emulsions and suspensions in the form of topical creams and ointments are perhaps the oldest sustained-release systems used in medicine. However, emulsions for parenteral administration were not developed until after World War II, when they were used to supply calories and essential fatty acids rapidly to patients who were unable to obtain oral nourishment. These lipid emulsions consisted of a vegetable oil dispersed in an aqueous solution, aided by egg phospholipids, and have now become the standard for the administration of total parenteral nutrition (TPN). The following brief historical background on IV emulsions is provided because they have been available for almost half a century. Moreover, product development issues facing a pharmaceutical scientist are similar regardless of the intended route, IV, subcutaneous (SC), or intramuscular (IM), as the three routes of administration are classified as parenteral.

All marketed IV emulsions have similar formulas: They are composed of soybean oil (10–20 percent) emulsified into an aqueous glycerol solution using egg phospholipids. Only one line of products (Liposyn™ II by Abbott) uses a 50:50 blend of safflower and soybean oils as the internal phase. The mean particle size of these TPN products ranges from 200 to 400 nm (Collins-Gold et al. 1990). The fat emulsions on the market are all isotonic since they are generally infused in large volumes intravenously.

TPN emulsions have a long history of clinical use and are therefore well understood. Relatively large doses of 2.5 to 3 g/kg per day are given to adult patients; this accumulated experience has established the overall safety of these emulsions. At the same time, many concerns about emulsions have arisen, such as their limited shelf life and their instability to added components (Washington and Davis 1987). This has stimulated research into the potential toxicities of intravenously administered emulsions. For example, it has been observed that serum triglycerides are elevated by lipid emulsions, leading to hyperlipidemia (PDR 1996). In addition, thrombocytopenia caused by phytosterolemia (Clayton et al. 1993) has been observed.

EMULSIONS AS COLLOIDAL DRUG CARRIERS

One of the challenges that has emerged in modern pharmaceutical development is the effective delivery of drugs to their targets. The use of emulsions as drug delivery systems began in the 1970s with experimentation into the incorporation of drugs such as barbituric acid (Jeppsson 1972) into fat emulsions. Subsequently drug delivery systems based on liposomes, emulsions, and micro- and nanoparticles have been explored for parenteral administration.

Both the particulate and chemical characteristics of emulsions can be exploited for drug delivery. When their particles are in the appropriate size range, emulsions are especially useful for the delivery of antifungal agents and immunomodulators because of their tendency for localization in the cells of the macrophage lineage after IV injection (Poste and Kirsh 1983).

The particulate nature of emulsions can be harnessed to deliver drugs to phagocytic cells of the reticuloendothelial system (RES), where they are taken up when injected intravenously. This property makes them useful for the delivery of agents against parasitic and infectious diseases. When injected intramuscularly or subcutaneously, emulsions migrate and accumulate at the lymph nodes before becoming available systemically.

Injection by these routes helps to target their payloads to the lymph nodes and greatly reduces delivery to the RES. These characteristics of emulsions are exploited in the delivery of vaccines, where emulsions act as good adjuvants for potentiating the immune response toward antigens. It is believed that emulsions allow slow release of antigens from the site of injection as well as presentation of antigens in a way that enhances both cellular and humoral response. Adjuvant emulsions such as MF59 (based on the oil squalene) may also act as immunostimulants, adding to their utility as antigen delivery systems (Ott et al. 1995).

In addition to harnessing the physical, particulate properties of emulsions to enhance the effectiveness of a drug, it is also possible to exploit the chemical properties of the emulsion. Depending on the type of emulsion, both hydrophilic and hydrophobic drugs can be incorporated into its inner phase. Hydrophilic drugs could be encapsulated in the aqueous phase of a w/o system, minimizing the degradation of drugs with limited hydrolytic stability. Hydrophobic drugs, which are either insoluble or sparingly soluble in water, could be incorporated into the oil phase of an o/w system. In this manner, a drug could be sequestered from direct contact with body fluids and tissues and released over a period of time. In principle, the release of a drug from an emulsion can be further prolonged by incorporation into a multiple emulsion, thus providing an extra barrier that the drug must cross before being exposed to the body compartment for release. The final advantage of drug incorporation into, and slow release from, emulsions is that only low concentrations of the drug contact body fluids and tissues at any time, reducing the toxicity of some drugs.

PREPARATION AND PROPERTIES

Manufacturing Processes

International regulatory authorities currently demand a consistent and reproducible manufacturing process for pharmaceutical products. This requirement is embodied in Good Manufacturing

Practices, or GMP. An important GMP requirement for parenteral products is that they be sterile. Other more general considerations for a manufacturing process include ease, low cost, and scalability.

A scheme demonstrating a manufacturing process for a typical emulsion is shown in Figure 6.1. The external phase solution is prepared by dissolving an emulsifier, a tonicity modifying agent, and various other excipients. The internal phase, which may include an emulsifier, is prepared as a separate solution. The internal and external phases are then mixed together by a high-speed mixer to form a coarse emulsion. The coarse emulsion must have adequate stability to last through the initial stages of the next step, which is a high-energy emulsification. The high-energy emulsification is the core step of the process; by reducing the particle size of the mixture, it imparts the desired product profile on the emulsion, such as a particular particle size distribution. This fine emulsion is passed through a filter with an appropriate pore size to eliminate any large oil droplets that still remain in the emulsion. The product is filled as large bulk aliquots that are subjected to a sterilization process. Finally, the sterilized bulk emulsion is filled into vials.

Two key steps in this generic process require elaboration: the high-energy emulsification and sterilization steps. Another critical step in the process for emulsions that are designed to deliver drugs is the drug incorporation step. These three processes will be discussed in more detail in the following sections.

High-Energy Emulsification

The goal of high-energy emulsification is to produce an emulsion with a submicron particle size distribution. Emulsions with this size distribution are kinetically more stable than those with larger particles (see "Stability"). The three most common scaleable methods described in the literature are conventional pressure homogenization, ultrasonic homogenization, and microfluidization. Table 6.1 lists these methods and the major manufacturers who supply the required emulsification equipment.

The conventional pressure homogenization technology has been employed for the production of many of the fat emulsions

Figure 6.1. *Schematic of a typical process for manufacturing an emulsion. Note that sterilization is accomplished by filtration through a 0.22 μm filter after emulsion formation.*

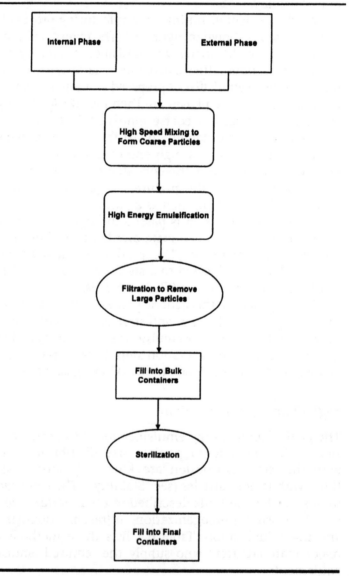

Table 6.1. *Emulsification Technologies*

Operation Principle	Commercial Equipment Manufacturers
Pressure Homogenization	APV Gaulin, Everett, Mass., USA
	APV Rannie, St. Paul, Minn., USA
	Avestin, Ottawa, Ontario, Canada
Ultrasonic Homogenization	Heat Systems, Ogden, Utah, USA
Microfluidization	Microfluidics, Inc., Newton, Mass., USA

currently on the market. As the coarse emulsion is passed through a valve under pressure, cavitational forces are created as a result of pressure and velocity changes. The dispersed phase is subjected to the shock waves generated by cavitation, which breaks up coarse, dispersed particles. Repeated passes of the coarse emulsion reduce the particle size further until an equilibrium value is reached. The equilibrium value is mainly dependent on the chemistry of the emulsion.

Ultrasonic homogenization is based on the disruption of the coarse, dispersed particles by applying short pulses of ultrasonic energy. The drawback of this method is that the throughput of the unit is too low to be suitable for large-scale operation.

The microfluidization technology is based on producing a fine emulsion by mutually impinging two streams of liquids at high pressure. The coarse emulsion is forced through an "interaction chamber," which consists of a system of channels within a ceramic block. The interaction chamber splits the coarse emulsion into two streams, which are then recombined at high speed to break the coarse particles into smaller particles. Typically, several passes of the coarse emulsion are needed to attain the desired particle size (Lidgate et al. 1992). The emulsion is cooled during the process to avoid prolonged exposure to the high temperatures associated with this process. In our experience, microfluidization is the most easily scaleable technology that reproducibly forms an emulsion with a mean particle size that is usually below 200 nm.

Sterilization

Conventional fat emulsions are generally sterilized by heat treatment. However, this sterilization method is not applicable to emulsions with components that degrade at high temperatures. For many drug-containing emulsions, the incorporated drugs may also be susceptible to degradation during sterilization. In such cases, a milder sterilization procedure may be desirable. For emulsions subject to alteration in the particle size distribution by heat treatment, individual emulsion components can be heat-sterilized separately, combined, and then emulsified under aseptic conditions. For heat-labile emulsions with particle sizes greater than 200 nm, the component liquids are presterilized by filtration through 0.22 μm filters, combined, and then emulsified under aseptic conditions (Hansrani et al. 1983). For emulsions with particle sizes of 200 nm or below, the processed emulsion can be sterilized by passing through a 0.22 μm filter (Lidgate et al. 1992; Ott et al. 1995).

Drug Incorporation

In order to obtain sustained release of a drug, it ideally should be incorporated into the internal phase of an emulsion. The drug is first solubilized in the internal phase (often with the help of cosolvents or other excipients) and is then combined with the external phase to form a coarse emulsion and processed as usual. If the drug is susceptible to degradation at high temperatures and shear forces (e.g., proteins and vaccine antigens), it is loaded after emulsion formation. For example, Ott et al. (1995) describe the addition of gD_2 and gB_2, the two glycoprotein herpes simplex viral antigens, to a preformed microfluidized squalene-in-water emulsion (MF59); the final vaccine preparation is then sterilized by passing through a 0.22 μm filter. Finally, in cases where the drug needs to be adsorbed to the internal phase or partially present in the external phase, it has been incorporated into a preformed, sterilized emulsion in a process called extemporaneous incorporation.

Characterization Techniques

In order to determine that an emulsion can be reproducibly manufactured and is appropriate for its intended use, it must be fully characterized. The basic techniques used to characterize emulsions measure properties such as the mean particle size, the particle size distribution, the zeta potential, viscosity, and chemical composition.

An emulsion is most readily characterized by its mean particle size and particle size distribution. Both light and electron microscopy have been used to assess the size, shape, and distribution of emulsion particles (Deitel et al. 1992). Particles in the micron and greater size range can be detected using light microscopy, but the method is blind to particles of submicron size. Electron microscopy can detect the submicron particles but is not suited to precise and accurate measurements of overall size and distribution. These drawbacks render microscopy of little practical use in a manufacturing environment, especially as an in-process control method.

Instead, photon correlation spectroscopy (dynamic light scattering, DLS) has become the technique of choice for determining the mean size and size distribution of submicron emulsions. This technique is capable of rapidly and directly analyzing emulsion particles suspended in solvent (Nicoli et al. 1991; Komatsu et al. 1995; Ott et al. 1995) and is thus ideal for in-process control of emulsion manufacture. In DLS, particle size and distribution are mathematically derived from a basic physical property of the emulsion—the rate of diffusion of its particles. The working range of this method is approximately 5 nm to several micrometers. The emulsion is simply diluted in the external phase until the particle concentration is below that necessary to prevent particle coincidence and is then analyzed directly. Typical analysis times are in the range of 10–30 min. The analysis yields data on mean particle size, the particle size distribution, and, if a Gaussian analysis of the data is performed, quality of the fit. The chi-squared value, which describes the goodness of fit of the analysis, has been adopted in the current USP draft as one of the characteristics for assessing size distribution (*Pharmacopeial Forum* 1991, 1994). Instruments such as the PSS Nicomp 370 have an additional algorithm for data analysis, which is well suited to analysis of non-Gaussian particle size distributions. Although DLS produces an idealized average of the emulsion particle size and distribution, it remains the best method for reproducible, precise characterization of an emulsion during manufacture.

Emulsions that are suitable for parenteral drug delivery usually have mean particle sizes in the 200–400 nm range, which are accurately detected using DLS. However, a small population of significantly larger (> 1 μm) particles is often found in the emulsion. Detection of this population of larger particles in submicron emulsions intended for SC or IM delivery is critical for two reasons: (1) Consistency of emulsion manufacture is ensured when the population of larger particles can be detected and minimized because these particles tend to slow the sterile filtration step of the process. (2) Emulsion stability can be monitored by detecting the larger particles, as their formation initiates the cascade of emulsion decomposition (see "Stability"). Thus, for submicron o/w emulsions intended to be in the 200–400 nm diameter range, the ability to detect and measure a relatively small population of large particles is pivotal in differentiating a "bad" emulsion from a "good" emulsion.

Optical particle sensing is an excellent method for counting and sizing a population of larger emulsion particles in the 1–500 μm diameter range. This technique involves the dilution of an emulsion into its external phase such that particle coincidence is insignificant in the size range of interest. The particles then flow through a sensor containing a photocell that is capable of counting and sizing the particles based on the principle of light obscuration; the particle passes in front of a light source and casts a shadow on the photocell, resulting in a signal that can be converted to a particle size. As with DLS, the light obscuration technique is an excellent tool for in-process control and product development. Typical analysis times are in the range of a few minutes, requiring sample volumes of only 10–100 μL.

The zeta potential can be directly correlated with emulsion stability. The zeta potential of an emulsion is the electrical potential in millivolts between the bulk solution and the shear plane around the emulsion particle. The shear plane is an imaginary sphere around the emulsion particle inside which the external phase moves with the particle as the particle moves through the solution. O/w lipid emulsions are typically stabilized by egg phosphatides and phospholipids, resulting, in general, in negatively charged emulsions with zeta potentials from -25 to -70 mV (Komatsu et al. 1997). However, positively charged emulsions can be prepared as described by Klang et al. (1994), who modified a typical lipid emulsion with polaxomer and stearylamine to produce a submicron emulsion with a zeta potential of 41 mV suitable for ocular drug delivery. The most popular technique for measuring the zeta potential is electrophoretic light scattering, a highly automated technique that combines electrophoresis with laser doppler velocimetry (Klang et al. 1994; Komatsu 1997). The charged emulsion particles are subjected to an applied electric field, accelerating until they reach a terminal velocity. The electrophoretic mobility depends on pH, ionic strength, viscosity, temperature, and the dielectric constant of the suspending liquid; therefore, if the emulsion needs to be diluted to perform this measurement, it must be diluted in a solvent that closely resembles the external phase of the emulsion. Inappropriate diluents could modify the particles' zeta potential, leading to spurious results (Komatsu et al. 1997).

The viscosity of the external phase is important with respect to emulsion stability and biocompatibility. Viscosity needs to be closely matched to that of plasma so as to minimize the possible adverse effects related to IV administration. Viscosity is a relatively simple property to measure. For example, Bakan et al. (1996) employed a simple, falling ball viscometer to measure the viscosity of a lipid emulsion used for contrast agents.

Both biocompatibility and consistency of emulsion manufacture depend on the chemical composition of the emulsion. As described above (see "Manufacturing Processes"), emulsions are typically produced by high-energy techniques that produce a heterogeneous distribution of the chemical components into the internal and external phases. The subsequent sterilization technique can also have an impact on composition, either through chemical decomposition due to extremes of temperature and/or pH or through selective component removal during filtration. The characterization of the chemical composition of an emulsion is a case-by-case problem dependent on the oils, surfactants, and excipients used to prepare the emulsion. For example, Herman and Groves (1992) employed high performance liquid chromatography (HPLC) to measure the phospholipids in an o/w emulsion and demonstrated that the pH drop resulting from heat sterilization was a consequence of phospholipid hydrolysis. Lidgate et al. (1992) employed gas chromatography (GC) and colorimetric methods for the component analysis of a Syntex adjuvant formulation (SAF) emulsion.

Stability

Emulsions are thermodynamically unstable because of their high interfacial surface tension at the oil-water interface. Accordingly, emulsions can have lifetimes of only a few minutes to several years. Many theoretical treatments of emulsion stability are discussed in the literature (Yamaguchi et al. 1995a, b).

The emulsion system reduces its free energy by progressively increasing the particle size until the dispersed phase separates out as a free liquid. The usual process of emulsion breakdown begins with creaming, followed by flocculation, and finally coalescence and phase separation. Creaming of an emulsion involves flotation of the less dense emulsion particles to the surface. In effect, this creates two emulsions, one of which is poorer in the internal phase, and the other richer, than the original emulsion. Gentle mixing of a creamed emulsion usually recreates the original homogeneous emulsion. Because creaming is readily reversible, it is usually not of consequence for the long-term storage of emulsions.

Flocculation, the next step of emulsion breakdown, involves aggregation of the emulsion particles such that there are strong, noncovalent interactions between the particles, yet the individual constituent particles have not lost their identity. Flocculation is of greater concern than creaming; flocculated particles are not as readily redispersed by agitation, although the association between particles is usually reversible under the appropriate conditions. It is likely that creaming accelerates flocculation, as creaming may result in emulsion particles being closer to each other. Conversely, flocculation may increase the rate of creaming in systems in which the aggregate and external phase densities are sufficiently different.

Although creaming and flocculation are reversible, once co-alescence of the emulsion has occurred, there is no reversal. In coalescence, the particles in each aggregate combine to form a single drop. The long-term consequence of coalescence is that the number of particles decreases, and the remaining particles increase in size, which finally leads to phase separation of the emulsion.

The individual factors that affect emulsion stability are complex, and it is not possible to readily predict stability. However, it has long been thought that emulsion stability is favored by small droplet radii, small density differences between the internal and external phases, and high viscosity of the external phase (Becher 1957). As discussed above, the emulsion manufacturing technique that is employed will also have a considerable impact on emulsion stability. For example, the addition of surfactants usually stabilizes emulsion particles by creating a mechanical or electrostatic barrier against coalescence and reducing interfacial free energy (Boyd et al. 1972).

Creaming and, ultimately, phase separation provide merely qualitative, visual evidence of emulsion instability. The underlying cause of these macroscopically observed characteristics is an increase in particle size and a change in the particle distribution, both of which are readily quantitated using the techniques described in above. For example, Lidgate et al. (1992) used DLS to measure the particle size range of the SAF emulsion and readily demonstrated that the particle size increased after 24 h. These data provided a sound explanation why the day-old emulsion was clogging the sterile filtration system. This slight particle size increase seen in the short term was not a predictor of long-term stability because this emulsion was stable for at least a 12-month period. Stress testing is the standard procedure for predicting long-term emulsion stability. Typical stress tests include heating, freeze-thaw cycling, centrifugation, and the addition of salts to emulsions. The commonality among these methods is that they perturb or accelerate the forces affecting the thermodynamics of the emulsion system. Yamaguchi et al. (1994) have evaluated the stability of a parenteral lipid emulsion used for carrying prostaglandin E_1 by studying the change in particle size during steam sterilization, while Dickinson et al. (1997) employed centrifugation at 16,000 rpm to study the reversibility of flocculation of o/w emulsions. Komatsu et al. (1997) examined the effects of salts on the coalescence of emulsions during freeze-thaw cycles.

APPLICATIONS TO DRUG DELIVERY

Although there is a long history of the use of topical drug emulsions and fat emulsions for TPN, emulsions have attracted relatively less attention as vehicles for parenteral drug administration. The field of colloidal drug carriers received a significant boost with the advent of liposomes and microparticles in the 1970s. Despite early pioneering work by Jeppsson (1975), it was first realized in the early 1980s that submicrometer, emulsions also belong to this class of drug carriers (Poste and Kircsh, 1983; Davis and Illum 1986; Singh and Ravin 1986). Since then, there has been much progress in the development of drug-carrying emulsions. Emulsions as vaccine adjuvants, where the emulsion itself is an active component in the elicitation of the immune response, have also been used quite extensively in recent years.

Carriers for Small Molecules

Theoretical considerations make emulsions an obvious choice for delivery of molecules that partition predominantly into an oil or aqueous phase. Early reports in the literature described substances as "oil-like" or lyophilic, but it has been found that if a substance is not soluble in water, it is not necessarily soluble in oils. Hence, this terminology has slowly given way to one in which substances are subdivided into those liking water (hydrophilic) or not liking water (hydrophobic). O/w emulsions have widely been used to deliver small molecules, while w/o emulsions have been used less often because they have inferior flow properties compared to o/w emulsions and are thus more difficult to inject. A few examples of commercially available emulsions for small molecular weight drugs are presented in Table 6.2 and are briefly discussed below in the context of the drugs with which they are formulated. Although all the current emulsion formulations cited in the table are used intravenously, their description is included here because the technology of IV emulsions (manufacturing, drug incorporation, in vitro release) is equally applicable to SC/IM emulsions.

Diazepam

The sedative diazepam is delivered as a solution formulation that requires cosolvents such as propylene glycol and ethanol. The formulation is associated with tissue irritation on injection, such as pain and thrombophlebitis. However, when diazepam was incorporated into the soybean oil phase of an Intralipid®-like emulsion, extensive clinical testing of this formulation (designated as Diazemuls®) in 2,435 patients showed that only 0.4 percent of the patients experienced any pain after administration (Von Dardel et al. 1983). Interestingly, the distribution and elimination phases after IV injection were the same for Diazemuls® as for diazepam in the solution formulation. Ninety-nine percent of the patients given the drug via emulsion exhibited sedation similar to that from a comparable strength solution of diazepam. The data clearly point to the superiority of the emulsion formulation. This product is now marketed in Europe and in the United States under the names Diazemuls® (manufactured by Pharmacia AB for Dumex, Denmark) and Dizac® (manufactured by Pharmacia AB for Ohmeda PPD Inc., USA), respectively.

Propofol

Propofol (or 2,6-diisopropylphenol) is used in the induction and maintenance of anesthesia or sedation and is sold in Europe and United States by Zeneca Pharmaceuticals. Propofol is very slightly soluble in water and is therefore formulated into the soybean oil phase of an Intralipid®-like emulsion. In this formulation, 0.005 percent disodium edetate has been added to retard the rate of growth of microorganisms in the event of accidental extrinsic contamination. It is also packaged under nitrogen to eliminate oxidative degradation (PDR 1996).

Table 6.2. *Commercially Available Drug Emulsions*

Emulsion	Manufacturer	Drug	Indication	Composition	Status	Other Comments
Diprivan®	Zeneca	propofol	sedative; hypnotic agent	10 mg propofol, 100 mg soybean oil, 12 mg egg phospholipids, 22.5 mg glycerol, 0.05 mg disodium edetate, NaOH to adjust pH	Marketed in the United States and Europe	Store between 4° and 22°C, refrigeration not recommended; shake well before use
Diazemuls®	Dumex	diazepam	sedation	5 mg diazepam, 150 mg soybean oil, 12 mg egg phospholipids, 22.5 mg glycerol, NaOH to adjust pH	Marketed in Europe	Shelf life: 18 months
Dizac®	Ohmeda	diazepam	management of anxiety disorders	5 mg diazepam, 150 mg soybean oil, 12 mg egg phospholipids, 22.5 mg glycerol, NaOH to adjust pH	Marketed in the United States	Storage at 25°C; can be exposed to temperature changes between 5° and 30°C for 4 h at least 20 times without deterioration of the emulsion quality

Table 6.2 continued on next page.

Table 6.2 continued from previous page.

Emulsion	Manufacturer	Drug	Indication	Composition	Status	Other Comments
Vitralipid®	Kabi	vitamins	nutrition	99 mg retinol palmitate, 0.5 mg ergocalciferol, 0.91 mg α-tocopherol, 15 mg vitamin K, 100 mg soybean oil, 12 mg phospholipids, 2.5 mg glycerol, NaOH to adjust pH	Marketed in Europe	Shelf life: 2 years at 2–8°C
Liple®	Green Cross	prostaglandin E_1	treatment of peripheral vascular disease	soybean oil, egg phospholipid based; exact composition not known	Marketed in Japan	

Vitamins

Vitamins A, E, and K are solubilized into the oil phase of an Intralipid®-like emulsion and are marketed in Europe. Low aqueous solubility of the vitamins is the primary motive behind the choice of an emulsion delivery system.

Prostaglandin E_1

Prostaglandin E_1 (PGE_1) is a potent vasodilator and inhibits platelet aggregation. Being hydrophobic, it has been solubilized in an aqueous solution with the aid of the host molecule α-cyclodextrin. PGE_1 in this formulation is found to be inactivated during passage through the lung. An alternative emulsion formulation comprising 3 μg PGE_1 in a soybean oil–egg phosphatidylcholine emulsion is marketed in Japan. The drug is present predominantly (93 percent) in the oil or interfacial phase of the emulsion. The drug is believed to be drawn out slowly from the inner phase of the emulsion by interaction with plasma proteins present in the bloodstream. The in vitro release profile from the emulsion has been shown to be dependent on the pH, dilution ratio, and temperature of the diluting buffer (Yamaguchi et al. 1994). Whereas the solution drug is available immediately, the emulsion formulation released PGE_1 over a period of at least 4 h. The emulsion formulation is associated with a lower side effect profile than the solution drug.

Amphotericin B

One of the early success stories with liposomes was the delivery of amphotericin B (AMB) to treat fungal infections (Lopez-Bernstein et al. 1985). AMB is practically insoluble in water and is given as a micellar solution with sodium deoxycholic acid. Dose-limiting nephrotoxicity prevents the use of higher doses of the drug in many treatment situations. Because the pathogens mostly localize in the organs of the RES in fungal infections, liposomes help to passively target the drug at the desired site of action (Lopez-Bernstein et al. 1985). A similar rationale was proposed for the use of emulsions for the delivery of AMB (Singh and Ravin 1986; Kirsh et al. 1988). In this work, AMB was incorporated into the Intralipid® emulsion by using deoxycholate and 3 percent dimethylacetamide. The use of emulsion AMB allowed administration of higher doses of AMB in animal models. Lower toxicity of emulsion AMB was demonstrated in vitro by reduced hemolytic activity and decreased lethality and nephrotoxiciy in vivo (Kirsh et al. 1988). A case was made that emulsion AMB was more practical than liposomal AMB because of the medical practitioner's familiarity with emulsions and the availability of a good safety record of emulsions, particularly of Intralipid®. Despite this seemingly strong argument and favorable animal toxicology data, no emulsion formulation has come on the market, while a liposomal formulation of AMB (Ambisome®, Nextar, USA) is now available commercially. It is unclear if the reasons for not pursuing further development of an AMB emulsion product are scientific or economic in nature, although such information would be highly valuable from the point of view of product development.

Other Small Molecules

Emulsions have also been investigated for the delivery of many other small molecules, such as many investigational anticancer compounds (Prankerd and Stella 1990). These compounds have low solubility in water and were incorporated into the oil phase of emulsions. Lin et al. (1997) recently reported the use of an ethiodized oil emulsion for the delivery of the anticancer drug doxorubicin hydrochloride in rats. Release of the drug was sustained for approximately 72 h in vitro, and the in vivo half-life could be tripled by incorporation of the drug into the emulsion. The drug concentration in the heart and kidney also decreased in the emulsion group. Thus, an emulsion formulation of doxorubicin hydrochloride appears promising for the sustained release of the drug and decreased cardiac toxicity. Emulsion formulations for taxol, another anticancer drug, are also under investigation (Lundberg 1997).

Carriers for Biologics

In general, considerably less is known about the sustained delivery of biological agents from emulsions. One reason for this is the relative newness of the biotechnology field. However, this is not the only explanation, as vaccine biologics have been around for many years. A more fundamental reason for less knowledge about the sustained delivery of biologics by emulsions is the diversity and complexity of the biological agents themselves. Therapeutic proteins and antibodies, vaccine antigens, and genes are all large macromolecular structures containing local regions of charge and hydrophilicity. These attributes make inclusion of these molecules within the internal phase of an emulsion difficult to achieve. The following subsections briefly describe the state of this field.

Vaccines

The use of emulsions as vaccine adjuvants has been investigated. Emulsions are believed to augment the immune response of a vaccine antigen through a combination of mechanisms. Sustained release, proper antigen presentation, and lymph targeting are all expected to play crucial roles in providing cellular and humoral immunity to the vaccinee. Freund's complete adjuvant (FCA), consisting of mineral oil and killed myobacteria in a crude w/o emulsion, was found to be a potent adjuvant. However, it was found to be too toxic for use in humans, as animal experiments revealed injection-site inflammation, pain, fever, and induction of autoimmune disorders. FCA minus the myobacteria, known as Freund's incomplete adjuvant (FIA), was significantly less toxic and was used in human studies (Gupta et al. 1993). Vaccination with FIA was also associated with side effects, but these were primarily limited to the injection site, because mineral oil is only slowly cleared from this site. Although FIA was found to be carcinogenic in rodents, long-term follow-up of humans failed to show increased mortality, tumors, or autoimmune diseases attributable to the adjuvant. Nevertheless, the "reactogenicity profile" of FIA was considered unacceptable for its further use in humans.

Recent work has centered around two, well-characterized, reproducible o/w emulsion adjuvants. These adjuvants, designated as SAF (manufactured by Syntex) and MF59 (manufactured by Chiron) have been widely investigated with a variety of vaccine antigens (Allison and Byars 1992; Byars et al. 1991; Hjorth et al. 1997; Ott et al. 1995). The Chiron adjuvant has been

Table 6.3. *Emulsions as Vaccine Adjuvants*

Emulsion	Manufacturer	Composition	Status
SAF	Syntex, Palo Alto, Calif., USA	5% squalane, 0.17% polysorbate 80, 2.5% Pluronic L-121 ± 250 µg/mL N-acetylmaramyl-L-alanyl-D-isoglutamine (or, muranyl dipeptide) in phosphate-buffered saline	Clinical trials
MF59	Chiron Corporation, Emeryville, Calif., USA	3.9% squalene, 0.47% polysorbate 80, 0.47% sorbitan trioleate in 10 mM sodium citrate or water	Marketed in Italy; clinical trials elsewhere

tested extensively in clinical trials and has proven safe in > 6,000 human subjects (Traquina et al. 1996). Clinical studies with vaccines against herpes simplex, cytomegalo-, hepatitis B, hepatitis C, human immunodeficiency, and influenza viruses have been conducted. An MF59-adjuvanted influenza vaccine (Martin 1997) has now been approved for sale in Italy. Table 6.3 provides further information on the two adjuvant emulsions.

Proteins and Genes

The use of emulsions has not been explored extensively for the delivery of proteins. Some work on insulin has been reported when the emulsion dosage form was intended for oral delivery (Matsuzawa et al. 1995). Insulin was encapsulated in a w/o/w emulsion in which the innermost phase consisted of 5 percent gelatin; the oil phase of 5 percent lecithin, 20 percent sorbitan trioleate, and 75 percent soybean oil; and the outermost phase of 3 percent polysorbate 80. The emulsion was stable for several days at 4°C but not at room temperature. A noticeable hypoglycemic effect in the ileum and colon loops was observed after direct administration of the emulsion to the stomach of rats. These data indicated that the emulsion might have protected insulin from proteolytic enzymes present in the gastrointestinal tract. Nevertheless, the evaluation of emulsions for parenteral sustained delivery of proteins is presently lacking.

Gene delivery has become the bottleneck in the development of genes as therapeutic agents. Liposomes and cationic lipids are presently the favorite nonviral methods for gene delivery (Lasic and Templeton 1996). Not surprisingly, interest is emerging in the use of emulsions as colloidal carriers for genes. Hara et al. (1997) recently reported the use of DNA (deoxyribonucleic acid)/cationic emulsion complexes for cell transfection. In this study, charged DNA was ionically coupled with 3β[N-(N',N'-dimethylaminoethane)-carbamoyl]cholesterol (DC-Chol), a charged variant of cholesterol, and incorporated into an o/w emulsion. The final emulsion contained 0.25 mg castor oil, 0.25 mg dioleoylphosphatidylethanolamine, 0.75 mg DC-Chol, and 0.125 mg polysorbate 80 in phosphate-buffered saline. Similarly, DNA was incorporated into reconstituted chylomicrons with the help of a quaternary ammonium derivative of DC-Chol. Emulsion formulations appeared to have more favorable physical and biological activities than cationic liposomes because they expressed similar or better transfection efficiencies in various cell lines and possessed no serum sensitivity. This preliminary study should pave the way for more careful examination of cationic emulsions for gene delivery.

OUTSTANDING ISSUES

Although emulsions have a long history of medical use, most commonly in drug delivery, they are probably one of the least understood sustained-release systems. We have attempted to outline a few issues that should be further investigated to improve our understanding of the mechanism of action of emulsions as vehicles. We believe that a deeper knowledge of emulsions would help the product development scientist to better exploit the delivery potential of these systems.

Characterization and Stability of Drug-Emulsion Formulations

As described above, there are many excellent techniques for characterizing emulsions. However, these techniques are of limited use in characterizing the molecule that is to be delivered by the emulsion. The methods typically employed to characterize small molecules or proteins are often subject to strong interferences from the emulsion components. Although the molecule is fully characterized before it is incorporated into the emulsion, for regulatory and scientific reasons, it must be fully characterized again after exposure to the emulsion. For example, emulsifiers present in emulsion formulations might cause denaturation of a biological macromolecule such as a protein, leading to a loss in its tertiary structure and thus its efficacy. The future challenge for the characterization of drug-emulsion formulations will be to develop techniques that result either in quantitative separation of the molecule from the emulsion for independent analysis or in the development of functional tests that allow direct characterization of the molecule in the emulsion.

Mode of Action

The biological fate of emulsions has not been fully explored. Most of the knowledge on emulsions is based on their analogy to colloidal particles. A good deal of pharmacodisposition data is derived from preclinical and human-testing experience using IV fat emulsions. Most preclinical studies have used a radioactively labeled oil to track the appearance of radioactivity in different tissues. From these studies, it cannot be distinguished whether the emulsions are transported as intact particles or as degraded fat compounds. Studies to characterize biological disposition of emulsions after SC or IM administration are also lacking. Similarly, mechanistic information on emulsion adjuvants is not available in the literature. A possible mechanism for the emulsion adjuvant effect is sustained release of the antigen by the emulsion. Another proposed mechanism is the ability of emulsions to carry the antigen to distant sites, as evidenced by observation of the extravasation of emulsions to the lymph nodes.

Finally, presentation of antigen on the emulsion particle surface is thought to be an important part of the action of emulsion adjuvants. Clearly, much more investigation is needed to fully understand the adjuvant activity of emulsions.

CONCLUSION

Emulsions represent a valuable system for the sustained delivery of drugs and biologics. Their simplicity and familiarity to pharmaceutical and medical scientists are significant assets for their development as carriers for new drugs and vaccines. Although the recognition of emulsions as particulate carriers has been slow to come, recent activity in the field attests to the growing interest in emulsions as sustained-release systems. With several emulsion drug-delivery systems on the market in the United States, Europe, and Japan, and a few emulsions in the late stages of clinical development as vaccine adjuvants, the field is showing signs of maturity and solid growth. This growth is likely to continue in the future, especially if critical research and development work is carried out in key areas of characterization, stability, and the biological mode of action of emulsions.

ACKNOWLEDGMENTS

We thank Dr. Gia Depillis for reviewing and editing this article.

REFERENCES

Allison, A. C., and N. E. Byars. 1992. Immunological adjuvants. In *Genetically engineered vaccines*, edited by J. E. Ciardi. New York: Plenum.

Bakan, D. A., M. A. Longino, J. P. Weichert, and R. A. Counsell. 1996. Physicochemical characterization of a synthetic lipid emulsion for hepatocyte-selective delivery of lipophilic compounds:

Application to polyiodinated triglycerrides as contrast agents for computed tomography. *J. Pharm. Sci.* 85:908–914.

Becher, P. 1957. Emulsions: theory and practice. ACS Monograph Series 135. New York: Reinhold.

Boyd J., C. Parkinson, and P. Sherman. 1972. Factors affecting emulsion stability, and the HLB concept. *J. Colloid Interface Sci.* 41: 359–370.

Byars, N. E., G. Nakano, M. Welch, D. Lehman, and A. C. Allison. 1991. Improvement of hepatitis B vaccine by the use of a new adjuvant. *Vaccine.* 9:309–318.

Clayton, P. T., A. Bowron, K. A. Mills, A. Massoud, M. Casteels, and P. J. Milla. 1993. Phytosterolemia in children with parenteral nutrition associated cholestatic liver disease. *Gastroenterology* 105: 1806–1818.

Collins-Gold, L. C., R. T. Lyons, and L. C. Batholow. 1990. Parenteral emulsions for drug delivery. *Adv. Drug Delivery Rev.* 5:189–208.

Davis, S. S., and L. Illum. 1986. Colloidal delivery systems–opportunities and challenges. In *Site-specific drug delivery*, edited by E. Tomlinson and S. S. Davis. New York: Wiley, pp. 210–224.

Deitel, M., K. L. Friedman, S. Cunnane, P. J. Lea, A. Chaiet, J. Chong, and B. Almeida. 1992. Emulsion stability in a total nutrient for total parenteral nutrition. *J. Am. Coll. Nutr.* 11:5–10.

Dickinson, E., and M. Golding. 1997. Rheology of sodium caseinate stabilized oil-in-water emulsions. *J. Colloid Interf. Sci.* 191:166–176.

Florence, A. T., and D. Whitehill. 1982. The formulation and stability of multiple emulsions. *Int. J. Pharm.* 11:277–308.

Gupta, R. K., E. H. Relyveld, E. B. Lindblad, B. Bizzini, S. Ben-Efraim, and C. K. Gupta. 1993. Adjuvants–a balance between toxicity and adjuvanticity. *Vaccine* 11:293–306.

Hansrani, P. K., S. S. Davis, and M. J. Groves. 1983. The preparation and properties of sterile intravenous emulsions. *J. Parent Sci Tech.* 37:145–150.

Hara, T., F. Liu, D. Liu, and L. Huang. 1997. Emulsion formulation as a vector for gene delivery in vitro and in vivo. *Adv. Drug Del. Rev.* 24:265–271.

Herman, C. J., and M. J. Groves. 1992. Hydrolysis kinetics of phospholipids in thermally stressed intravenous lipid emulsion formulations. *J. Pharm. Pharmacol.* 44:539–542.

Hjorth, R. N., G. M. Bonde, E. D. Piner, K. M. Goldberg, and M. H. Levner. 1997. The effect of Syntex adjuvant formulation (SAF-m) on humoral immunity to the influenza virus in the mouse. *Vaccine* 15:541–546.

Jeppsson, R. 1972. Effects of barbituric acid using an emulsion form intravenously. *Acta Pharm Suec.* 9:81–90.

Jeppsson, R. 1975. Comparison of pharmacological effects of some local anaesthetic agents when using water and lipid emulsion as injection vehicles. *Acta Pharmacol. et Toxicol.* 36:299–312.

Kirsh, R., R. Goldstein, J. Tarloff, D. Parris, J. Hook, N. Hanna, P. Bugelski, and G. Poste. 1988. An emulsion formulation of amphotericin B improves the therapeutic index when treating systemic murine candidiasis. *J. Infect. Dis.* 158:1065–1070.

Klang, S. H., J. Frucht-Perry, A. Hoffman, and S. Benita. 1994. Physico-chemical characterization and acute toxicity evaluation of a positively-charged submicron emulsion vehicle. *J. Pharm. Pharmacol.* 46:986–993.

Komatsu, H., A. Kitajima, and S. Okada. 1995. Pharmaceutical characterization of commercially available intravenous fat emulsions: Estimation of average particle size, size distribution and surface potential using photon correlation spectroscopy. *Chem. Pharm. Bull.* 43:1412–1415.

Komatsu, H., S. Okada, and T. Handa. 1997. Suppressive effects of salts on droplet coalescence on commercially available fat emulsion during freezing for storage. *J. Pharm. Sci.* 86:497–502.

Lasic, D. D., and N. S. Templeton. 1996. Liposomes in gene therapy. *Adv. Drug Delivery Rev.* 20:221–266.

Lin, S-Y., W-H. Wu, and W-Y. Lui. 1997. In vitro release, pharmacokinetic and tissue distribution studies of doxorubicin hydrochloride (Adriamycin-HCl®) encapsulated in lipiodolized w/o emulsion and w/o/w multiple emulsions. *Pharmazie* 47:439–443.

Lidgate, D. M., T. Trattner, R. M. Shultz, and R. Maskiewicz. 1992. Sterile filtration of a parenteral emulsion. *Pharm. Res.* 9:860–863.

Lopez-Bernstein, G., V. Fainstein, R. Hopfer, K. Mehta, M. P. Sullivan, M. Keating, M. G. Rosenblum, R. Mehta, M. Luna, E. M. Hersh, J. Reuben, R. L. Juliano, and G. P. Bodey. 1985. Liposomal amphotericin B for the treatment of systemic fungal infections in patients with cancer: A preliminary study. *J. Infect. Dis.* 151:704–710.

Lundberg, B. B. 1997. A submicron lipid emulsion coated with amphiphilic polyethylene glycol for parenteral administration of paclitaxel (Taxol). *J. Pharm. Pharmacol.* 49:16–21.

Martin, J. T., 1997. Development of an adjuvant to enhance the immune response to influenza vaccine in the elderly. *Biologics* 25:209–213.

Matsuzawa, A., M. Morishita, K. Takayama, and T. Nagai. 1995. Absorption of insulin using water-in-oil-in-water emulsion from an enteral loop in rats. *Biol. Pharm. Bull.* 18:1718–1723.

Nicoli, D. F., D. C. McKenzie, and J.-S. Wu. 1991. Application of dynamic light scattering to particle size analysis of macromolecules. *Am. Laboratory.* November.

Ott, G., G. L. Barchfeld, D. Chernoff, R. Radhakrishnan, P. van Hoogevest, and G. Van Nest. 1995. MF59: Design and evaluation of a safe and potent adjuvant for human vaccines. In *Vaccine design: The subunit and adjuvant approach*, edited by M. F. Powell and M. J. Newman. New York: Plenum Press, pp. 277–296.

PDR. 1996. Diprivan prescribing information. In *Physicians' Desk Reference*. Montvale, N. J., USA: Medical Economics Company.

Poste, G., and R. Kirsh. 1983. Site-specific (targeted) drug delivery in cancer therapy. *Biotechnology* 1:869–873.

Prankerd, R. J. and V. J. Stella. 1990. The use of oil-in-water emulsions as a vehicle for parenteral drug administration. *J. Parenter. Sci. Tech.* 44:139–149.

Singh, M., and L. J. Ravin. 1986. Parenteral emulsions as drug carrier systems. *J. Parent. Sci. Technol.* 40:34–41.

Traquina, P., M. Morandi, M. Contorni, and G. Van Nest. 1996. MF59 adjuvant enhances the antibody response to recombinant hepatitis B surface antigen vaccine in primates. *J. Infect. Dis.* 174:1168–1175.

USP. 1991. *Pharmacopeial Forum* 17:2219.

USP. 1994. *Pharmacopeial Forum* 20:7170.

Von Dardel, O. C. Mebius, T. Mossberg, and B. Svensson. 1983. Fat emulsions as a vehicle for diazepam. A study of 9492 patients. *Br. J. Anaesth.* 55:41–47.

Washington, C., and S. S. Davis. 1987. Ageing effects in parenteral fat emulsions: The role of fatty acids. *Int. J. Pharm.* 39:33–37.

Yamaguchi, T., N. Tanabe, Y. Fukushima, T. Nasu, and H. Hayashi. 1994. Distribution of prostaglandin E_1 in lipid emulsion in relation to release rate from lipid particles. *Chem. Pharm. Bull.* 42:646–650.

Yamaguchi, T., K. Nishizaki, S. Itai, H. Hayashi, and H. Oshima. 1995a. Physicochemical characterization for a parenteral lipid emulsion: Determination of Hameker constants and activation energy of coalescence. *Pharm. Res.* 12:342–347.

Yamaguchi, T., K. Nishizaki, S. Itai, H. Hayashi, and H. Oshima. 1995b. Physicochemical characterization for a parenteral lipid emulsion: Influence of cosurfactants on flocculation and coalescence. *Pharm. Res.* 12:1273–1278.

7

Liposomes for Local Sustained Drug Release

Christien Oussoren
Gert Storm
Daan J. A. Crommelin
Utrecht Institute for Pharmaceutical Sciences,
Utrecht University
Utrecht, The Netherlands

Judy H. Senior
Maxim Pharmaceuticals Inc.
San Diego, California

Liposomes have been successfully employed for the controlled release and site specific delivery of drugs in several therapeutic fields. Liposomal formulations are versatile drug carrier systems because their physicochemical properties can be relatively easily altered, and both water- and lipid-soluble drugs can be incorporated in relatively high concentrations. Moreover, liposomes are biodegradable and possess low inherent toxicity (Storm et al. 1993). Large-scale production of liposomes is possible, and liposomes with pharmaceutically acceptable shelf lives can be produced (Crommelin and Schreier 1994). Several liposomal products for parenteral administration are already on the market. Evidently, convincing data was produced regarding chemical, physical, and biological characterization of the product, reproducibility of the production process, sterility, and apyrogenicity. To date, marketed liposome and lipid-based formulations are mostly intended for intravenous (IV) use. The main reason for the application of marketed formulations is to increase the therapeutic index by virtue of increased delivery to target tissue and/or reduced delivery to nontarget tissue sensitive to the toxicity of the drug.

Another attractive characteristic of liposomes is their ability to act as a sustained-release system after local administration. Pharmaceutical research and development of liposomes as a sustained-release system after local administration is mainly focused on favorably altering the pharmacokinetics of the drug so that the therapeutic agent is delivered over an extended period of time. Thus, local sustained-release systems aim for prolonged drug concentrations at either the local site of injection or within the blood compartment.

This chapter will focus on the use of liposomes in the formulation of depot preparations for prolonged drug release after local injection. Particular emphasis will be given to the subcutaneous (SC) and intramuscular (IM) routes.

LIPOSOMES AS PHARMACEUTICAL DOSAGE FORMS

Liposomes consist of one or more phospholipid bilayers separated by one or more internal aqueous compartments. They come in many varieties, encompassing differences in chemical composition and physical characteristics. Liposomes are heterogeneous in size and range from about 0.02 μm for small unilamellar vesicles (SUVs) to several μm for multilamellar vesicles (MLVs) or more for multivesicular liposomes (MVLs) (Figure 7.1). As they comprise both an aqueous and a lipophilic phase, liposomes can be used as carriers for hydrophilic, lipophilic, and amphiphilic drugs.

Figure 7.1. *Two-dimensional representations of major liposome types and phospholipid bilayers.*

SUV	MLV	LUV	MVV	DepoFoam™

| 0.02–0.1 μm | 0.1–5 μm | 0.1–1 μm | 1–10 μm | 1–100 μm |

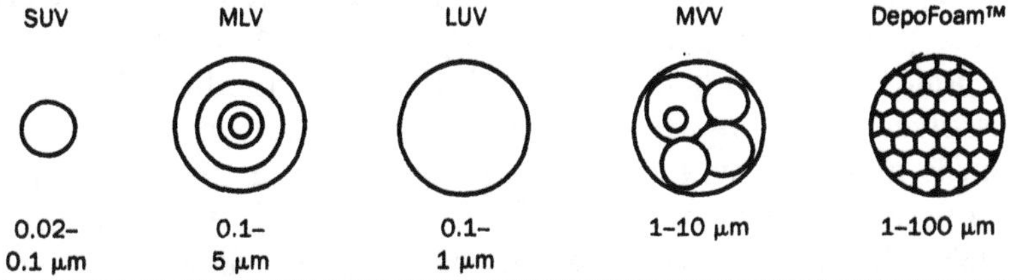

SUV: small unilamellar vesicles; MLV: multilamellar vesicles; LUV: large unilamellar vesicles; MVV: multivesicular vesicles.

Liposome characteristics, such as size, bilayer fluidity, and surface charge, depend, in part, on their lipid composition. A wide variety of different types of saturated or unsaturated, naturally occurring, and synthetic lipids have been used in the preparation of liposomes. The physicochemical characteristics of the bilayers of liposomes also highly depend on other variables, such as temperature, pH, ionic strength, and the presence of divalent cations. The number of phospholipid bilayers depends mainly on the type of liposome and method of preparation. For instance, larger liposomes prepared by the thin-film hydration method consist of several aqueous compartments and lipid bilayers, whereas small unilamellar liposomes (known as SUVs) consist of an aqueous core surrounded by one lipid bilayer (Figure 7.1).

Typical Methods for the Preparation of Liposomes

Several techniques have been successfully developed for the large-scale preparation of liposomes. In general, two steps can be discerned in the preparation of liposomes: (1) hydration of the lipids resulting in vesicle formation and drug encapsulation, and (2) size reduction to the desired vesicle size. In some production processes, these steps are linked together. In this section, we will give some examples of the large-scale preparation of liposomes. More specific details on liposome preparation for parenteral administration and their scale-up feasibility are given elsewhere (Barenholz and Crommelin 1994; Chapter 17 of this book).

Lipids can be hydrated by adding an aqueous solvent and subsequent mixing. The most simple and widely used method for lipid hydration is the so-called "thin-film" method. Lipids are dissolved in an organic solvent in a round-bottomed flask. The film is hydrated by adding an aqueous solution above the transition temperature of the lipids. Shaking yields a dispersion of MLVs. The use of lipids that are freeze-dried followed by rehydration with an aqueous solution of the drug allows fast hydration because of the high porosity of the lyophilized powder (Van Hoogevest and Frankhauser 1989). Another way to hydrate lipids is the injection of a solution of lipid in a large volume of solvent that is miscible with water (e.g., ethanol and dimethylsulfoxide [DMSO]) into an aqueous phase. The resulting dilution of the organic solvent induces vesicle formation (Martin 1990). Liposome formation can also be achieved by dispersing phospholipids in a mixture of an aqueous and organic solvent followed by removal of the organic solvent. During the gradual removal of the organic phase, liposomes are formed (Allen 1984). For most methods, the resulting vesicle size depends on factors such as the stirring rate, vesicle charge, and ionic strength.

Methods for the encapsulation of a drug into liposomes should preferably result in a high encapsulation efficiency, minimizing the amount of extraliposomal drug that has to be removed from the dispersions, and a high drug/phospholipid ratio, in order to limit the dose of phospholipids administered to the patient. Hydrophilic drugs can be passively encapsulated in liposomes during the preparation of the liposome dispersion. Upon hydration of phospholipids in an aqueous medium, bilayers are formed. During the formation of the bilayers, a portion of the aqueous phase, including dissolved drug, is entrapped within the aqueous compartments of the liposomes. Entrapment of hydrophilic drugs depends on the solubility of the drug in water, the phospholipid concentration, and the method of preparation. A relatively high encapsulation efficiency is obtained by using lyophilized lipids due to the increased intimacy of the association between the dehydrated lipid and the drug (Kirby and Gregoriadis 1984). A strategy to improve the encapsulation efficiency of hydrophilic drugs is to subject liposomes to freezing/thawing cycles in the aqueous hydration medium containing the drug. During freezing and thawing, a relatively high concentration of the drug is built up between the lipid bilayers (Chapman et al. 1990). Other approaches to improve the encapsulation efficiency of hydrophilic drugs involve the active loading of the drug into preformed liposomes. For example, high encapsulation of doxorubicin can be achieved by the pH gradient method, which is achieved by a transmembrane pH gradient (Haran et al. 1993). Accumulation of doxorubicin is the result of diffusion of unprotonated doxorubicin through the liposomal bilayer into the liposomes, where it becomes protonated and entrapped inside the aqueous interior of the liposomes. Such active loading methods have the advantage of not only a high encapsulation efficiency and capacity but may also result in reduced leakage of the encapsulated compound as compared to passive entrapment (Bolotin et al. 1994).

Lipophilic or amphiphatic drugs can intercalate into the lipophilic phase, the liposomal bilayers. The encapsulation efficiency of these molecules mainly depends on the liposome composition and the type of aqueous hydration medium. The interaction of such compounds with the bilayer can result in alteration of vesicle properties, such as permeability and stability of the bilayer structure. For lipophilic drugs, maximal incorporation into the liposomal bilayer mainly depends on the amount of lipid and the solubility of the drug in lipid bilayers.

If drug encapsulation during liposome preparation is not complete, it is usually desirable to remove free, nonencapsulated drug from the liposome dispersion, especially in the case of toxic agents or drugs that affect liposome stability. Therefore, the final step in the preparation of liposomes deals with the removal of free, nonencapsulated drug from the liposome dispersion. Classical bench-scale techniques are dialysis and gel filtration. For some liposome dispersions, e.g., doxorubicin-containing liposomes, an ion exchange resin offers a good alternative (Storm et al. 1985). Free drug can also be removed from the liposomal dispersion by ultracentrifugal sedimentation or by ultrafiltration (Barenholz and Crommelin 1994).

Particle size is a critical factor for the in vivo behavior of liposomes (Senior 1987). Therefore, the size of liposomes designed to be used for pharmaceutical applications should be well defined and reproducible when preparing different batches. A standard procedure for obtaining liposomes in every defined size range on commercial scale is not available presently. Selecting the technique for large-scale sizing of liposomes depends mainly on the desired liposome size. For the large-scale production of small unilamellar liposomes, the most promising techniques are high-pressure homogenizers (Brandle et al. 1993). Dispersions of larger-sized liposomes with narrow size distributions are more difficult to scale-up. Extrusion through a filter with a defined pore size under low or medium pressure is possible. Clogging of the filter can be prevented by using continuous flow through a high-pressure extruder (Schneider et al. 1994).

Liposomal products that are designed for parenteral administration require either a terminal sterilization step or aseptic processing. Terminal sterilization is preferred if this can be done without destroying the integrity of the product. Typical methods are heat sterilization by autoclaving or filtration. Heat sterilization by autoclaving (121°C, 15 min) may damage liposomal products. However, damage can be prevented if the proper conditions are chosen (Zuidam et al. 1993, 1996). If the liposomal product is heat labile, sterilization by filtration through 0.2 μm pores before aseptic filling into the final vial may be a good alternative. Heat labile large liposomes, such as MLVs, should be manufactured aseptically.

Stability of Liposome Formulations During Storage

Liposomal products must be stable during storage and transport. For most pharmaceuticals, a shelf life of at least one year is a minimum prerequisite. Factors affecting liposomal stability are of a physical and/or a chemical nature.

Liposomal shelf life can be largely limited by the leakage of drug from the vesicles so that the percentage of free drug increases relative to encapsulated drug on storage. Another limiting factor for liposomal shelf life can be changes in particle size, although this rarely occurs without concurrent loss of encapsulated drug. Increases in the average size and size distribution occurs by aggregation and/or fusion of liposomes. These processes are strongly dependent on phospholipid composition, temperature during storage, and the aqueous medium. The physical instability of liposomes may also become manifest by leakage of drug molecules from the liposomes. The leakage rate strongly depends on the physicochemical nature of the drug and the bilayer composition (Crommelin et al. 1993). Leakage of encapsulated highly lipophilic, bilayer-interacting drugs in storage is usually not a major problem. These substances stay in the bilayer even in freezing-thawing or freeze-drying–rehydration cycles when the proper conditions are chosen (Van Winden et al. 1997b). Water-soluble drugs may leak from the liposomes more easily. With hydrophilic, nonbilayer-interacting encapsulated material, the retention on storage depends much more on bilayer characteristics. In response to temperature changes, bilayers may undergo a phase transition, shifting at the phase transition temperature from a "gel" state, in which the fatty acid side chains of the membrane are closely packed and relatively ordered, to a "fluid" state in which the acyl side chains of the bilayers have more rotational freedom or vice versa (Figure 7.2). In general, liposomes with rigid bilayers are more stable with regard to the leakage of drugs than liposomes with more fluid

Figure 7.2. *Representation of the bilayer transition of liposomes.*

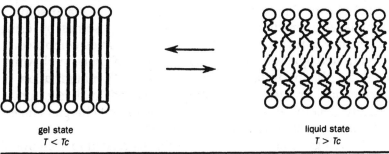

gel state
$T < Tc$

liquid state
$T > Tc$

Depending on phospholipid composition and temperature (T), bilayers are in a gel state or in liquid state. Phase transition occurs at the phase transition temperature (Tc). Above the Tc, liposomes have more fluid bilayers. The Tc is determined by the length and the degree of saturation of the two phospholipid acyl chains. Generally, the Tc increases with longer acyl chains and a higher degree of saturation.

bilayers, provided that the rigid liposomes are not stored around the transition temperature of the bilayer.

Chemical destabilization may be the result of the peroxidation of acyl chains of unsaturated phospholipids and hydrolysis of the ester bonds of the liposomal phospholipids. Peroxidation of the liposomal lipids can be minimized by using an inert atmosphere (e.g., nitrogen), metal-complexing agents (e.g., EDTA [ethylenediamine tetraacetic acid]), and antioxidants (e.g., α-tocopherol) (Frøkjaer et al. 1984). If lipid peroxidation is not prevented, it may cause an increase in the permeability characteristics of the bilayer. The impact of lipid peroxidation on liposome behavior and safety on parenteral administration has not been studied in detail. Hydrolysis of the acyl chains of phospholipids is more difficult to control. In an aqueous liposome dispersion, the liposomal phospholipids can hydrolyze to free fatty acids and lyso-phospholipids. The kinetics of phospholipid hydrolysis were extensively investigated by Grit et al. (1993). Storage conditions such as temperature, pH, bilayer rigidity, and the composition of the aqueous medium were found to influence the hydrolysis rate (Grit and Crommelin 1992).

Strategies to improve the shelf life of pharmaceutical liposomes include (1) the careful selection of liposome composition and the composition of the aqueous medium to ensure chemical and physical stability of the liposomes; (2) storage at low temperatures to reduce chemical destabilization of the liposomal phospholipids; (3) reduction of leakage of water-soluble, nonbilayer-interacting active agents from liposomes by using either liposomes with rigid bilayers or liposomes composed of unsaturated lipids and substantial fractions of cholesterol, which rigidifies the phospholipid bilayers; and (4) minimization of aggregation or fusion by adding a charged lipid (e.g., phosphatidylglycerol [PG] or phosphatidic acid) and charged cholesterol esters, such as cholesteryl hemisuccinate.

A strategy to circumvent the problems related to the limited shelf lives of liposomes in aqueous media is to freeze-dry liposomes. Freeze-drying of liposomes generally results in a cake with a large surface area. The process consists of a freezing step and subsequent sublimation of the water. The presence of a cryoprotectant, the nature of the bilayer, liposome size, and several technological parameters regarding the freezing process are important variables to optimize the freeze-drying process of liposome dispersions. A number of studies demonstrate that when the proper conditions and drug are selected, freeze-drying of drug-containing liposomes for pharmaceutical production is feasible (Van Winden and Crommelin 1997a). Currently, relatively little information on the large-scale freeze-drying of liposomes is available.

LIPOSOMES INJECTED SUBCUTANEOUSLY OR INTRAMUSCULARLY AS SUSTAINED– RELEASE SYSTEMS

SC and IM injection of liposomes may serve two purposes: maintain prolonged drug concentrations in the blood compartment or maintain prolonged concentrations at the local site of injection. In this section, we will focus on the potential of subcutaneously and intramuscularly injected liposomes to act as a sustained-release system for maintaining prolonged therapeutic drug concentrations in the blood compartment. Examples of sustained-drug release systems for maintaining prolonged drug concentrations at the injection site are described at the end of this chapter.

Following SC and IM injection, liposomes do not have direct access to the bloodstream. Instead, liposomes remain at the injection site or are taken up by lymphatic capillaries draining the injection site (Figure 7.3). Liposomes that remain at the injection site will slowly destabilize and degrade in time, concurrently losing their drug content. The actual concentration and duration of drug concentrations in the blood compartment depend mainly on liposome retention at the injection site and the rate of release of drug from the liposomes. Factors influencing liposome retention, drug release, and other aspects related to SC and IM injection of liposomes are discussed in this section.

Figure 7.3. *Representation of drug release and absorption of liposomes from the SC and IM injection site.*

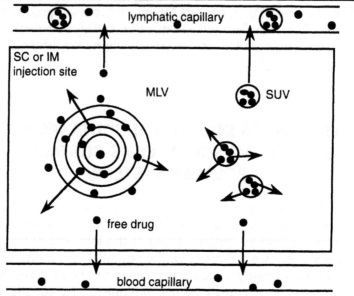

Following SC and IM injection, large liposomes remain at the injection site, whereas small liposomes (roughly < 0.1 μm) may enter the lymphatic capillaries and reach the blood circulation. Sustained release from locally injected liposomes occurs by the gradual release of drug from liposomes remaining at the injection site. After release, free drug enters the blood circulation. Small molecular weight drugs (< 16 kD) will enter the blood compartment by passing through the pores in the blood capillary walls, whereas larger molecules are mainly transported by the lymphatics.

Factors Influencing Liposome Retention at the Injection Site

Liposome-Related Factors

Liposome size is the most important factor that determines liposome retention at the injection site after SC and IM injection. Several reports refer to a cutoff value of 0.1 μm, above which the liposomes do not appear in the blood to any substantial extent and remain at the injection site (Allen et al. 1993a; Patel 1988; Tümer et al. 1983). A more detailed study on the fate and behavior of SC administered liposomes showed that a substantial fraction of dispersions with a larger mean diameter is absorbed from the SC injection site (Oussoren et al. 1997b) (Figure 7.4A and B). As liposome dispersions are heterogeneous with respect to size, it is likely that the relatively small liposomes from the dispersion are taken up by lymphatic capillaries, whereas the larger liposomes remain at the injection site. This size dependent retention at the injection site is likely to be related to the process of particle transport through the interstitial tissue. The structural organization of the interstitium dictates that larger particles will have more difficulty passing through the interstitium and will largely remain at the site of injection. Smaller liposomes can migrate through the aqueous channels in the interstitium and may be transported to the blood compartment via lymphatic capillaries draining the injection site. However, although size is a crucial factor for liposome retention at the injection site, even after injection of small liposomes, a substantial fraction of the injected dose remains at the injection site. Typically, after the injection of small liposomes (mean size about 0.07 μm), about 40 percent of the injected dose remains at the SC injection site. Considering the relatively high and prolonged retention of large liposomes compared to small liposomes, it appears that if the main goal is to obtain local sustained release for a long period of time, local injection of large, nonsized liposomes that remain at the site of injection is preferred above local injection of small liposomes.

Other liposome-related factors do not appear to influence liposome retention at the SC injection site substantially. Retention of small (0.07 μm) liposomes is independent of the presence of charged lipid or cholesterol in the liposomal bilayer, bilayer fluidity, and the presence of a hydrophilic polyethylene glycol (PEG) coating on the liposome surface. Moreover, the injected lipid dose does not appear to influence the absorption of liposomes from the injection site at doses ranging between 0.01 and 10 μmol total lipid (Oussoren and Storm 1997; Oussoren et al. 1997b).

Figure 7.4a. *Influence of liposome size on liposome absorption after SC administration: Percentage of injected dose absorbed from the SC injection site.*

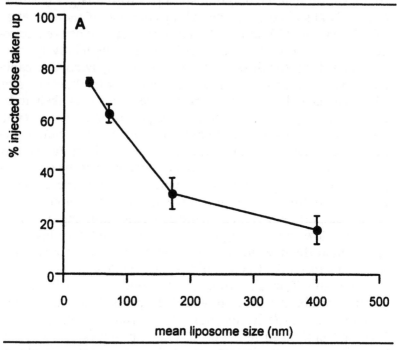

A single dose of radiolabeled liposomes (EPC:EPG:Chol, 10:1:4 molar ratio, 2.5 μmol TL [total lipid]) of varying sizes was injected subcutaneously on the dorsal side of the foot of rats. Levels of liposomal label were determined 52 h after injection. Values represent the mean percentage ± SD of 4 rats.

Influence of the Anatomical Site of Injection

Another important factor determining the fate of liposomes after SC administration is the anatomical site of injection. After SC injection of small (0.1 μm) liposomes into the flank of rats, disappearance from the site of injection is lower than when liposomes are injected into the foot pad or into the dorsal side of the foot (Oussoren et al. 1997a). For example, after SC injection into the flank of rats, the injected liposomes remain almost completely at the site of injection, whereas 40 to 50 percent of the injected

Figure 7.4b. *Influence of liposome size on liposome absorption after SC administration: Percentage of injected dose circulating in the total blood volume.*

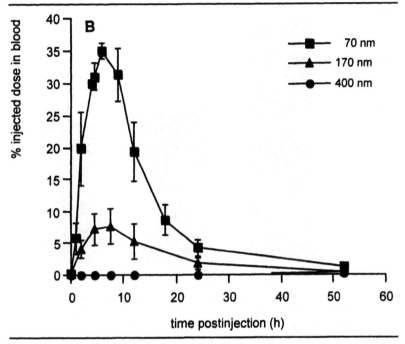

A single dose of radiolabeled liposomes (EPC:EPG:Chol, 10:1:4 molar ratio, 2.5 µmol TL [total lipid]) of varying sizes was injected subcutaneously on the dorsal side of the foot of rats. Levels of liposomal label were determined 52 h after injection. Values represent the mean percentage ± SD of 4 rats.

dose is taken up from the injection site after injection into the foot (Figure 7.5). The observed site dependent disposition is attributed to differences in the structural organization of the subcutaneous tissue at the different sites of injection. These results demonstrate that the anatomical site of SC injection should be considered carefully when designing sustained-release formulations based on the SC administration of liposomes. The observed site dependent disposition may be of less importance if large liposomes are injected, since retention of large liposomes at the SC injection site is invariably high.

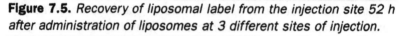

Figure 7.5. *Recovery of liposomal label from the injection site 52 h after administration of liposomes at 3 different sites of injection.*

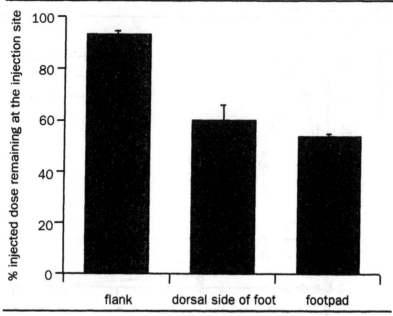

A single dose of radiolabeled liposomes (EPC:EPG:Chol, 10:1:4 molar ratio, mean diameter 0.1 μm, 2 μmol TL) was injected into the flank, into the dorsal side of the foot and into the footpad of rats. Values represent the mean percentage of injected dose remaining at the site of injection ± SD of 4 rats.

Massage of the local injection site induces lymphatic absorption of interstitial fluid, including injected material. Accordingly, blood levels of drug encapsulated in 0.2 μm liposomes increased substantially (about 10-fold) when the SC injection site is manually massaged. A possible utilization of this approach can be found in the development of self-controlled drug depots with which the patient can regulate drug release by massaging the subcutaneous depot of liposome-encapsulated drugs (Trubetskoy et al. 1998).

Drug Release from Liposomes After Subcutaneous and Intramuscular Injection

In contrast to free drugs that are generally rapidly cleared from the interstitium, high concentrations of liposome-encapsulated drugs remain at the injection site for a prolonged period of time (Postma et al. 1999) (Figure 7.6). Sustained drug release from

Figure 7.6. *Percentage of the injected dose of [111]Indium-labeled desferal remaining at the injection site after SC injection of free or liposome-encapsulated desferal.*

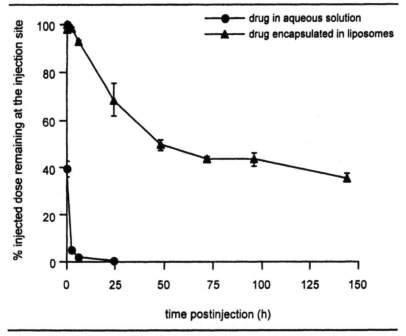

A single dose of large, unsized liposomes (DSPC:DPPG:Chol, 10:1:4 molar ratio) was injected subcutaneously in the neck region of mice. The amount of radioactivity at the SC injection site was determined by drawing regions of interest using a gamma camera and expressed as the percentage of the injected dose [111]Indium-labeled desferal. Values represent the mean percentage injected dose remaining at the injection site ± SD of 3 to 6 mice (Source: Postma et al. 1999).

liposomes is the result of leakage from locally destabilized liposome bilayers or due to degradation of the liposomes and subsequent release of the encapsulated drug. Destabilization and degradation of liposomes at the injection site is a complex process and is influenced by several factors. Serum proteins, phagocytosing cells, shear stress, and liposome aggregation at the injection site may induce leakage of the liposomal contents. The exact mechanism and factors influencing drug release from locally injected liposomes is not known at present. However, the rate and duration of release from liposomes depends strongly on liposome stability. The influence of liposome characteristics on the release of liposome-encapsulated drugs in vivo has been mainly studied after IV administration. Less attention has been directed toward the role of such parameters after local injection. Release characteristics after IV administration may be predictive for the relative release characteristics of liposomes after local administration. Stable liposomes composed of saturated lipids are known to release their content slower than liposomes with higher membrane fluidity. The same relationship was reported between liposomal stability and drug release after IM administration of liposomes (Schreier et al. 1987). The rate of release of encapsulated drug as well as the erosion of liposomes at the injection site were found to be a function of the fluidity of the lipid membranes. The release rate of gentamicin from egg phosphatidylcholine (EPC) liposomes was about 7 times slower than from soy phosphatidylcholine (SPC) liposomes, with more fluid bilayers when injected intramuscularly. Additionally, after intratumoral injection of [111]Indium-labeled desferal encapsulated in liposomes, the fraction of label remaining at the injection site was about 10-fold higher when encapsulated in liposomes with rigid bilayers than when encapsulated in liposomes with more fluid bilayers (Koppenhagen 1997). From these observations, it may be concluded that drug absorption rates may be controlled (within certain limits) by using lipids yielding different degrees of bilayer fluidity.

Other liposome-related factors are less important for drug release. In several studies, retention of differently charged liposomes at the injection site after IM injection has been described. Results suggest that retention of negatively charged liposomes at the IM injection site is somewhat less than the retention of neutral and positively charged liposomes (Kim and Han 1995;

Eppstein 1982). However, plasma concentrations of methotrexate are not substantially influenced by liposome charge (Kim and Han 1995). Studies on the influence of injection volume after SC and IM injection of chloroquine-containing liposomes showed that release rates of chloroquine are slightly lower when small volumes were injected than when large volumes were used (Kadir et al. 1991). The authors speculate that the effect may be related to the formation of a stable aggregate at the injection site. Smaller injection volumes result in a smaller and more compact depot from which drug release will be slower than from a larger depot. Lowering the drug-to-lipid ratio appears to improve the antimalarial activity of liposome-encapsulated desferal, probably as a result of even more prolonged sustained release of the drug (Postma et al. 1998).

Pharmacokinetics of the Released Drug

After release from locally injected liposomes, a drug can be absorbed into the blood compartment. Molecules that are small enough to pass through the blood endothelial capillary (smaller than about 16 kDa) enter the blood compartment directly. Larger molecules enter the blood compartment mainly via the lymphatic capillaries (Supersaxo et al. 1990). Drug concentrations in the blood compartment are usually low since drug release and subsequent absorption from the injection site is slow. Particularly rapidly cleared drugs, such as proteins and peptides, do not accumulate in the blood compartment, as clearance from the blood is faster than absorption from the injection site. Therefore, sustained drug delivery by SC, IM, or local injection of liposomes is of particular interest for drugs that require low but prolonged plasma concentrations for optimal efficacy. For example, methotrexate is an antitumor drug with a short half-life in vivo and requires repeated administration for optimal efficacy. SC administration of methotrexate encapsulated into a lipid-based drug delivery system resulted in 120-fold lower but prolonged plasma concentrations than SC administration of the free drug, whereas the AUCs (areas under concentration-time curve) were similar (Table 7.1) (Bonetti et al. 1994). As a consequence of the altered pharmacokinetics, the drug's potency was substantially increased (130-fold), whereas toxicity (LD50) was substantially decreased (110-fold). Similar observations were made after local

Table 7.1. *Pharmacokinetic Parameters of Methotrexate in Case of Subcutaneous Injection of Methotrexate in Aqueous Solution and Encapsulated in a Lipid-Based Sustained-Release Formulation*

	Methotrexate In In Aqueous Solution	Methotrexate Encapsulated In a Lipid-Based Sustained-Release Formulation
$t_{1/2abs}$ (h)	0.16 ($\pm$0.01)	50 ($\pm$4)
C_{max} (μmol/L)	17 ($\pm$5)	0.14 ($\pm$0.06)
$t_{1/2el}$ (h)	0.53 ($\pm$0.01)	100 ($\pm$5)
AUC (μmol.h/L)	18 ($\pm$4)	28 ($\pm$9)

$t_{1/2abs}$: absorption half-life from the SC injection site; C_{max}: maximum plasma level; $t_{1/2}$ elim: elimination half-life in plasma; AUC: area under the curve (from Bonetti 1994).

(intrathecal) administration of morphine encapsulated into the lipid-based sustained-release system. Low cerebrospinal fluid (CSF) levels were maintained for up to 1 week after intrathecal injection of the encapsulated drug, whereas intrathecal injection of unencapsulated morphine yielded a high CSF concentration within 30 min of injection, which then rapidly declined (Figure 7.7). In some cases, plasma concentrations after SC administration of drugs encapsulated in large liposomes are even below the detection limit, whereas therapeutic effects are observed. Apparently, for some drugs, very low but prolonged plasma concentrations are sufficient for exerting therapeutic effects (Kim and Howell 1987; Roy and Kim 1991).

Lymphatic Absorption of Liposomes

Following SC and IM injection, small liposomes (< 0.1 μm) may be taken up by lymphatic capillaries draining the injection site. Once liposomes have entered the lymphatic capillaries, they will pass through a system of lymphatic vessels and will encounter one or more lymph nodes, where a fraction will be retained. Relatively high concentrations of liposomes are found in lymph nodes after SC administration (Oussoren et al. 1997b). The ability of the lymphatic system to take up liposomes from interstitial spaces has been exploited for the targeting of drugs to regional lymph nodes. Both the SC and IM routes of administration have been investigated for the delivery of liposomes to regional lymph nodes. Delivery of liposomes to regional lymph nodes has a number of applications, including diagnosis and treatment of diseases with lymphatic involvement, such as tumor metastases, viral and bacterial infections, and immunization.

Liposome-encapsulated antitumor drugs that have been administered subcutaneously or intramuscularly as a strategy for the treatment of lymphatic metastases inhibit tumor cell growth in lymphatic metastases more effectively than the free, nonencapsulated drug (Yaguchi et al. 1990; Konno et al. 1990). The

Figure 7.7. *Free (triangle, open) and total (circle, open) morphine concentrations in cerebrospinal fluid (CSF) following intrathecal administration of 150 µg DepoMorphine or morphine sulfate (circle, closed).*

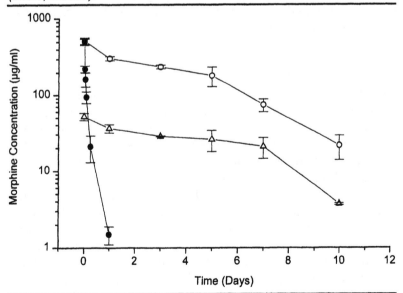

Each point represents the average and standard error of the mean (SEM) for 3 animals. (Courtesy of R. C. Willis, SkyePharma plc.)

relatively high therapeutic efficacy may be attributable to two different mechanisms: (1) The sustained release of the drug from liposomes remaining at the injection site may provide low drug concentrations in interstitial tissue for prolonged periods of time. Part of the released drug may be absorbed by lymphatic vessels and be transported to regional lymph nodes, thus providing low but lasting concentrations in lymph node tissue. (2) Liposomes are efficiently captured in regional lymph nodes after being taken up from the SC injection site by lymphatic vessels. Sustained release of a drug from intranodal liposomes may provide relatively high and prolonged concentrations in lymph node tissue.

Liposomes that have been taken up by the lymphatic capillaries and passed through the lymph nodes reach the general circulation, where they behave as if administered by the IV route. Peak levels in blood are reached within about 6 to 12 h after injection, after which plasma concentrations decline. Once liposomes have reached the blood compartment, they tend to accumulate in organs rich in cells of the mononuclear phagocyte system (MPS). Due to a rich blood supply and on abundance of MPS cells, the major sites of accumulation are the liver and the spleen (Oussoren et al. 1997b).

Tissue Protective Effect of Liposomes

Apart from the sustained release of drugs, the administration of liposome-encapsulated drugs may offer another important advantage over the administration of the free drug. SC and IM administration of highly irritating drugs, such as several antitumor drugs, often results in tissue irritation and necrosis at the site of injection. Several studies have demonstrated that liposomes can protect surrounding tissue from the damaging effects of irritating drugs after SC and IM injection. Forssen and Tökes (1983) were the first to report on the protective effect of liposomes against the dermal toxicity of intradermally administered doxorubicin. More recently, several studies have confirmed the protective effect of liposomes against the damaging effects of doxorubicin (Gabizon et al. 1993; Balazsovits et al. 1989) and other irritating drugs such as novaminsulfon (Kadir et al. 1992) and the antitumor drugs vincristine and mitoxantrone (Boman et al. 1996; Oussoren et al.

1998). The protective effect of liposomes depends on the route of administration and the liposomal lipid composition (Oussoren et al. 1998). After IM injection, the protective effect against local toxicity is much stronger than after SC administration. The differences may be related to faster clearance of the injected material after IM injection than after SC injection. Additionally, liposomes with rigid bilayers protect surrounding tissue better than liposomes with more fluid bilayers. Although fluid liposomes initially protect surrounding tissue from direct dermal toxicity of the encapsulated drug, slow release may result in prolonged and persistent exposure of the injection site to low doses of the toxic drug, which may be a serious cause of local tissue damage (Oussoren et al. 1998).

THERAPEUTIC APPLICATIONS OF LIPOSOMES INJECTED SUBCUTANEOUSLY AND INTRAMUSCULARLY

Sustained-release systems aim to provide drug release at rates that are sufficiently controlled to provide periods of prolonged therapeutic action. Pharmacotherapeutic reasons for developing sustained-release injectable products include maintenance of an optimal drug therapeutic concentration profile of rapidly cleared drugs and reduction of adverse effects due to avoidance of peak concentrations. Table 7.2 provides an overview of literature reports on the SC and IM injection of liposomes for sustained release. The following section highlights selected examples.

Small Molecular Weight Drugs

Antitumor Drugs

The objective for developing sustained-release formulations of antitumor drugs is to treat some malignancies by prolonging their exposure to antitumor drugs at a therapeutic concentration. Improvements in safety may also be observed when sustained low-dose administration of antitumor drugs is less toxic

Table 7.2. *Overview of Studies on the Application of Subcutaneous and Intramuscular Application of Liposomes for Sustained Drug Release*

Therapeutic Agent	Liposome Characteristics	Treatment	Animal/Model	Effect	Reference
amikacin	DOPC:DPPG:Chol:triolein (9.3:2.1:15:1.8) mean diameter between 15 and 35 µm	single dose SC infiltration	*Staphylococcus aureus* infected mice	Sterilization of the injection site and reduced systemic toxicity compared to the free drug	Roerhborn et al. (1995)
arabinofuranosyl-cytosine	varying composition and varying mean diameter	single or multiple SC injections	murine L1210 leukemia model	Increased survival time equivalent or in some cases superior to results obtained after IV administration of liposomes and superior to free drug.	Allen et al. (1993a)
bleomycin	DOPC:DPPG:Chol:triolein (9.3:2.1:15:1.8) mean diameter between 14 and 27 µm	single SC injection	murine B-16 melanoma model	dose-responsive inhibition of tumor growth and increased survival time compared to free drug.	Roy and Kim (1991)
calcitonin	EPC:Chol (1:1) or EPC sonicated	single IM injection	rat	Liposomal calcitonin produced a greater degree of hypocalcemia and prolonged hypocalcemia than an equal dose-free drug.	Fukunaga et al. (1984)

Table 7.2 continued on next page.

Table 7.2 continued from previous page.

Therapeutic Agent	Liposome Characteristics	Treatment	Animal/Model	Effect	Reference
chloroquine	DSPC:DPPG:Chol (10:1:10) mean diameter about 0.3 µm	single IM and SC injection	mouse	Slightly higher chloroquine concentrations in blood after injection of chloroquine liposomes compared to free drug over a 72 h observation period.	Titulaer et al. (1990)
chloroquine	DSPC:DPPG:Chol (10:1:10) mean diameter about 0.3 µm	single SC or IM injection 5, 8 or 10 days before infection	murine malaria	Protection against a challenge of parasitised erythrocytes for a period of at least 10 days.	Crommelin et al. (1991)
cytarabine	DOPC:DPPG:Chol:triolein 9.3:2.1:15:1.8 mean diameter between 15 and 35 µm	single SC injection	murine L1210 leukemia model	Increased survival time compared to free drug.	Kim and Howell (1987)
desferal	EPC:EPG (10:1) EPC:EPG:Chol (10:1:4) DSPC:DPPG:Chol (10:1:4) mean diameter between 0.3 and 0.5 µm	2 SC injections prior to infection or 2 SC injections 6 and 8 days after infection	*Plasmodium vinckei* infected mice	Suppression of parasitaemia and increased long-term survival.	Postma et al. (1998)
enrofloxacin	PC:Chol (1:1) unsized MLV mean diameter about 2.4 µm	single IM injection	rabbit	Prolonged and therapeutic plasma concentrations of enrofloxacin compared to free drug.	Cabanes et al. (1995)

Table 7.2 continued on next page.

Table 7.2 continued from previous page.

Therapeutic Agent	Liposome Characteristics	Treatment	Animal/Model	Effect	Reference
gentamicin	DOPC:DPPG:Chol:triolein (9.3:2.1:15:1.8) mean diameter between 15 and 35 μm	single SC injection	*Staphylococcus aureus* infected mice	Dose-responsive inhibition of infection.	Grayson et al. (1995)
GM-CSF	DMPC unsized	single SC injection	mouse	Sustained and stable levels of cytokine in the serum for 24 h.	Anderson et al. (1994)
insulin	EPC:Chol:DCP or EPC:Chol:SA varying ratio unsized	single IM injection	rat	Prolonged absorption half-life after injection of liposomal drug, which was more pronounced when higher amounts of cholesterol were incorporated.	Arakawa et al. (1975)
insulin	EPC:Chol:DCP or EPC:Chol (weight ratio 10:2:1 and 10:2, resp) mean diameter about 0.1 μm	single SC injection	diabetic dog	Insulin levels still detectable 24 h after injection; decrease in serum glucose level for at least 7 h.	Stevenson et al. (1982)
interferon-α	varying composition and varying size	single IM injection	mouse	Increased localization at the site of injection for at least 3 days as compared to the free drug, depending on liposome composition	Eppstein (1982)

Table 7.2 continued on next page.

Table 7.2 continued from previous page.

Therapeutic Agent	Liposome Characteristics	Treatment	Animal/Model	Effect	Reference
interferon-β	DSPC:DPPG (9:1) unsized	IM injection at day 1 and 6 after infection	simian varicella virus infection in monkey	Liposomal IFN-β was effective in reducing infection and preventing death, whereas free IFN-β at the same dosing schema was not efficacious.	Eppstein et al. (1989)
interferon-γ	DMPC:Chol:DCP (9:9:2) mean diameter about 0.3 μm	single SC injection	mouse	Increased bioavailability and prolonged serum titers as compared to the free drug.	Rutenfranz et al. (1990)
interleukin-2	DMPC mean diameter about 1.7 μm	single SC injection	mouse	Sustained levels in serum for 48 and 72 h.	Anderson et al. (1992)
interleukin-2	EPC:Chol (2:1) unsized	2 peritumor injections weekly for 2 weeks	rat hepatoma tumor model	Inhibition of solid tumor growth and prolonged survival time compared to the free drug.	Konno et al. (1991)
interleukin-7	EPC:Chol (8:2) unsized	1 SC injection weekly for 2 weeks	guinea pig	Lower but sustained blood concentrations and enhanced effects on the blood lymphocyte numbers as compared to the free drug.	Bui et al. (1994)

Table 7.2 continued on next page.

Table 7.2 continued from previous page.

Therapeutic Agent	Liposome Characteristics	Treatment	Animal/Model	Effect	Reference
methotrexate	DOPC:DPPG:Chol:triolein 9.3:2.1:15:1.8 mean diameter about 14 μm	single SC injection	mice	Increased plasma half-life and lower plasma peak concentrations, resulting in increased single-dose potency without producing significant changes in the therapeutic index.	Bonetti et al. (1994)
morphine	DOPC:DPPG:Chol:triolein 9.3:2.1:15:1.8 mean diameter between 15 and 35 μm	single SC injection	mice	Low plasma concentration for up to 1 week after injection.	Kim et al. (1993)
praziquantel	DPPC:Chol (7:6) unsized	single SC injection 1 or 2 weeks before infection	mouse	Prophylactic action toward schistosomiasis, demonstrated by increased survival time and hepatic worm count as compared to free drug.	Ammar et al. (1994)

DOPC: dioleoyl/phosphatidylcholine
Chol: cholesterol
EPG: egg phosphatidylglycerol
GM-CSF: Granulocyte macrophage colony-stimulating factor
SA: stearylamine
SC: subcutaneous
IFN: interferon

DPPG: diphosphatidylglycerol
EPC: egg phosphatidylcholine
PC: phosphatidylcholine
DCP: dicetylphosphatase
DSPC: distearoylphosphatidylcholine
IM: intramuscular
IV: intravenous

than intermittent high-dose bolus injection. Kim and Howell (1987) were the first to report on the therapeutic efficacy of SC–injected MVL containing antitumor drugs. The therapeutic efficacy of SC injected liposomes containing cytarabine was studied in mice intravenously inoculated with L1210 leukemia cells. The results show that SC administration of cytarabine entrapped in MVLs results in increased survival compared to the mice receiving free cytarabine. Cytarabine was not detected in the plasma after SC administration of cytarabine liposomes. The authors hypothesize that there is a low but therapeutic concentration of the drug in the plasma for a sufficient time to increase the survival of the mice. Further details of clinical studies are discussed later in this chapter (see section "Medical Applications of Multivesicular Liposomes").

Therapeutic efficacy of SC injected liposomes containing bleomycin in the B16 melanoma mouse model showed that MVLs (mean diameter about 19 μm) containing bleomycin are more effective in reducing tumor growth than the free drug at an equivalent dose (Roy and Kim 1991). At higher doses, the therapeutic efficacy of liposomal bleomycin increased, whereas free bleomycin was found to be severely toxic. Analysis of blood from mice that had been injected with liposomal bleomycin did not reveal detectable levels of bleomycin. As the liposomes were large and therefore expected to remain at the SC injection site, the observed therapeutic efficacy is likely attributable to sustained release of bleomycin from the liposomes.

The therapeutic efficacy of 1-β-D-arabinofuranosylcytosine (ara-C) encapsulated in PEG–coated liposomes was studied after SC administration in mice inoculated intravenously or intraperitoneally with L1210 leukemia cells (Allen et al. 1993b). The therapeutic efficacy of free ara-C is negligible when given as a single SC, IV, and IP (intraperitoneal) injection. Administration of free ara-C by infusion or as repeated injections improves the efficacy of free ara-C substantially. When ara-C was encapsulated in liposomes and administered subcutaneously, the therapeutic efficacy of a single injection was in all cases considerable at a relatively low dose (25 mg/kg) against both IV– and IP–inoculated leukemia. For most liposome formulations, a single SC injection was as effective and in some cases even more efficacious than a single IV injection. The therapeutic effect of small liposomes

(0.09 μm) may be the result of drug released from liposomes reaching the blood circulation. However, larger liposomes (mean diameter about 0.17 μm) were as effective as small liposomes, despite a substantial decrease in the amount of liposomes that reached the circulation. These results confirm that an important component of the mechanism of action of SC–administered liposomes is through the sustained release of the drug from the site of injection.

Other Small Molecular Weight Drugs

Several examples have been reported on the application of liposomal sustained-release systems for small molecular weight drugs other than antitumor drugs (see Table 7.2). Generally, SC and IM administration of these drugs encapsulated in liposomes results in low but prolonged plasma concentrations of the drug and enhanced therapeutic efficacy as compared to the free drug. A relatively new application of SC injected liposomes as a sustained drug release system is the use of liposome-encapsulated desferal in the treatment of malaria (Postma et al. 1999). Single SC injections of unencapsulated desferal are not effective in the treatment of *Plasmodium vinkei* infection in mice. Continuous IP infusion of desferal can clear parasitemia in *Plasmodium vinkei* infected mice even at a late stage of malaria. To optimize desferal delivery, SC–administered liposomes containing desferal were used for treatment. Suppression of parasitemia and long-term survival of mice was obtained after 2 SC injections of liposomal desferal prior to infection or 2 injections given in an advanced stage of infection (Postma et al. 1998).

Proteins and Peptides

Several protein and peptide-based therapeutic drugs, such as insulin and growth hormones are routinely administered subcutaneously or intramuscularly in order to obtain therapeutic plasma concentrations for a prolonged period of time. However, they are generally rapidly cleared from the blood and are short lived in their action. In many situations, the therapeutic activity of such drugs can be increased by utilizing a sustained-release system (Storm et al. 1991). Peptide and protein-based products are particularly well suited for liposomal-sustained drug delivery, since they are all water-soluble compounds and very potent. Encapsulation of peptides and proteins in liposomes may raise some difficulties, as the protein activity should be preserved. However, when the conditions are carefully chosen, several methods are available for the encapsulation of proteins within liposomes without detectable protein degradation (Meyer et al. 1994; Anderson et al. 1994; Heeremans et al. 1995; Corvo 1998).

Insulin

Several studies have addressed the administration of insulin-containing liposomes after SC and IM administration. Arakawa et al. (1975) originally reported on the sustained release of insulin from nonsized liposomes after IM administration. Almost 50 percent of the injected dose of insulin was still present at the IM injection site 8 h after injection of insulin containing liposomes, whereas no insulin was detectable at this time-point when administered in the free form. Stevenson et al. (1982) investigated the potential of liposomes as sustained release systems for insulin in diabetic dogs after SC administration. When encapsulated in small liposomes (0.08 μm), insulin could still be detected in plasma 24 h after injection and resulted in a prolonged hypoglycemic response. The authors attributed this extended response to the results of a slower metabolic clearance of liposome-encapsulated insulin. A review on the application of liposomes for delivery of insulin by SC injection is given by Sprangler (1990).

Cytokines

Although during the last decade many cytokines have been developed for therapeutic applications, only a few are used clinically. This is due to their relatively short half-life, which requires either continuous infusion or frequent injections to produce a therapeutic effect. Sustained-release formulations of cytokines that prolong blood concentrations while retaining biologic activity in vivo may add significantly to the utility of cytokines as therapeutic agents.

Sustained release of IL-2 (interleukin-2) from SC–injected liposomes may facilitate the effectiveness of IL-2–based immunotherapy. The depot characteristics and biodistribution of IL-2 entrapped in large liposomes (mean diameter 1–2 μm) was studied after SC, IV, IP, and intrathoracic administration (Anderson et al. 1992). Administration by the three different local routes produced depots of IL-2 liposomes. SC administration of IL-2 liposomes resulted in substantially higher serum concentrations than equal SC doses of free IL-2 at all time-points tested during the 48 h time period and was detected in the serum 48 and 72 h after injection. Therapeutic effectiveness of SC–injected recombinant IL-2 entrapped in large liposomes was studied by Konno et al. (1991). Repeated peritumor injection inhibited solid tumor growth and prolonged the survival time of rats. IL-2 serum concentrations and tumor tissue histology suggest that the therapeutic effect of liposomal IL-2 is related to the slow release of IL-2 into the circulation, which enhances the effective induction of lymphokine activated killer (LAK) cells, and local activation of tumor-infiltrating macrophages.

Interleukin-7 (IL-7) is of interest as a therapeutic cytokine because of its lymphopoietic activity. Free soluble IL-7 has been shown to induce proliferation of bone marrow cells and T and B lymphocytes in vivo. To achieve a lymphopoietic effect, frequent SC injections of soluble IL-7 are required. Liposome-formulated IL-7 reduced the frequency of SC injections required to achieve a similar effect. The need for twice-daily SC administration of soluble IL-7 was reduced to a weekly SC injection of the IL-7 liposome formulation. Pharmacokinetic data suggest that the liposome-containing formulation provides a sustained blood level for more than 7 days and extends the IL-7 residence time in the body (Bui et al. 1994).

Reports in the early 1980s described a significant higher retention of interferon-α (IFN-α) at the IM injection site after injection of liposome-encapsulated IFN-α than after injection of free IFN-α (Eppstein 1982). Later the therapeutic efficacy was examined for IM injected interferon-β (IFN-β) encapsulated in liposomes in African green monkeys infected with simian varicella virus. The antiviral activity of liposome-encapsulated IFN-β was substantially higher than when aqueous IFN-β was given in the same dosing regimen. Thus, liposomal IFN-β formulations release IFN-β in a biologically active form with apparent favorably altered pharmacokinetics compared with aqueous, non-encapsulated IFN-β. Previous results showed that IFN-β was gradually released from the IM injection site over a 3–5 day period when administered intramuscularly in liposomal form to mice (Eppstein et al. *1989*).

DRUG PRODUCTS USING MULTIVESICULAR LIPOSOMES

MVLs, commercially known as DepoFoam™ formulations (Skye-Pharma plc), are distinguishable from other lipid-based drug delivery systems by their structure. Each microscopic particle encloses multiple nonconcentric aqueous chambers bounded by a single bilayer lipid membrane with a foamlike appearance under the microscope (Figure 7.1 and Figure 7.8). This drug delivery system has advantages of good stability during storage as an aqueous suspension, control over the drug release rate, and highly efficient entrapment of hydrophilic molecules. The type of water-stable drugs that can be entrapped ranges from small

Figure 7.8. *Typical image of the fracture plane through a Depo-Foam™ vesicle that shows the bilayer walls of the multiple interior compartments tightly packed into a roughly 10 μm diameter sphere. (Courtesy of M. B. Sankaram, SkyePharma plc.)*

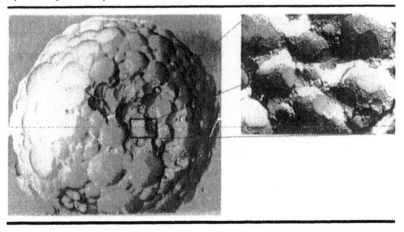

molecules, oligonucleotides, and peptides, to macromolecules such as proteins and nucleic acids. Like other liposomes and lipid-based systems, the particles are made from lipids commonly found in biological membranes and appear to be biodegradable by the usual lipid metabolic pathways in vivo (Brownson et al. 1998). MVL are suited for applications involving local drug administration and release by virtue of the relatively large size of the particles (approximately in the range 5–100 μm). Routes of administration being used in the clinic include intrathecal, SC, and epidural routes. IM, intraocular, intra-articular, and IP routes, along with direct injection at the diseased site, are also feasible.

Preparation and Stability of Multivesicular Liposomes

The manufacture of MVLs begins by emulsifying a mixture of an aqueous phase containing the drug to be encapsulated with the organic phase containing the lipids in chloroform to form a water-in-oil (w/o) emulsion. A lipid combination commonly used is dioleoyl-phosphatidylcholine (DOPC), dipalmitoyl-phosphatidylglycerol (DPPG), cholesterol, and triolein. The first w/o emulsion is dispersed and emulsified in a second aqueous phase (containing glucose and lysine) to make a water-in-oil-in-water (w/o/w) emulsion (Kim et al. 1983). The w/o/w emulsion is sparged with nitrogen to remove the chloroform, at which time numerous submicrometer to micrometer-sized water compartments, separated by lipid bilayers, take on a close-packed polyhedral structure (Spector et al. 1996). The resulting vesicles are then diafiltered by cross-flow filtration to exchange buffers and remove unencapsulated drug. Aseptic production procedures are used to manufacture MVLs for pharmaceutical applications, since the products are unsuited to terminal sterilization.

MVLs are relatively stable in aqueous media. They are stored under refrigeration in a ready-to-use, injectable form and are stable under the recommended storage conditions of 2–6°C for at least 24 months. The stability of the vesicles may be related to their three-dimensional structure. The presence of internal aqueous chambers may serve to confer increased mechanical strength to the vesicle (New 1995). The role of triglyceride in the particles is incompletely understood. Recent ^{13}C nuclear magnetic resonance (NMR) studies with triolein in MVL-encapsulating cytarabine (DepoCyt®) show that the triolein is not detected residing in the bilayer regions of the lipid dispersions (Ellena 1998a, b), yet the triglyceride component is required to achieve the unique appearance and properties of the vesicles. The multivesicular nature of the particles also means that, unlike conventional liposomes, a single breach in the external membrane of the vesicle will not result in total release of the internal aqueous contents.

Medical Applications of Multivesicular Liposomes

Small Molecular Weight Drugs

An MVL formulation of cytarabine (DepoCyt®) is being used in the treatment of neoplastic meningitis, which results from metastatic infiltration of the meninges by malignant cells most commonly arising from acute leukemias, lymphomas, or carcinomas. The drug product is administered locally into the cerebrospinal fluid (CSF) via intracerebral spinal fluid (I-CSF) injection. The formulation requires fewer injections to be administered than current therapies using free or unencapsulated drug. Less frequent injections should reduce the risk of infection, discomfort, and cost, while being more effective at maintaining therapeutic concentrations of drug in the CSF over an extended period of time. DepoCyt® has been evaluated in Phase I/II and Phase III clinical trials against current therapy, and a Phase IV trial is ongoing. The administration of cytarabine in the form of MVL directly into the CSF is beneficial for targeting the meninges, with minimal systemic exposure of cytarabine. Because of the sustained release of cytarabine from the particles, drug exposure is prolonged over time, resulting in lower peak cytarabine levels compared with standard dosing with cytotoxic agents. In patients with meningitis, intrathecally administered DepoCyt® provides a more convenient dosing schedule than methotrexate standard therapy, with response rates that may be at least comparable with methotrexate. DepoCyt administration may also result in a longer time to the progression of key signs and symptoms of the disease and perhaps longer survival than conventional intrathecal chemotherapy (reviewed by Senior 1998).

Peptides, Proteins, and Oligonucleotides

A number of protein-, peptide-, and oligonucleotide-based drugs have been successfully encapsulated in MVLs (Katre et al. 1996; Ye et al. 1997). Insulin-like growth factor-I (IGF-I), IL-2, the antisense oligonucleotide ISIS 2922, and various antigens for use in vaccine development have been formulated successfully and are being tested in model systems.

IGF-I has been encapsulated in MVL (DepoIGF-I™) with high encapsulation yield and high loading. DepoIGF-1™ has clinical potential in several disease states. The parent drug, IGF-I, is being evaluated for its therapeutic benefit in several disease states, including amyotrophic lateral sclerosis (ALS), diabetes, and acute renal failure. DepoIGF-I™ has been successfully scaled up to allow cGMP production at a small scale. The product showed high protein loading of up to 300 mg per mL of encapsulated volume (Ye et al. 1997), along with a high encapsulation efficiency of up to 85 percent, and a low content of free protein (< 1 percent) in the final dispersion (Ye et al. 1997; Katre et al. 1998). Little chemical change in the protein was detectable as a result of the formulation process, and full bioactivity of the encapsulated protein was retained (Ye et al. 1997). Drug release assays conducted in vitro in biological suspending media such as human plasma and artificial interstitial fluid indicate that these formulations provide sustained release of encapsulated proteins over a period of more than a week (Katre et al. 1998). Pharmacokinetic studies in a rat model where a single SC injection of the product showed a sustained level of IGF-1 in serum maintained over a 1-week period with DepoIGF-I™ formulations, while free IGF-I is cleared in 1 day (Katre et al. 1998).

Other Applications

Morphine encapsulated in MVL (DepoMorphine™) has completed Phase I clinical studies, and a Phase II study is being conducted in patients for postoperative pain management. In preclinical studies, a rat model was used to show that administration of a single epidural dose could provide sustained analgesia (Kim et al. 1996). The pharmacokinetics of epidural morphine encapsulated in MVL™ were conducted in a dog model (Yaksh et al. 1999) to determine the minimum drug exposure intervals, the maximum tolerable dose, and the safety of repeated delivery of the maximum tolerable dose. The studies indicated that Depo-Morphine™ was capable of sustained release of morphine sulfate over a 3–4 day period when administered epidurally. Previously, sustained release of morphine was observed after intrathecal administration in a rat model (Figure 7.7).

An MVL suspension of the aminoglycoside antibiotic amikacin (DepoAmikacin™) is intended for the local treatment of soft tissue and closed body compartment infections caused by microorganisms susceptible to amikacin. DepoAmikacin™ provides sustained release of amikacin over a 7- to 14-day period after SC administration. These parameters were tested in preclinical toxicity testing/ pharmacokinetics (PK) and efficacy studies and in a Phase I clinical study. DepoAmikacin™ was well tolerated at all dose levels tested (15–240 mg amikacin). Efficacy findings in animal models show potential therapeutic benefit from sustained release of amikacin at a site of local infection while reducing overall systemic exposure (Roerhborn et al. 1995).

Although only liposome-encapsulated cytarabine, amikacin, and morphine have entered the clinic as locally administered depot formulations as of this writing, many other drugs are possible candidates. For example, MVLs are also being tested using bupivacaine for local infiltration. The system is being evaluated in patients undergoing surgical procedures under local anesthesia. Other small molecules tested in the feasibility stage include clonidine, an alpha-2 adrenergic agonist, for the management of intractable chronic pain for a defined period of time following intrathecal injection and low molecular weight heparin (LMW-Heparin). Local delivery to the eye for treatment of cytomegalovirus (CMV)-induced retinopathy and proliferative vitreoretinopathy, for example, have been evaluated in rabbit and other models (Assil and Weinreb 1987; Gariano et al. 1994; Besen et al. 1995)

CONCLUSION

Liposomes are being used extensively in medical products and cosmetics. Marketed liposomal products focus on IV application of relatively toxic drugs (Storm and Crommelin 1998) or topical application as cosmetics. Liposomes can also provide sustained-release drug delivery after local injection. In a depot form, they are capable of sustained release of a variety of drugs over several days or more as illustrated by MVLs given intrathecally or epidurally. Following SC and IM injection, larger-sized liposomes provide the longest residence times at the injection site, thus providing a depot for sustained systemic drug delivery. The rate and extent of release of encapsulated drug may be controlled by manipulating the physical and chemical characteristics of the liposomes. Locally injected liposomes have shown promise and are being evaluated for the treatment of a variety of diseases in preclinical models and clinical trials.

REFERENCES

Allen, T. M. 1984. Removal of detergent and solvent traces from liposomes. In *Liposome technology*, vol. I, edited by G. Gregoriadis. Boca Raton, Fla., USA: CRC Press, pp. 109–122.

Allen, T. M., C. B. Hansen, and L. S. S. Guo. 1993a. Subcutaneous administration of liposomes: A comparison with the intravenous and intraperitoneal routes of injection. *Biochim. Biophys. Acta* 1150:9–16.

Allen, T. M., C. B. Hansen, and A. Peliowski, 1993b. Subcutaneous administration of sterically stabilized (Stealth) liposomes is an effective sustained release system for 1-β-D-arabinofuranosylcytosine. *Drug Delivery* 1:55–60.

Ammar, H. O., M. S. El-Ridy, M. Ghorab, and M. M. Ghorab, 1994. Evaluation of the antischistosomal effect of praziquantel in a liposomal delivery system in mice. *Int. J. Pharm.* 103:237–241.

Anderson, P. M., D. C. Hanson, D. E. Hasz, M. R. Halet, B. R. Blazar, and A. C. Ochoa, 1994. Cytokines in liposomes: Preliminary studies with IL-1, IL-2, GM-CSF and interferon-γ. *Cytokine* 6:92–101.

Anderson, P. M., E. Katsanis, S. F. Sencer, D. Hasz, A. C. Ochoa, and B. Bostrom. 1992. Depot characteristics and biodistribution of interleukin-2 liposomes: Importance of route of administration. *J. Immunother.* 12:19–31.

Arakawa, E., Y. Imai, H. Kobayashi, K. Okumura, and H. Sezaki. 1975. Application of drug-containing liposomes to the duration of the intramuscular absorption of water-soluble drugs in rats. *Chem. Pharm. Bull.* 23:2218–2222.

Assil, K. K., and R. N. Weinreb. 1987. Multivesicular liposomes: Sustained release of the antimetabolite cytarabine in the eye. *Arch. Ophthalmol.* 105:400–403.

Balazsovits, J. A. E., L. D. Mayer, M. B. Bally, P. R. Cullis, M. McDonell, R. S. Ginsberg, and R. E. Falk. 1989. Analysis of the effect of liposome encapsulation on the vesicant properties, acute and cardiac toxicities, and antitumor efficacy of doxorubicin. *Cancer Chemother. and Pharmacol.* 23:81–86.

Barenholz, Y. and D. J. A. Crommelin. 1994. Liposomes as pharmaceutical dosage forms. In *Encyclopedia of pharmaceutical technology* edited by J. Swarbrick and J. C. Boylan. New York: Marcel Dexker, 9:1–39.

Besen, G., M. Flores-Aguilar, K. K. Assil, B. D. Kupperman, P. Gangan, M. Pursley, D. Munguia, C. Vuong, E. De Clercq, G. Bergeron-Lynn, S. P. Azen, and W. Freeman. 1995. Long-term therapy for Herpes retinitis in an animal model with high-concentrated liposome-encapsulated HPMPC. *Arch. Ophthalmol.* 113:661–668.

Bolotin, E. M., R. Cohen, L. K. Bar, N. Emmaneul, S. Ninio, D. D. Lasic, and Y. Barenholz. 1994. Ammonium sulfate gradients for efficient and stable remote loading of amphipatic weak bases into liposomes and ligando liposomes. *J. Liposome Res.* 4:455–479.

Boman, N.L., V. A. Tron, M. B. Bally, and P. R. Cullis. 1996. Vincristine-induced dermal toxicity is significantly reduced when the drug is given in liposomes. *Cancer Chemother. and Pharmacol.* 37:351–355.

Bonetti, A., E. Chatelut, and S. Kim. 1994. An extended-release formulation of methotrexate for subcutaneous administration. *Cancer Chemother. and Pharmacol.* 33:303–306.

Brandle, M. M., D. Bachmann, M. Drechsler, and K. H. Bauer. 1993. Liposome preparation using high-pressure homogenizers. In *Liposome technology*, vol. I, edited by G. Gregoriadis. Boca Raton, Fla., USA: CRC Press, pp. 49–65.

Brownson, E.A., M. Langston, A. G. Tsai, T. Gillespie, T. P. Davis, M. Intaglietta, M. B. Sankaram. 1998. Biodistribution during sustained release from DepoFoam, a lipid-based parenteral drug delivery system. *Proc. Int. Symp. Controlled Release Bioact. Mat.* 25:42–43.

Bui, T., C. Faltynek, and J. Y. Ho. 1994. Differential disposition of soluble and liposome-formulated human recombinant interleukin-7: Effect on blood lymphocyte population in guinea pigs. *Pharm. Res.* 11:633–641.

Cabanes, A., F. Reig, J. M. Garcia Anton, and M. Arboix. 1995. Sustained release of liposome-containing enrofloxacin following i.m. administration. *Am. J. Vet. Res.* 56:1498–1501.

Chapman, C. J., W. L. Erdahl, R. W. Taylor, and D. R. Pfeiffer. 1990. Factors affecting solute entrapment in phospholipid vesicles prepared by freeze-thaw extrusion method: a possible general method for improving the effiency of entrapment. *Chem. & Phys. of Lipids* 55:73–83.

Corvo, L. (1998) Liposomes as delivery systems for superoxide dismutase in experimental arthritis. PhD thesis, Utrecht University, Utrecht, The Netherlands.

Crommelin, D. J. A. and H. Schreier. 1994. Liposomes. In *Colloidal drug delivery systems,* edited by J. Kreuter. Marcel Dekker, Inc., pp. 73–190.

Crommelin, D. J. A., W. M. C. Eling, P. A. Steerenberg, U. K. Nässander, G. Storm, W. H. De Jong, Q. G. C. M. Van Hoesel, and J. Zuidema. 1991. Liposomes and immunoliposomes for controlled release or site specific delivery of anti-parasitic drugs and cytostatics. *J. Controlled Release* 16:147–154.

Crommelin, D. J. A., H. Talsma, M. Grit, and N. J. Zuidam. 1993. The physical stability of liposomes on long term storage. In *Phospholipids handbook,* edited by G. Cevc. New York: Marcel Dekker, Inc. pp. 335–348.

Ellena, J. F., R. Solis, D. S. Cafiso, and M. S. Sankarm. 1998a. DepoCyt®, a liposomal anticancer formulation, undergoes a bilayer to non-bilayer transition near room temperature. *Biophysical J.* 74:A372.

Ellena, J. F., R. Solis, D. S. Cafiso, and M. S. Sankarm. 1998b. The triolein in DepoCyt®, is completely or almost completely excluded from bilayers. *Biophysical J.* 74:A374.

Eppstein, D. A. 1982. Altered pharmacologic properties of liposome-associated human interferon alpha II. *J. Interferon Res.* 2:117–125.

Eppstein, D. A., M. A. van der Pas, C. A. Gloff, and K. F. Soike. 1989. Liposomal interferon-β: Sustained release treatment of simian varicella virus infection in monkeys. *J. Infectious Diseases* 159:616–620.

Forssen, E. A., and Z. A. Tökes. 1983. Attentuation of dermal toxicity of doxorubicin by liposome encapsulation. *Cancer Treat. Rep.* 67 (5):481–484.

Frøkjaer, S., E. L. Hjorth, and O. Wørts. 1984. Stability and storage of liposomes. In *Optimization of drug delivery,* edited by H. Bungaard, A. Bagger, and H. Kofod. Copenhagen, Denmark: Munksgaard, pp. 384–404.

Fukunaga, M., M. M. Miller, K. Y. Hostetler, and L. J. Deftos. 1984. Liposome entrapment enhances the hypoglycemic action of parenterally administered calcitonin. *Endocrinology* 115:757–61.

Gabizon, A. A., O. Pappo, D. Goren, M. Chemla, D. Tzemach, and A. T. Horowitz. 1993. Preclinical studies with doxorubicin encapsulated in polyethlene glycol-coated liposomes. *J. Liposome Res.* 3:517–528.

Gariano, R. F., K. K. Assil, C. A. Wiley, D. Munguia, R. N. Weinreb, and W. R. Freeman. 1994. Retinal toxicity of the antimetabolite 5-fluorouridine 5-monophosphate administered intravitreally using multivesicular liposomes. *Retina* 14:75–80.

Grit, M. and D. J. A. Crommelin. 1992. The effect of ageing on the physical stability of liposome dispersions. *Chem. & Phys. of Lipids* 1167:49–55.

Grit, M., N. J. Zuidam, and D. J. A. Crommelin. 1993. Analysis and hydrolysis kinetics of phospholipids in aqueous dispersions. In *Liposome technology,* vol. 1, edited by G. Gregoriadis. Boca Raton, Fla., USA: CRC Press, pp. 455–486.

Grayson, L., J. Hansborough, R. Zapata-Sirvent, A. Roehrborn, T. Kim, and S. Kim. 1995. Soft tissue infection prophylaxis with gentamicin encapsulated in multivesicular liposomes: Results from a prospective, randomized trial. *Crit. Care Med.* 23:84–91.

Haran, G., R. Cohen, L. K. Bar, and Y. Barenholz. 1993. Transmembrane ammonium sulfate gradients in liposomes produce efficient and stable entrapment of amphiphatic weak bases. *Biochim. Biophys. Acta* 1151:201–205.

Heeremans, J. L. M., H. R. Gerritsen, S. P. Meusen, F. W. Mijnheer, R. S. Gangaram Panday, R. Prevost, C. Kluft, and D. J. A. Crommelin. 1995. The preparation of tissue-type plasminogen activator (t-PA) containing liposomes: entrapment efficiency and ultracentrifugation damage. *J. Drug Targeting* 3:301–310.

Kadir, F., W. C. M. Eling, D. Abrahams, J. Zuidema, and D. J. A. Crommelin. 1992. Tissue reaction after intramuscular injection of liposomes in mice. *Int. J. Clin. Pharm. Ther. Toxicol.* 30:374–382.

Kadir, F., W. C. M. Eling, D. J. A. Crommelin, and J. Zuidema. 1991. Influence of injection volume on the release kinetics of liposomal chloroquine administered subcutaneously or intramusculary to mice. *J. Control Release* 17:277–284.

Katre, N. V., J. Asherman, H. Schaefer, and M. Hora. 1996. A multivesicular lipid-based sustained-release system for the delivery of therapeutic proteins. *Proc. 8th Inter. Pharm. Technol. Symp.* (IPTS-96):20–21.

Katre, N. V., J. Asherman, H. Schaefer, and M. Hora. 1998. Multivesicular liposome (DepoFoam™) technology for the sustained delivery of insulin-like growth factor-I (IGF-I). *J. Pharm. Sci.* 87 (11): 1341–1346.

Kim, C. K., and J. L. Han. 1995. Lymphatic delivery and pharmacokinetics of methotrexate after intramuscular injection of differently charged liposome-entrapped methotrexate to rats. *J. Microencapsulation* 12:437–446.

Kim, S., and S. B. Howell. 1987. Multivesicular liposomes containing cytarabine for slow release sc administration. *Cancer Treat. Rep.* 71:447–450.

Kim, S., Turker, M. S., Chi, E. Y., Sela, S., and Martin, G. 1983. Preparation of multivesicular liposomes. *Biochim. Biophys. Acta* 728:339–348.

Kim, T., J. Kim, and S. Kim. 1993. Extended-release formulation of morphine for subcutaneous administration. *Cancer Chemother. and Pharmacol.* 333:187–190.

Kim, T., S. Murdande, A. Gruber, S. Kim. 1996. Sustained-release morphine for epidural analgesia in rats. *Anesthesiology* 85:331–338.

Kirby, C., and G. Gregoriadis. 1984. Dehydration-rehydration vesicles: A simple method for high yield drug entrapment in liposomes. *Biotechnology* 2:979–984.

Konno, H., T. Tadakuma, K. Kumai, T. Takahashi, K. Ishibiki, O. Abe, and S. Sakagughi. 1990. The antitumor effects of Adriamycin entrapped in liposomes on lymph node metastases. *Japanese J. Surgery* 20:424–428.

Konno, H., A. Yamashita, T. Tadakuma, and S. Sakaguchi. 1991. Inhibition of growth of rat hepatoma by local injection of liposomes containing recombinant interleukin-2. *Biotherapy* 3:211–218.

Koppenhagen, F. J. 1997. Liposomes as delivery system for recombinant interleukin-2 in anticancer immunotherapy. PhD thesis, Utrecht Univeristy, Utrecht, The Netherlands.

Martin, F. J. 1990. Pharmaceutical manufacturing of liposomes. In *Specialized drug delivery systems: Manufacturing and production technology,* edited by P. Tyle. New York: Marcel Dekker, pp. 267–316.

Meyer, J., L. Whitcomb, and D. Collins. 1994. Efficient encapsulation of proteins within liposomes for slow release in vivo. *Biochem. Biophys. Res. Comm.* 99:433–438.

New, R. R. C. 1995. Influence of liposome characteristics on their properties and fate. In *Liposomes as tools in basic research and industry,* edited by J. R. Philipott and F. Schuber. Boca Raton, Fla., USA: CRC Press, pp. 3–20.

Oussoren, C., and G. Storm. 1997. Lymphatic uptake and biodistribution of liposomes after subcutaneous injection. III. Influence of surface modification with poly(ethyleneglycol). *Pharm. Res.* 14:1479–1484.

Oussoren, C., J. Zuidema, D. J. A. Crommelin, and G. Storm. 1997a. Lymphatic uptake and biodistribution of liposomes after subcutaneous injection. I. Influence of the anatomical site of injection. *J. Liposome Res.* 7:85–99.

Oussoren, C., J. Zuidema, D. J. A. Crommelin, and G. Storm. 1997b. Lymphatic uptake and biodistribution of liposomes after subcutaneous injection. II. Influence of liposomal size, lipid composition and lipid dose. *Biochim. Biophys. Acta.* 7:227–240.

Oussoren, C., W. M. C. Eling, D. J. A. Crommelin, G. Storm, and J. Zuidema. 1998. The influence of the route of administration and liposome composition on the potential of liposomes to protect tissue against local toxicity of two antitumor drugs. *Biochim. Biophys. Acta* 1369:159–172.

Patel, H. M. 1988. Fate of liposomes in the lymphatics. In *Liposomes as drug carriers,* edited by G. Gregoriadis. New York: John Wiley & Sons, Ltd. pp. 51–61.

Postma, N. S., O. C. Boerman, W. J. G. Oyen, J. Quidema, and G. Storm. 1999. Absorption and biodistribution of [111]indium-labelled desferrioxamine (^{111}In-DFO) after subcutaneous injection of ^{111}In-DFO liposomes. *J. Controlled Release* 58:51–60.

Postma, N. S., C. C. Hermsen, J. Zuidema, and W. M. C. Eling. 1998. *Plasmodium vinckei:* Optimization of deferrioxamine B in the treatment of murine malaria. *J. Exp. Parasitology.* 89:323–330.

Roy, R., and S. Kim. 1991. Multivesicular liposomes containing bleomycin for subcutaneous administration. *Cancer Chemother. and Pharmacol.* 28:105–108.

Roerhborn, A., J. F. Hansbrough, B. Gualdoni, and S. Kim. 1995. Lipid-based slow-release formulation of amikacin sulfate reduces foreign body-associated infections in mice. *Antimicrobial Agents Chemotherapy* 39:1752–1755.

Rutenfranz, I., A. Bauer, and H. Kirchner. 1990. Pharmacokinetic study of liposome-encapsulated human interferon-γ after intravenous and intramuscular injection in mice. *J. Interferon Res.* 10:337–341.

Schneider, T., A. Sachse, G. Rössling, and M. M. Brandle. 1994. Large scale production of liposomes of defined size by a new high pressure extrusion device. *Drug Develop. Industr. Pharm.* 20:2787–2807.

Schreier, H., M. Levy, and P. Mihalko. 1987. Sustained release of liposome-encapsulated gentamicin and the fate of phospholpid following intramuscular injection in mice. *J. Controlled Release* 5:187–192.

Senior, J. H. 1987. Fate and behaviour of liposomes in vivo: A review of controlling factors. *Crit. Rev. Ther. Drug Carrier Syst.* 3:123–193.

Senior, J. H. 1998. Medical applications of multivesicular lipid-based particles. In *Medical applications of liposomes,* edited by D. Lasic and D. Paphadjopoulos. Amsterdam: Elsevier, pp. 733–750.

Spector, M. S., J. A. Zasadzinski, M. B. Sankaram. 1996. Topology of multivesicular liposomes, a model biliquid foam, Langmuir 12:4704–4708.

Sprangler, R. S. 1990. Insulin administration via liposomes. *Diabetes Care* 13:911–922.

Stevenson, R. W., H. M. Patel, J. A. Parson, and B. E. Ryman. 1982. Prolonged hypoglycemic effect in diabetic dogs due to subcutaneous administration of insulin in liposomes. *Diabetes* 31:506–511.

Storm, G., and D. J. A. Crommelin. 1998. Liposomes: Quo vadis? *PSTT* 1:19–31.

Storm, G., H. P. Wilms, and D. J. A. Crommelin. 1991. Liposomes and bio-therapeutics. *Biotherapy* 3:25–42.

Storm, G., C. Oussoren, P. A. M. Peeters, and Y. Barenholz. 1993. Tolerability of liposome in vivo. In *Liposome technology,* vol. III, edited by G. Gregoriadis. Boca Raton, Fla., USA: CRC press, pp. 345–381.

Storm, G., L. Van Blooys, M. Brouwer, and D. J. A. Crommelin. 1985. The interaction of cytostatic drugs with adsorbents in aqueous media. *Biochim. Biophys. Acta* 818:343–351.

Supersaxo, A., W. R. Heine, and H. Steffen. 1990. Effect of molecular weight on the lymphatic absorption of water-soluble compounds following subcutaneous administration. *Pharm. Res.* 7:167–169.

Titulaer, H. A. C., W. M. C. Eling, D. J. A. Crommelin, P. A. M. Peeters, and J. Zuidema. 1990. The parenteral controlled release of liposome encapsulated chloroquine in mice. *J. Pharm. Pharmacol.* 42:529–532.

Trubetskoy, V. S., K. R. Whiteman, V. P. Torchilin, and, G. L. Wolf. 1998. Massage-induced release of subcutaneously injected liposome-encapsulated drugs to the blood. *J. Controlled Release* 50:13–19.

Tümer, A., C. Kirby, J. Senior, and G. Gregoriadis. 1983. Fate of cholesterol-rich liposomes after subcutaneous injection into rats. *Biochim. Biophys. Acta* 760:119–125.

Van Hoogevest, P., and P. Frankhauser. 1989. An industrial liposome dosage form for muramyl-tripeptide-phosphatidylethanolamine (MTP-PE). In *Liposomes in the therapy of infectious diseases and cancer,* edited by G. Lopez-Berestein and I. J. Fidler. New York. Alan R. Liss, pp. 453–466.

Van Winden, E. C. A., and D. J. A. Crommelin. 1997a. Long term stability of freeze-dried lyoprotected doxorubicin liposomes. *Eur. J. Pharm. Biopharm.* 43:295–307.

Van Winden, E. C. A., W. Zhang, and D. J. A. Crommelin. 1997b. Effect of freezing rate on the stability of liposomes during freeze-drying and rehydration. *Pharm. Res.* 14:1151–1160.

Yaksh, T., J. Provender, M. Rathburn, R. Myers, P. Richter, and F. Kohn. 1999. Evaluation of safety of epidural sustained release encapsulated morphine in dogs. Submitted for publication.

Yaguchi, T., M. Yamauchi, H. Takagi, N. Ohishi, and K. Yagi. 1990. Effect of sulfatide-inserted liposomes containing entrapped adriamycin on metastasised cells in lymph nodes. *J. Clin. Biochem. Nutr.* 9:79–85.

Ye, Q., M. Stevenson, J. Asherman, S. Chen, B. Shirley, and N. V. Katre. 1997. A sustained-release system for efficient encapsulation with high loading of insulin-like growth factor-I (IGF-I). *Pharm. Res.* 14, Abstract 3020:S-469.

Zuidam, N. J., S. S. L. Lee, and D. J. A. Crommelin. 1993. Sterilization of liposomes by heat treatment. *Pharm. Res.* 10:1591–1596.

Zuidam, N. J., H. Talsma, and D. J. A. Crommelin. 1996. Sterilization of liposomes. In *Handbook of nonmedical applications of liposomes. From design to microreactors,* vol. III, edited by Y. Barenholz and D.D. Lasic. Boca Raton, Fla., USA: CRC Press, pp. 71–80.

8

Synthetic Polymers for Nanosphere and Microsphere Products

Michael Radomsky
Epilctye Pharmaceutical Inc.
San Diego, California

LinShu Liu
Orquest, Inc.
Mountain View, California

Taro Iwamoto
Otsuka America Pharmaceuticals, Inc.
Palo Alto, California

Although the development of the first controlled drug delivery system can be traced back to the late 1940s, considerable progress in a rapidly increasing number of novel drug delivery systems was made in the 1970s and 1980s. During this period, the basic principles of drug release from solid matrices were established, mathematical modeling of these systems was developed, new polymers were synthesized, the parameters affecting the chemical and physical properties of matrices were identified, and the release profiles of incorporated substances were determined. There are many reviews on drug delivery systems using synthetic polymers that have been published to describe the progress and challenges that occurred during this period (Baker 1980; Kydoniens 1980; Chien 1982; Das 1983; Langer and Peppas 1981; Siegal and Langer 1984; Langer 1990; Couvreur and Puisieux 1993; Okada and Toguchi 1995). In the 1980s, as the result of biotechnology, large-scale, affordable production of potent, bioactive proteins became possible. However, the production of a new peptide or protein drug is only one step in developing a therapeutically useful commercial product. Administration of these proteins from an efficacious delivery system became a major challenge. Microspheres were a potential strategy that was proposed and investigated.

Microspheres have been successfully developed from the bench scale to products approved by the Food and Drug Administration (FDA). One of the most commercially successful applications of injectable microspheres to date has been the encapsulation of peptides for treating hormonal diseases (Okada et al. 1989, 1991; Okada and Toguchi 1995; Sanders et al. 1984). Microspheres and nanospheres have been investigated as materials for the delivery of pharmaceuticals with a variety of polymers, fabrication techniques, and encapsulation methods by a number of researchers in academic and industrial settings. This chapter reviews and summarizes the most critical aspects of microsphere technology. The rationale for microsphere use, polymer types, fabrication techniques, the limitations of the technology, and future trends are discussed.

Microspheres are micrometer-sized particles, typically about 100 μm in diameter, but larger and smaller spheres are also used. A property of microspheres is that they can be suspended in an aqueous vehicle and injected percutaneously through a small gauge needle without the need for anesthesia. Nanospheres are smaller than microspheres (typically less than 1 μm) but can be fabricated with the same materials and methods as microspheres (see equation 1 for the relationship between process parameters and micro/nanosphere size). The authors do not make a distinction between the fabrication or potential applications of these different-sized particles and in this chapter refer to microspheres and nanospheres as nano/microspheres.

RATIONALE FOR MICROSPHERE USE

The rationale for sustained-release injectable products was presented in Chapter 1. However, the specific advantages of nano/microsphere technology are summarized in Table 8.1 and extensively reviewed elsewhere (Baker 1980; Langer 1990; Langer and Peppas 1981; Siegal and Langer 1984; Couvreur and Puisieux 1993; Okada and Toguchi 1995).

The rate of drug release from a polymeric system intended for parenteral use can be controlled by the delivery system (i.e., the polymer delivery system determines the rate of drug absorption). In contrast, the oral delivery of drugs can be highly

Table 8.1. *Advantages and Disadvantages of Sustained-Release Microsphere Formulations*

Advantages	Disadvantages
• Well-controlled drug absorption controlled by the delivery system rather than fluctuating biological factors.	• Customized formulation or manufacturing process required for each application.
• A wide range of product characteristics, such as rate of drug release, duration of release, and type of drug, are possible for treatment of a variety of diseases.	• Sterilization: Terminal sterilization is not typically feasible, aseptic processing is usually required.
	• Acceptable residual solvent concentrations can be difficult to achieve.
	• Stability of drugs during manufacture or stability of polymers during long-term storage may be problematic.
• Local delivery of compounds may result in a reduction of side effects.	• Availability of high quality, FDA-approved polymers is limited.
• Drug stability may be improved because of protection from in vivo environment.	• Drug losses during manufacturing can be as high as 25 to 50 percent or more.
• Patient compliance is not an issue once the formulation is injected.	• The drug cannot be "removed" easily following injection into patient.

dependent on absorption through the gastrointestinal wall, which can vary from patient to patient and can be dependent on factors such as a fed versus fasted state. However, in polymeric systems, these variabilities of delivery can be eliminated by the proper design of the delivery system. With microspheres that are injected percutaneously, the rate of drug absorption is controlled by the polymeric delivery system rather than the potentially highly variable process of transport across a biological barrier.

Many different clinical indications of microspheres are possible because a wide range of drug dosages of many classes of compounds (e.g., small molecules, peptides, proteins, oligonucleotides, etc.) can be achieved with these types of delivery systems.

MICROSPHERE TECHNOLOGY

Many varieties of microsphere formulations have been investigated. In the following sections, the polymers used in nano/microsphere systems and the corresponding fabrication techniques are discussed.

Polymer Types

Table 8.2 shows the various classes of polymers that have been investigated for use in drug delivery systems. A wide range of polymers is now available. When selecting a polymer for use in a drug delivery system, the following properties of the polymer should be considered:

- Chemical, physical, and biological properties
- Availability of pure, well-characterized material in sufficient quantities
- Degradation rate
- Safety/toxicity profile (if known)
- Regulatory history (if any)

Table 8.2. *Classes of Polymers Useful for Nanosphere and Microsphere Fabrication*

Polymer Class	Example	Features
Polyesters	polylactide, polyglycolide, poly(lactide-coglycolide) polycaprolactones	Medical grade source available, well-established safety profile, long history of use in medical products
Polyanhydrides	poly(biscarboxyphenoxy propane-cosebacic acid)	Polymer surface erosion, drug polymer linkage possible
Polyphosphazenes	Poly(dichlorophos phazene) reacted with nucleophile	Control degradation rate by side group substitution
Polymer blends	Blend of polycaprolactone with poly(hydroxybutyric acid and poly(lactide-coglycolide)	New biomaterials with improved physical, mechanical, and degradation properties

This section discusses the only two classes of biodegradable, synthetic polymers presently used in FDA–approved medical products (i.e., polyesters and polyanhydrides) and also discusses polyphosphazenes and blended polymers because of their enormous potential. Poly(amino acids), poly(ortho esters), polyiminocarbonates, and other infrequently studied polymers are not included in this discussion because they are not appropriate for nano/microsphere use, not appropriate for injectable formulations, or are not commercially available. However, as new biomaterials are developed for drug delivery applications, the necessary safety data generated, and commercial production begins, these and other novel polymers may gain widespread use for sustained-release injectable applications.

For each of the polymer classes reviewed in this chapter, the following properties of each polymer are discussed (when available):

- Chemical description
- Synthesis

- Solubility properties

- Degradation (in vitro and in vivo)

- Safety and toxicity implications

- Any approved commercial products or products in development

Polyesters

Lactide/Glycolide Polymers. Lactide/glycolide homo-polymers, poly(lactic acid) (PLA) or poly(glycolic acid) (PGA), or the copolymer, poly(lactic acid-co-glycolic acid) (PLGA) are the most common types of synthetic, degradable polymers investigated for use in pharmaceutical drug delivery systems. These polymers have been used for decades in medical products (e.g., degradable sutures) and have been used in FDA–approved microsphere products (e.g., Lupron Depot®).

PLA, PGA, and PLGA are typically synthesized by a ring-opening condensation of their cyclic dimers, carried out at 120°C for several hours that produces medical grade, commercially available material. The knowledge of this polymerization reaction has advanced to the stage such that polymers with broad characteristics can be obtained by varying the stereochemistry of monomers, the polymer molecular weight and distribution, and the crystallinity of polymers.

Because microspheres are not fabricated from solid polymers and must be dissolved in a suitable solvent, solubility is an important property of these polymers as drug delivery vehicles. PLA and PLGA containing less than 50 percent glycolic acid are soluble in common solvents, such as chloroform, dichloromethane, dioxane, tetrahydrofuran, and ethyl acetate. PGA and copolymers containing less than 50 percent lactic acid are insoluble in common solvents that are utilized in a pharmaceutical manufacturing process. When selecting a polymer, it is important to also consider the properties of the solvent that will be utilized (e.g., solvent availability, solvent price, safety profile of the solvent, and acceptable residual solvent levels).

The degradation mechanism of lactide/glycolide polymers occurs through hydrolysis of the ester bonds, which can be affected by significant changes in environmental temperature, pH, or the presence of catalyst. Crystallinity and water uptake are critical factors in determining the in vitro and in vivo degradation rates. PLA, due to the methyl group in the β carbon of lactic acid, is more hydrophobic than PGA. The segment length and contribution of block or random structure in the copolymer can also affect the water uptake. The rate of water uptake is proportional to the degradation rate. Therefore, the time to complete degradation can be modified by adjusting the lactide to glycolide ratio in PLGA. The degradation of semicrystalline polymers proceeds in two phases: the amorphous regions are first hydrolyzed followed by degradation of the crystalline regions. These polymers degrade by bulk hydrolysis of the ester bonds, which initially results in a decrease in molecular weight with no polymer weight loss.

A number of pharmaceutical agents have been formulated in lactide/glycolide polymer matrices in the form of injectable microspheres and tested for their biocompatibility. No evidence of an inflammatory response, irritation, or other adverse effect has been reported upon implantation, and the response to injected lactide/glycolide polymers has been well described (Miller et al. 1977; Chu 1985; Cutrigh et al. 1971; Craig et al. 1975). Chapter 15 also describes the toxicity and biocompatibility of polymeric delivery devices.

Polycaprolactones. The most commonly used polymer in this class is poly(ε-caprolactone), which can be synthesized from anionic or cationic polymerization, coordination polymerization, and radical polymerization. Coordination polymerization enables the formation of a high molecular weight polymer, while anionic polymerization is the most convenient method for the synthesis of low molecular weight polymers with hydroxyl terminals.

Polycaprolactone is not soluble in aliphatic hydrocarbons, alcohol, and diethyl ether, while it is soluble in aromatic hydrocarbons, cyclohexane, and 2-nitropropane. This presents some difficulty in processing microspheres because aliphatic hydrocarbons are more desirable solvents for manufacturing microspheres. Aromatic hydrocarbons are much harder to remove during processing than aliphatic hydrocarbons.

Like lactide/glycolide polymers, polycaprolactone and its co-polymers degrade both in vivo and in vitro by bulk hydrolysis. The degradation rate is affected by the size and shape of the matrix, additives, molecular weight, pH, temperature, and the presence of a catalyst. The degradation rate of polycaprolactone can be increased by copolymerization or blending with lactide or glycolide, as well as the addition of oleic acid or tertiary amines.

Safety studies on polycaprolactone have been conducted with a polycaprolactone contraceptive delivery system. The drug incorporated in microspheres under the trade name Capronor™ has been implanted and extensively evaluated (Pitt 1990).

Poly(β-hydroxybutyrate). Poly(β-hydroxybutyrate) is a naturally occurring polymer that can be produced from propionic or patanoic acid by bacterial fermentation. The copolymer poly(hydroxybutyrate-cohydroxyvalerate) can be obtained from the mixture of propionic acid and patanoic acid by the action of *Alcaligenes eutrophus* (Doi et al. 1988). The compositions of copolymers and, subsequently, their characteristics, such as melting point, glass transition temperature, and susceptibility to hydrolysis, can be controlled by altering the ratio of starting materials. In general, copolymers with higher hydroxyvalerate content or lower molecular weights produce a faster degrading polymer (Doi et al. 1988).

Poly(β-hydroxybutyrate) or blends with other polyesters have been used to fabricate membranes and microspheres. Poly(β-hydroxybutyrate) microspheres are quite stable in aqueous solution, only 5 percent to 20 percent of the polymer was eroded when incubated with phosphate buffer at 85°C for 5 months (Domb 1994). Currently, there are no published safety data available for this polymer.

Poly(phosphate esters). Poly(phosphate esters) can be synthesized from the reaction of ethyl or phenyl phosphorodichloridates and various dialcohols by interfacial condensation using a phase-transfer catalyst with bisphenol A as a co-monomer. Polyethylene glycol segments of various molecular weights can also be inserted into the main chains of poly(phosphate esters). This biodegradable polymer has been considered as a potential drug carrier for controlled drug delivery as well as a matrix for tissue regeneration (Leong 1991).

The degradation and solubility properties of these polymers are dependent on the side chains. Poly(phosphate esters) from 4,4'-isopropylidenediphenol have a relatively slower degradation rate and a longer drug release duration; after 3 months, 20 percent of the incorporated cortisone was released from a polymer matrix into a buffer solution (Leong 1991).

This class of polymer is still in the early stage of investigation and development. Although the degradation rate of poly(phosphate esters) has been tested in vivo using a rabbit model, few published data exist on their toxicity and biocompatibility (Li et al. 1989; Leong 1991).

Polyanhydrides

Polyanhydrides are polymers gaining popularity in the drug delivery field. The degradation profile of these polymers present a distinct advantage over the more common polyesters and is described in detail below. A polyanhydride was first synthesized in 1909 (Bucher and Slade 1909) but has received attention in this field for only the past two decades.

Early work with polyanhydrides provided low molecular weight polymers, making them impractical for many applications. High molecular weight polyanhydrides became available following a systematic study on the mechanism of polymerization by Langer and his colleagues (Domb 1992; 1993; Domb and Langer 1987, 1988; Tamada and Langer 1992; Leong et al. 1985; Mathiowitz et al. 1988, 1990; Mathiowitz and Langer 1987, 1992). Factors affecting the polymer molecular weight are the purity of the starting materials, reaction time and temperature, and the removal of reaction products or by-products. Since both polymerization and depolymerization are involved in anhydride interchange, rapid removal of the condensation product acetic anhydride (polymerization) and internal ring formation (depolymerization) leads to a high molecular weight polymer. The products can be removed by maintaining a vacuum during the polymerization reaction (Domb and Langer 1988).

Higher molecular weight polyanhydrides were obtained using a melt polymerization technique by increasing the polymerization process while minimizing the depolymerization process. For example, by reacting pure, individually prepared prepolymers to produce a copolymer of biscarboxyphenoxy propane and sebacic acid [P(CPP-SA)] in a 80:20 molar ratio, a molecular weight of 116,800 was achieved. This is about 10 times higher than the molecular weight obtained by using unisolated and unpurified prepolymer mixtures. Catalysts such as cadmium acetate, earth metal oxides, and $ZnEt_2$-H_2O increase the electron deficiency of the carbonyl carbon, thus facilitating polycondensation. The molecular weight of polyanhydrides can be doubled by the use of catalysts in P(CPP-SA, 20:80) synthesis (Domb and Langer 1987).

The majority of polyanhydrides are soluble in solvents such as dichloromethane and chloroform. Aromatic polyanhydrides [e.g., poly(bis(p-carboxyphenoxy)methane] and poly[1,3-bis(p-carboxyphenoxy)propane] have a much lower solubility than aliphatic polyanhydrides [e.g., poly(sebacic acid)] (Tamada and Langer 1992).

Polyesters bulk erode while polyanhydrides surface erode, which can result in constant drug delivery rates from properly designed systems from polyanhydride systems. Polyanhydrides are highly hydrophobic polymers linked by a very sensitive water-labile anhydride bond. Since the rate of degradation of these bonds occurs by hydrolysis and is much faster than the rate of water penetrating through the surface into the center of the device, polyanhydrides only erode at the surface (i.e., at the polymer-water interface). This surface erosion provides maximum control of the release process and aids in the stability of incorporated water-labile drugs by shielding them from contact with water. The rate of drug release can be controlled by the constant degradation of the polymer and not by diffusion, dissolution, water penetration, or some other nonlinear transport process.

The degradation rate of polyanhydride can be varied from days to years by altering the ratio of the aromatic segment (i.e., CPP) to the aliphatic segments (i.e., SA). When the SA concentration in the polymer reached 80 percent, an increase in degradation rate of 800 times was observed, in comparison with the P(CPP) homo-polymer (Leong et al. 1985). A study on the effect of solution pH revealed that the degradation rate increased markedly as the pH rises (Leong et al. 1985).

The safety of several polyanhydrides and their degradation products have been safely evaluated in terms of biocompatibility and cytotoxicity and additionally is FDA approved for use in humans (Leong et al. 1986; Laurencin et al. 1987b; Tamargo et al. 1989; Brem et al. 1989, 1992, 1995).

Polyphosphazenes

Polyphosphazenes are another type of potential candidates as erodible biostructural materials for the controlled delivery of drugs. Polyphosphazenes containing amino acid ester side groups were the first bioerodible polyphosphazene synthesized (Grolleman et al. 1986a, b).

Polyphosphazenes can be synthesized by a substitution reaction of the reactive poly(dichlorophosphazene) with a wide range of reactive nucleophiles, such as amines, alkoxides, and organometallic molecules. The reaction is carried out at room temperature in tetrahydrofuran or aromatic hydrocarbon solutions (Allcock and Kugel 1965; Allcock et al. 1966; Allcock 1985). These polymers are typically soluble in the most polar organic solvents.

In contrast with most polyesters and polyanhydrides, the hydrolytic stability of polyphosphazenes is determined not by changes in the backbone structure but by changes in the side chains. Typically, amino acids and imidazole derivatives are used in the synthesis of hydrolytically degradable polyphosphazene. The rate of hydrolysis can be slowed by the incorporation of hydrophobic side groups such as phenoxy or methylphenoxy groups (Laurencin et al. 1987a). Polyphosphazenes with the appropriate side groups are capable of undergoing facile hydrolysis to phosphate and ammonia that can be safely excreted. These polymers typically erode hydrolytically to compounds that are generally regarded as safe: ethanol, glycine, phosphate, and ammonia.

Blended Polymers

A simple mixture of polymers may create a biomaterial with novel properties. Several bioerodible polymer blends have been developed for the purpose of altering the degradation and drug release properties of the polymers: blends of polycaprolactone with poly(hydroxybutyric acid) and lactide-glycolide copolymers (Holland et al. 1990; Yasin et al. 1992), blends of hydrophobic polymers with hydrophilic polymers such as polysaccharides or polyethylene glycol (Park et al. 1992; Holland et al. 1987), PLGA with polyethylene glycol (Cleek et al. 1997), and PLGA with polyvinyl alcohol (Pitt et al. 1992).

Blended polymers can be prepared simply by mechanical mixing of the two polymers to form a single material. A number of mixing processes have been studied and applied to different types of polymers for various applications. The degree of miscibility of a blended polymer can be measured by determining the glass transition temperature. For miscible polymer blends, there is only one glass transition temperature with a value between that of the two parent polymers; for immiscible polymer blends, there are two glass transition temperatures. By examining the glass transition temperature of polymer blends, the homogeneity and miscibility of the blend can be determined.

A systematic study of the miscibility of various biodegradable polymers to form new useful degradable materials has been completed (Domb 1993). Materials with unique physical, chemical, and biological properties can be created from blended polymers. The degradation and drug delivery rate can be designed for a specific application. If well-characterized materials are used, a comprehensive, experimental investigation into the safety profile of the polymers may be unnecessary.

Fabrication Methods and Techniques

There are dozens of methods that have been developed to fabricate nano/microspheres from synthetic polymers, polyesters, and polyanhydrides and are summarized in Table 8.3. Some of these methods are only slightly different from one another. It is impossible to determine one optimal fabrication method; selection of the most appropriate method must consider a variety of factors. The selection of the method is dependent on the nature of polymers, biological activity, stability, dose, and the optimal release profile of the drug.

These methods are typically multistep processes. The polymer is first dissolved in an appropriate organic solvent followed

Table 8.3. *Types of Fabrication Methods and Techniques for Nanospheres and Microspheres*

Fabrication Technique	Comment
Solvent extraction	Useful for entrapment of water-insoluble agents
Solvent evaporation	Produces less porous microspheres with slower drug release profile than those produced by extraction
Phase separation	Useful for entrapping water-soluble agents
Holt-melt encapsulation	Successfully applied to polyanhydride microspheres
Spray drying	Simple method that scales up easily but with a low efficiency

by dispersion of core materials containing the drug (either in an aqueous solution or in the form of small solid particles) into the organic phase. Next the organic phase is added to a water phase or another nonsolvent organic phase containing surfactants. Then the organic solvent in the first phase is removed by extraction, evaporation, or phase separation. Since dissolution of the polymer in organic solvent is required, deactivation of labile drugs (e.g., proteins), cells, or viruses may occur.

Microsphere Formation and Size Control

When a two-phase polymer solution is mixed by constant stirring, droplets or particles are formed. The size and size distribution of the droplets are determined immediately after mixing of the two phases. The droplet size is directly proportional to the final size of the nano/microspheres. There is a series of parameters that may affect particle size and size distribution: The speed and uniformity of stirring, the ratio and nature of the two phases, the ratio of core materials to organic phase, the nature of core materials, and the design of apparatus can all affect the nano/microsphere size. Presently, no universal quantitative relationship between these parameters and the average size of droplets has been established. However, an understanding of this relationship should be helpful in the design and synthesis of nano/microspheres. A relationship proposed by Fornusek and Vetvicka is a simple and useful example (Fornusek and Vetvicka 1986):

$$d \propto \frac{K(D_v R V_d \gamma)}{D_s N V_m C_s} \tag{1}$$

where d is the average droplet size; D_v is the diameter of the reactor; D_s is the diameter of stirrer; R is the ratio of the two phases; V_d and V_m are the viscosity of the droplet phase and suspension medium phase, respectively; γ is the surface tension between the two immiscible phases; N is the stirring speed; C_s is the concentration of the stabilizer; and K is a systematic parameter dependent on the apparatus design. In general, the average size of droplets increases as the mixing force decreases.

A key element in forming consistent droplets is the uniformity of the mixing force. The more uniform the mixing process, the more uniform the size of the droplets. Droplet formation in a two-phase suspension system is a dynamic process. Therefore, it is essential that the individuality of the initially formed droplets is maintained throughout the course of nano/microsphere formation, which can be achieved by the presence of a suitable droplet stabilizer. The function of a suspension agent is to form a thin film around the droplets (or particles) and thus prevent their coalescence. An ideal suspension stabilizer is usually a polymeric compound that is insoluble in the droplets and slightly soluble in the aqueous phase. Polyvinyl alcohol (PVA), polyvinyl pyrrolidone (PVP), polysorbate 80, sodium oleate, methylcellulose, alginate, and gelatin are often used for this purpose. Further discussions can be found in Arshady (1990, 1991), Nixon (1985), Deasy (1984), Chemtob (1984), and Kydonieus (1980).

Solvent Removal Methods

Nano/microspheres are formed by removing the solvent in the droplets that are formed as described above. There are three methods employed to remove the solvent from the droplet phase: solvent extraction, solvent evaporation, and phase separation.

1. **Solvent Extraction.** To utilize the solvent extraction method, the suspension medium must be miscible with the solvent. For example, acetone can be utilized as the polymer solvent and water used as the suspension medium. This procedure is useful for entrapping water-insoluble agents in biodegradable polymers.

2. **Solvent Evaporation.** When utilizing this method, the organic solvent is evaporated rather than extracted from the emulsion. Following formation of the droplet phase, this phase is emulsified with water, and the solvent is finally removed under vacuum to form discrete and hardened microspheres. The characteristics of the resulting nano/microspheres are strongly affected by the rate of solvent evaporation, which is dependent on pressure and temperature.

    The solvent must diffuse from the droplet phase into the medium before it can be removed by evaporation; therefore, solvent removal by extraction is generally faster than the evaporation method. Consequently, nano/microspheres produced by solvent evaporation are less porous than those prepared by solvent extraction, resulting in nano/microspheres having a slower drug release profile.

3. **Phase Separation.** This process involves the addition of the droplet phase into a nonsolvent phase such as silicon oil or olive oil. Under these conditions, coacervation of polymers encapsulated with or without core materials is possible. The resulting microspheres are separated by filtration, washed, and dried. This procedure is useful for entrapping water-soluble agents such as peptides and macromolecules in the polymers.

The methods described above are based on the different ways in which the solvent is removed. In some of the published literature, methods to fabricate polymeric nano/microspheres are classified by the method in which different solvents are combined: oil-in water (o/w), oil-in-oil (o/o), water-in-oil-in-water (w/o/w), and water-in-oil-in-oil (w/o/o). O/w is an emulsion technique followed by solvent evaporation to entrap hydrophobic substances. For hydrophilic substances, o/o and w/o/o can achieve a high encapsulation efficiency. W/o/w techniques have been developed to entrap water soluble compounds and those that are labile in organic solvents such as proteins and peptides.

Hot Melt Microencapsulation

Hot melt encapsulation has been successfully applied in the preparation of nano/microspheres of polyanhydrides. In this process, small, solid particles of drug are dispersed into melted polyanhydride. This mixture is suspended in an immiscible solvent, such as silicon oil or olive oil and preheated to 5°C above the melting point of the polymer. The system is cooled, the solvent drained, and the microsphere washed with petroleum ether to give free-flowing particles. By carefully controlling the stirring rate, a narrow size distribution can be achieved with excellent process reproducibility.

Spray Drying Microencapsulation

Spray drying is an easy and rapid method for preparing large quantities of microspheres. In this method, the polymer is dissolved in a solvent with a low boiling point; drug can be preloaded into the polymer solution either by dissolution or dispersion of small particles. The solution is then sprayed through an atomizer and simultaneously dried by an upward flow of nitrogen. Polymers with a low glass transition temperature may not be suitable to this technology due to the aggregation of particles during the process, resulting in an uneven particle morphology and size distribution.

TECHNOLOGY LIMITATIONS

Microsphere technology has many advantages, which has resulted in several commercially successful products (Chapter 3). However, there are limitations to this technology that must be considered before initiating a nano/microsphere development program. Most of the disadvantages of microspheres are related to the many challenges of manufacturing on a commercial scale and are outlined below and summarized in Table 8.1.

- *Manufacturing process:* The manufacturing process for microspheres is very complex, typically resulting in the need for a customized, aseptic manufacturing facility.

- *Sterilization:* Terminal sterilization is typically not feasible due to degradation of drug and/or polymer. Costly and difficult aseptic processing must usually be employed.

- *Residual solvent:* Producing product with low levels of residual solvent is usually difficult. Alternative less toxic solvents may need to be utilized.

- *Stability:* The stability of drugs during manufacture and the shelf life of the product can present technical obstacles. Labile drugs may degrade (or lose activity) during processing, and polymer and/or drugs may degrade during long-term storage.

- *Biomaterials:* Procurement of consistent polymeric materials is important to the manufacture of quality product. The availability of degradable, synthetic polymers is limited at the present time; only polyesters and polyanhydrides are in products that are approved for use in the United States.

- *Encapsulation efficiency:* It can be difficult to develop a process that utilizes drug efficiently. Drug losses of 25 to 50 percent or more are common.

- *Complex system:* There does not appear to be a formulation and fabrication process that is appropriate for all drug compounds and indications. Therefore, a custom-made formulation must be developed for each particular compound or application.

In spite of these limitations, microsphere technology has found its place in today's pharmacy. Developing these complex products is challenging, typically limited by financial and/or time constraints. However, throughout this book the development of a microsphere product, Lupron Depot®, is discussed, and it appears that the increased time and development costs were a wise investment.

CONCLUSION AND FUTURE TRENDS

Successful microsphere products have been commercialized, and many more products are presently in development. Microsphere technology can result in products that provide benefits to patients and greatly improve health care; however, the authors could not agree on the magnitude of the use of microspheres in the future. Will microspheres have widespread use as a generic drug delivery system in the pharmaceutical industry or will the technology be utilized only for niche applications? We have outlined some very significant advantages of microspheres for drug delivery applications, but it remains to be seen in this cost-conscious healthcare environment to what extent the technical limitations will be overcome as this technology moves forward into the 21st century.

REFERENCES

Allcock, H. R. 1985. Inorganic macromolecules. *Chem. Eng. News* 63:22.

Allcock, H. R., and R. L. Kugel. 1965. Synthesis of high polymeric alkoxy- and aryloxyphosphonitriles. *J. Am. Chem. Soc.* 87:4216–4217.

Allcock, H. R., R. L. Kugel, and K. J. Valan. 1966. Phosphonitrillic compounds. VII. High molecular weight poly(alkoxy- and aryloxyphosphazenes). *Inorg. Chem.* 5:1709–1715.

Arshady, R. 1990. Microspheres and microcapsules: A survey of manufacturing techniques, Part 2: Coacervation. *Poly. Eng. Sci.* 30:905–914.

Arshady, R. 1991. Preparation of biodegradable microspheres and microcapsules: 2. Polylactides and related polyesters. *J. Control. Rel.* 17:1–12.

Baker, R., ed. 1980. *Controlled release of bioactive materials.* New York: Academic Press.

Brem, H., A. Domb, D. Lenartz, C. Dureza, A. Olivi, and J. I. Epstein. 1992. Brain biocompatibility of a biodegradable controlled release polymer consisting on anhydride copolymer of fatty acid dimer and sebacic acid. *J. Control. Rel.* 19:325–330.

Brem, H., M. Sisti, P. C. Burger, J. Greenhoot, S. Piantodosi, and M. G. Ewend. 1995. The safety of interstitial chemotherapy with BCNU-loaded polymer followed by radiation therapy in the treatment of newly diagnosed malignant gliomas: Phase I trial. *J. Neurooncol.* 26:111–123.

Brem, H., A. Kader, J. I. Epstein, R. Tamargo, A. Domb, R. Langer, and K. W. Leong. 1989. Biocompatibility of bioerodible controlled release polymers in rabbit brain. *Sel. Cancer Ther.* 5:55–65.

Bucher, J. E., and W. C. Slade. 1909. The anhydrides of isophthalic and terephtalic acids. *J. Am. Chem. Soc.* 31:1319–1321.

Chemtob, C. 1984. Microencapsulation by coacervation. *Labo-Pharma. Probol. Tech.* 32:702–709.

Chien, Y. W., ed. 1982. *Novel drug delivery systems, fundementals, developmental, concepts, biomedical assessments.* New York: Marcel Dekker.

Chu, C. C. 1985. Degradation phenomena of two linear aliphatic polyester fibre used in medicine and surgery. *Polymerization* 26:591–594.

Cleek, R. T., K. C. Ting, S. G. Eskin, and A. G. Mikos. 1997. Microparticles of poly(DL-lactic acid-co-glycolic acid)/poly(ethylene glycol) blends for controlled drug delivery. *J. Control. Rel.* 48:259–268.

Couvreur, P., and F. Puisieux. 1993. Nano- and microparticles for the delivery of polypeptides and proteins. *Adv. Drug Delivery Rev.* 10:141–162.

Craig, P. H., J. A. Williams, K. W. Davis, A. D. Magoun, A. J. Levy, S. Bogdansky, and J. P. Jones. 1975. A biologic comparison of polyglactin 910 and polyglycolic acid synthetic absorbable sutures. *Surg. Gynecol. Obstet.* 141:1–10.

Cutrigh, D. E., J. D. Beasley, and B. Perez. 1971. Histologic comparison of polylactic and polyglycolic acid sutures. *Oral Surg.* 32:165–173.

Das, K. G., ed. 1983. *Controlled release technology: Bioengineering aspects.* New York: John Wiley & Sons.

Deasy, P. 1984. *Microencapsulation and related drug processes.* Basel, Switzerland: Marcel Dekker Verlag, chapters 3–4.

Doi, Y., A. Tamaki, M. Kunioka, and K. Soga. 1988. Production of copolyesters of 3-hydroxybutyrate and 3-hydroxyvalerate by alcaligenes eutrophus from butyric and pentanoic acids. *Appl. Microbiol. Biotech.* 28:330–334.

Domb, A. J. 1992. Synthesis and characterization of biodegradable arometic anhydride copolymers. *Macromolecules* 25:12–17.

Domb, A. J. 1993. Degradable polymer blends. I. Screening of miscible polymers. *J. Polym. Sci. Part A: Polymer Chemistry* 31:1973–1981.

Domb, A. 1994. Implantable biodegradable polymers for site-specific drug delivery. In *Polymeric site-specific pharmacotherapy,* edited by A. J. Domb. New York: John Wiley and Sons Ltd., pp. 1–27.

Domb, A., and R. Langer. 1987. Polyanhydrides. I. Preparation of high molecular weight polyanhydrides. *J. Polym. Sci.* 25:3373–3386.

Domb, A., and R. Langer. 1988. Solid state and solution stability of poly(anhydrides) and poly(esters). *Macromolecules* 21:1925–1919.

Fornusek, L., and V. Vetvicka. 1986. Polymeric microspheres as diagnostic tools for cell surface marker tracing. *CRC Crit. Rev. Therap. Drug Carrier Sys.* 2:137–174.

Grolleman, C. W. J., A. C. de Visser, J. G. C. Wolke, C. P. A. T. Klein, H. van der Goot, and H. Timmerman. 1986a. Studies on a bioerodible drug carrier system based on a polyphosphazene. Part II. Experiments in vitro. *J. Control. Rel.* 4:119–131.

Grolleman, C. W. J., A. C. de Visser, J. G. C. Wolke, C. P. A. T. Klein, H. van der Goot, and H. Timmerman. 1986b. Studies on a bioerodible drug carrier system based on a polyphosphazene. Part III: Experiments in vivo. *J. Control. Rel.* 4:133–142.

Holland, S. J., A. M. Jolly, M. Yasin, and B. J. Tighe. 1987. Polymers for biodegradable medical devices. II. Hydroxybutyrate-hydroxyvalerate copolymers: Hydrolytic degradation studies. *Biomaterials* 8:289–295.

Holland, S. J., M. Yasin, and B. J. Tighe. 1990. Polymers for biodegradable medical devices. VII. Hydroxybutyrate-hydroxyvalerate copolymers: Degradation of copolymers and their blends with polysaccharides under in vitro physiological conditions. *Biomaterials* 11:206–215.

Kydoniens, A. F., ed. 1980. *Controlled release technologies: Methods, theory and application.* Boca Raton, Fla., USA: CRC Press.

Langer, R. S. 1990. New methods of drug delivery. *Science* 249:1527–1533.

Langer, R. S., and N. A. Peppas. 1981. Present and future applications of biomaterials in controlled drug delivery systems. *Biomaterials* 2:201–214.

Laurencin, C., J. H. Koh, T. X. Neenan, H. R. Allcock, and R. S. Langer. 1987a. Controlled release using a new biodegradable polyphosphagene matrix system. *J. Biomed Mater. Res.* 21:1231–1246.

Laurencin, C. T., A. J. Domb, C. D. Morris, V. I. Brown, M. Chasin, R. F. McConnell, and R. Langer. 1987b. High dosage administration of polyanhydrides in vivo: Studies of biocompatibility and toxicicology. *Proc. Int. Symp. Control. Rel. Bioact. Mater.* 1:140–141.

Leong, K. W. 1991. Synthetic bioerodible polymer drug delivery systems. In *Polymers for controlled drug delivery*, edited by P. J. Tarcha. Boca Raton, Fla., USA: CRC Press.

Leong, K. W., B. C. Brott, and R. Langer. 1985. Bioerodible polyanhydrides as drug-carrier matrices. I. Characterization, degradation, and release characteristics. *J. Biomed. Mater. Res.* 19:941–955.

Leong, K. W., P. D'Amore, M. Marletta, and R. Langer. 1986. Bioerodible polyanhydrides as drug-carrier matrices: II. Biocompatibility and chemical reactivity. *J. Biomed. Mat. Res.* 20:51–64.

Li, N. H., M. Richards, K. Brandt, and K. W. Leong. 1989. Poly(phosphate esters) as drug carriers. *Poly. Preprints* 30:454–455.

Mathiowitz, E., and R. Langer. 1987. Polyanhydride microsphere drug carriers: Hot melt microencapsulation. *J. Control. Rel.* 5:13–22.

Mathiowitz, E., and R. Langer. 1992. Polyanhydride microspheres as drug delivery systems. In *Microcapsules and nanoparticles in medicine and pharmacy*, edited by M. Donbrow. Boca Raton, Fla., USA: CRC Press, pp. 99–123.

Mathiowitz, E., P. Dor, C. Amato, and R. Langer. 1990. Polyanhydride microspheres as drug carriers. III. Morphological characterization of microspheres by solvent removal. *Polymer* 31:547–555.

Mathiowitz, E., W. M. Saltzman, A. J. Domb, P. Dor, and R. Langer. 1988. Polyanhydride microspheres as drug carriers. II. Microencapsulation by solvent removal. *J. Appl. Poly. Sci.* 35:755–774.

Miller, R. A., J. M. Brady and D. E. Cutright. 1977. Degradation rates of oral reabsorbable implants (PLA-PGA), rate modified with change in poly(lactide-co-glycolide) copolymer ratios. *J. Biomed. Mater. Res.* 14:470–478.

Nixon, J. R. 1985. Preperation of microcapsules with possible pharmaceutical use. *Endeavour New Ser.* 9:123–128.

Okada, H. and H. Toguchi. 1995. Biodegradable microspheres in drug delivery. *Crit. Rev. Therap. Drug Carrier Sys.* 12:1–99.

Okada, H., T. Heya, Y. Igari, Y. Ogawa, H. Toguchi, and T. Shimamoto. 1989. One-month release injectable microspheres of leuprolide acetate inhibit steroidogenesis and genital organ growth in rats. *Int. J. Pharm.* 45:231–239.

Okada, H., Y. Inoue, T. Heya, H. Ueno, Y. Ogawa, and H. Toguchi. 1991. Pharmacokinetics of once a month injectable microspheres of leuprolide acetate. *Pharm. Res.* 8:787–791.

Oldshue, J. Y. 1983. *Fluid mixing technology, chemical engineering.* *New York:* McGraw-Hill.

Park, T. G., S. Cohen, and R. Langer. 1992. Poly(L-lactic acid)/pluronic blends: Characterization of phase separation behavior, degradation, and morphology and use as protein release matrices. *Macromolecules* 25:116–122.

Pitt, C. G. 1990. Polymers–therapeutic uses. In *Biodegradable polymers as drug delivery systems,* edited by M. Chasin and R. Langer. New York: Marcel Dekker.

Pitt, C. G., Y. Cha, S. S. Shah, and K. J. Zhu. 1992. Blends of PVA and PLGA: Control of the permeability and degradability of hydrogels by blending. *J. Control. Rel.* 19:189–200.

Sanders, L. M., J. S. Kent, G. I. McRae, B. H. Vickery, T. R. Tice, and D. H. Lewis. 1984. Controlled release of a leuteinizing hormone-releasing hormone analogue from poly(d,l-lactide-co-glycolide) microspheres. *J. Pharm. Sci.* 73:1294–1297.

Siegal, R. A., and R. Langer. 1984. Controlled release of polypeptides and other macromolecules. *Pharm Res.* 1:1–10.

Tamada, J., and R. Langer. 1992. The development of polyanhydrides for drug delivery applications. *J. Biomater. Sci. Polymer Edn.* 3:315–353.

Tamargo, R. J., J. I. Epstein, C. S. Reinhard, M. Chasin, and H. Brem. 1989. Brain biocompatibility of a biodegradable controlled release polymer in rats. *J. Biomed. Mat. Res.* 23:253–266.

Yasin, M., and B. J. Tighe. 1992. Polymers for biodegradable medical devices. VIII. Hydroxybutyrate-hydroxyvalerate copolymers: Physical and degradative properties of blends with polycaprolactone. *Biomaterials* 13:9–16.

9

Nano/Microspheres from Natural Polymers

LinShu Liu
Orquest, Inc.
Mountain View, California

Joseph Kost
Ben-Gurion University
Beer-Sheva, Israel

Natural polymers are the subject of significant research in drug delivery systems. Studies on natural polymers as drug carriers have concentrated on materials of two types: proteins and polysaccharides. Proteins (e.g., collagen, gelatin, albumin, and fibrinogen) are polymers of natural amino acids and vary in sequence and molecular weight. Polysaccharides, including cellulose, hyaluronate, chitosan, starch, xanthan gum, and alginate, differ from each other in the identity of their recurring monosaccharide units, the length of the polymer chain, the type of O-glycosidic bonds linking the monomers, and the degree of branching. Natural polymers are found in both plants and animals and provide several advantages over synthetic polymers: They are products of living organisms, readily available, nontoxic, and capable of further chemical modifications to meet the requirements of specific applications.

The primary limitation of natural polymers in drug delivery systems is the difficulty in controlling the degradation rate because most natural polymers degrade by enzymatic hydrolysis. The degradation of these polymers is highly tissue specific and difficult to predict from in vitro studies. An additional concern is the antigenicity of these polymers. However, with recent advances in biotechnology, recombinant and immune-inert biopolymers may be possible for future medical applications.

It is not surprising that the list of natural polymers approved for medical applications is quite long. For example, both Kelco in the United States and Protan in Norway produce medical grade alginates. The first medical application of sodium hyaluronate was in ophthalmic surgery and has since been used for the treatment of arthritis by direct injection into the joint. In addition, hyaluronate has been utilized as a carrier for drug delivery applications. A commercial product under the trademark Healon® was introduced by Kabi Pharmacia (Upsala, Sweden), and several other companies have developed similar formulations. In the late 1980s, additional hyaluronan preparations for the treatment of arthritis were approved in Japan (Arzt®, Seikagaku and Kaken, Tokyo) and in Italy (Hyalagan®, Fidia). In 1973, collagen sutures manufactured by Sherwood Medical were approved for use, and additional collagen products have been approved for medical applications in plastic surgery, orthopedics, dentistry, and cardiovascular surgery. Labopharm introduced an amylose derivative for controlled-release matrices under the trade name Contramid™, obtained by treating amylose with cross-linking agents. Gelatin (i.e., denatured collagen) is used in the pharmaceutical industry for the manufacture of drug capsules and in surgery as a hemostatic sponge. Fats and oils have also been employed for the preparation of liposomes and emulsions for drug delivery and are discussed in Chapters 5 and 7.

This chapter reviews the processing steps and applications of nano- and microspheres from natural polymers and focuses on the use of proteins and polysaccharides employed in drug delivery systems. The sources and suppliers of medical grade natural polymers discussed in this chapter are listed in Table 9.1.

Table 9.1. *Sources and Suppliers of Medical Grade Natural Polymers*

Polymers	Source	Provider
Albumin	Blood	Baxter Heathcare Corp.
Collagen	Dermal tissue	Collagen Corp. Integra Lifescience Company Kensey-Nash Inc.
Fibrinogen	Blood	Baxter Healthcare Corp.
Gelatin	Denatured collagen	Vyse Gelatin Company
Alginate	Seaweed extracts	Pronova Biopolymer Chemical MFG Corp.
Cellulose	Plant cell wall	Eastman Chemical company
Chitosan	Fungi wall membrane, commercially derived from chitin by chemical conversion with alkali	Pronova Biopolymer
Dextran	Bacteria	Medisan Pharmaceutical
Hyaluronate	Rooster combs bacteria	Anika Inc. Lifecore Biomedical Corp. Genzyme Corp. Fidia
Starch	Plant seed	National Starch and Chemical Surjikos Laing-National

METHODOLOGY OF NANO/MICROSPHERE PREPARATION

Polymeric nano/microspheres can be synthesized from monomers or existing polymers. The preparation from monomers must include a polymerization step. For example, poly(alkyl cyanoacrylate) nanospheres are obtained by emulsion polymerization, and reservoir type polyamide microspheres can be prepared by interfacial polymerization. Nano/microspheres can be also prepared from polymers by inducing aggregation and further stabilization by heat denaturation or chemical cross-linking. For hydrophobic polymers, such as poly(lactide), poly(lactide-co-glycolide), and water-insoluble polymers, such as esterified hyaluronate derivatives, solvent evaporation/extraction is a simple and well-documented method (Chasin and Langer 1990).

Most natural polymers are hydrophilic, requiring the formation of the appropriate-sized polymer droplets followed by cross-linking to reduce their solubility in an aqueous system. Suspension cross-linking techniques are employed to prepare of small particles for most natural polymers and consist of the formation of droplets with the desired size, cross-linking, and recovery of the resulting nano/microspheres. Another common, effective, and convenient approach is coacervation, by which ionic bonds between two compounds are formed in one solvent mixture. Spray drying is also used to process nano/microspheres and can be continually and easily scaled up for large-scale production.

Suspension Cross-Linking

Droplet Formation

The most important characteristic of nano/microspheres is the size of the particles and their distribution. Particle size is primarily determined during droplet formation in an oil-water emulsion and can be controlled by adjusting the stirring speed, duration, and uniformity; the concentration of materials in the water phase; the type of the oil phase; the ratio of water to oil phase; and the design of the apparatus. These relationships can be described by equation 1 (Kafarov and Babanov 1959; Scully 1976; Arshady and Ledwith 1983; Arshady 1990).

$$d \propto \frac{K D_v R v_d \gamma}{D_s N v_m C_s} \tag{1}$$

The average droplet size at steady state (d) is proportional to the surface tension between the two immiscible phases (γ), the volume ratio of the dispersion phase to the continuous phase (R), and the ratio of viscosities of the dispersed phase (droplets) to the continous medium phase (v_d/v_m), as well as the ratio of

D_v/D_s, where D_v is the diameter of the reactor and D_s is the diameter of stirrer, N represents the mixing force, C_s is the concentration of the stabilizer, and K respresents the apparatus design parameters.

In general, the average size of droplets increases as the mixing force decreases. The key element in the uniformity of the the size distribution of the droplets is the uniformity of the mixing force throughout the mixing process. Uniform mixing will produce a uniform droplet size.

Droplet formation in a two-phase suspension system is a dynamic process; the small aqueous droplets in the oil phase are unstable and therefore tend to coalesce. Coalescence is usually inhibited or stabilized by the presence of a surfactant forming a thin film around the individual droplets (or particles). The concentration and solubility of the stabilizer in both phases is the most important characteristic in choosing a surfactant. An ideal suspension stabilizer is usually a polymeric compound that is insoluble in the droplets (dispersed phase) and slightly soluble in the aqueous phase (continuous phase). Polyvinyl alcohol (PVA), polyvinyl pyrrolidone (PVP), sorbitan trioleate, and sodium oleate are the most utilized stabilizers. In addition, methylcellulose, alginate, and gelatin are frequently employed.

Temperature also affects the droplet size, through its effect on viscosity and surface tension. Dehydration caused by a temperature increase can also result in inter/intramolecular cross-linking (Yapel 1979).

Ultrasound irradiation (sonication) is an efficient means of mixing and dispersion. When ultrasound is applied for droplet generation, the droplets size depends on the ultrasound irradiation intensity, frequency and duration. To protect the polymers from degradation during sonication, pulsed, low intensities of short duration are usually used.

Cross-Linking Chemistry

In many cases, protein/polysaccharide-based nano/microspheres formed in the liquid phase require further chemical cross-linking to enhance their stability. The cross-linking process can be accomplished through the addition of bi- or multifunctional active reagents (the cross-linking agent) that react with at least two polymeric chains to generate a three-dimensional structure. Cross-linking can also occur without the addition of cross-linking agents (i.e., self-cross-linking) by dehydration between two macromolecules with a primary amine and a carboxyl functional group. Regardless of the type of cross-linking, active functional groups attached to the sugar rings or the polypeptide chains are necessary. These active groups, are not part of the main chain construction and are thus free for further reaction. The most common functional side groups for protein cross-linking are the primary amine in lysine, the carboxyl group in glutamine, and the hydroxyl group in serine. For polysaccharide cross-linking, hydroxyl and carboxyl functional groups are the most frequently involved. The primary amine in deacetyl chitin can also be utilized. Highly reactive groups can be also generated from functional groups with relatively low reactivity. For instance, active sulfhydryls can be generated by a selective reduction reaction of a cysteine or a disulfide residue in proteins, or highly active formyl groups can be generated by oxidizing the sugar rings of most polysaccharides.

Many of the chemical reagents applied as cross-linkers for the preparation of polysaccharide/protein-based nano/microspheres were originally developed for chromatography supports in the coupling of affinity ligands and also for the modification and conjugation of proteins and polysaccharides. Usually, these reagents are electrophilic.

For most drug delivery applications, the incorporation of drugs during the cross-linking process affects the cross-linking efficiency. Therefore, many approaches load drug into the matrix after the cross-linking reaction is complete.

The following is a brief description of the most frequently used cross-linking reagents.

Bis-epoxides. Biofunctional compounds containing epoxide groups on both ends, such as 1,4-butainediol diglycidyl ether and diepoxyoctane that have a low toxicity and are well-documented reagents, can be used to cross-link biopolymers. Epoxide groups can react with a nucleophile-containing compound, such as hydroxyls, primary amines, and sulfhydryls (reaction 1). The reaction with hydroxyls requires a high pH, the reaction with amine can be performed at more moderate alkaline pH, and the reaction with sulfhydryl requires a pH at close to physiological conditions.

$$R-OH \quad + \quad \overset{O}{\triangle}\diagdown\diagup R'\diagdown\diagup\overset{O}{\triangle} \quad \longrightarrow \quad R-O\overset{OH}{\diagdown\diagup}\diagdown R'\diagdown\diagup\overset{OH}{\diagdown\diagup}O-R$$

Divinyl sulfone. Divinyl sulfone is a highly active reagent that requires an aqueous alkaline solution (reaction 2). The reaction is very fast in most cases, even when carried out at room temperature.

$$R-OH \quad + \quad H_2C{=}HC{-}\overset{O}{\underset{O}{\overset{\parallel}{S}}}{-}CH{=}CH_2 \quad \longrightarrow \quad R{-}O{-}H_2C{-}\overset{O}{\underset{O}{\overset{\parallel}{S}}}{-}CH_2{-}O{-}R$$

Isocyanates. Isocyanate-containing reagents can be used to cross-link hydroxyl containing molecules, including polysaccharides and proteins (reaction 3). This reaction occurs preferably in alkaline solutions.

$$R{\cdot}OH \quad + \quad O{=}C{=}N\diagup{}^{'R}\diagdown N{=}C{=}O \quad \longrightarrow \quad R{-}O\overset{O}{\overset{\parallel}{\diagup}}\underset{H}{N}\diagdown{}^{'R}\diagdown\underset{H}{N}\overset{O}{\overset{\parallel}{\diagup}}O{-}R$$

Isocyanates also react with amine-containing molecules. The reaction involves an attack of the nucleophile onto the electrophilic carbon atom of the isocyanate groups. The resulting electron shift and proton loss creates a stable isourea linkage between the isocyanate-containing compound and the amine, without any leaving group involved. The stability of isocyanate reagents is limited, since their stability is very sensitive to moisture, and the reagent may rapidly decompose, resulting in the formation of carbon dioxide and an aromatic amine.

Aldehyde. Aldehyde groups can react with hydroxyl groups to form hemiacetals and acetals, effectively reacting with amino groups to form Schiff bases. Aldehyde groups can also react with sulfhydryl groups. The most frequently used aldehyde-containing reagent is glutaraldehyde. The reaction with glutaraldehyde is pH dependent, efficient at low pH, and especially efficient at high pH. The rather labile Schiff base interaction can be chemically stabilized by reductive amination in the presence of sodium borohydride or sodium cyanoborohydride, creating a secondary amine linkage between the two molecules (reaction 4). Glutaraldehyde is the most popular bis-aldehyde homofunctional cross-linker.

Carbodiimides. 1-ethyl-3-(3-dimethylaminopropyl)carbodiimide hydrochloride (EDC) is the most popular carbodiimide for cross-linking polysaccharides and proteins. It mediates the formation of an amide bone between carboxylates and amines. The reaction process includes the formation of a highly active intermediate, which is short lived in aqueous environments. Subsequently, the carbonyl group of this ester is attacked by a nucleophilic amine, resulting in the loss of an isourea derivative and the formation of an amide bond. There are no additional chemical structures introduced in forming these bonds, thus carbodiimides are also called zero-length reagents (reaction 5). Carbodiimide-mediated amide bond formation effectively occurs between pH 4.7 and 7.5. The major competing reaction in water is hydrolysis.

Periodates. Periodate cleaves adjacent hydroxyl groups in sugar residues to create highly reactive aldehyde functionality (reaction 6). The reaction can be performed in mild conditions. The amount of periodate added can be adjusted to cleave only certain sugars in the polysaccharide chains and mucoproteins. The created aldehydes can be further reacted with, for example, diaminocompounds via Schiff base formation and reductive amination to complete cross-linking.

Other reagents such as alkyl halogens or heterobifunctional cross-linkers such as epichlorohydrin and reagents that contain amine-reactive or carbonyl-reactive groups on one end, and sulfhydryl-reactive, maleimide or photoreactive groups on the other end can also be used for cross-linking proteins and polysaccharides.

Self-Cross-Linking. Self-cross-linking is also called thermal cross-linking or thermaldehydration. Nano/microspheres derived from proteins are often stabilized by incubation at certain temperatures, which results in intermolecular and/or intramolecular cross-linking accompanied by water loss. To avoid oxidation, thermaldehydration is sometimes carried out under high vacuum.

In drug delivery applications, the relatively high temperatures needed for thermal cross-linking exclude this method for the incorporation of temperature-sensitive drugs such as proteins and peptides. Temperature can affect the degree of cross-linking in the resulting nano/microspheres and, consequently, their swellability, rate of degradation, and amount of incorporated drug. For example, albumin microspheres were prepared using the suspension method (Sugibayashi et al. 1979) and were further treated at 100, 150, and 180°C. As the reaction temperature increased, the albumin cross-linking density increased, and the cumulative amount of released core substance, 5-fluorouracil, decreased from 40 percent to 10 percent, and to 5 percent when incubated in phosphate-buffered saline for 1 week.

Protein-based nano/microspheres obtained by thermal cross-linking result in relatively short cross-linkages compared to those formed with chemical cross-linking agents. The less porous morphology may exclude drugs from the nano/microspheres during the heat treatment process. It has been reported (Gupta et al. 1986a, b, c; Arshady 1990) that albumin nanospheres produced at higher temperatures contain less drug than those obtained similarly but at a lower temperature. Drug exclusion from the albumin nanospheres could account for the lower drug loading, although drug decomposition at higher temperatures may be partially responsible for the differences in drug loading efficiency.

Coacervation

Coacervation can be defined as a phenomenon or process associated with colloids, where dispersed particles separate from the solution to form a second phase. Coacervation is an important method in the preparation of microparticles from natural polymers. Coacervation of proteins, polysaccharides, or other polyelectrolyte solutions can be induced by changing the solution's physical conditions, such as temperature, ionic strength, pH, or by the addition of counterions. Coacervation has been extensively used to prepare nano/microspheres derived from gelatin, alginate, chitosan, and related products. The following section is a brief discussion of the parameters affecting the coacervation process.

Counterions

The polyelectrolyte complex is formed from two compounds carrying counterions, such as collagen with chondroitin sulfate, alginate with calcium chloride, alginate with chitosan, chitosan with sodium sulfate, or chitosan with gelatin. The selection of the core and shell materials is essential. Usually, the core material should have a relatively high molecular weight compared with that of the shell material. Shell material with a high molecular weight may cause aggregation of the formed microparticles. One recent example is the collagen/chondroitin sulfate microsphere system developed by Shao and Leong (1995). When a collagen solution is extruded into a chondroitin sulfate bath, complexation occurs, and a coacervate is formed. Alternatively, when the chondroitin sulfate solution is extruded into the collagen bath, the microcapsules obtained are fragile and have no mechanical stability. Collagen is a high molecular weight fibrous protein (300 kDa) with a triple helical structure, while chondroitin sulfate is a much smaller molecule. Thus, to construct stable microspheres, collagen should be chosen as the internal component. A similar phenomenon was observed in preparing of alginate/calcium chloride microspheres (Dainty et al. 1986).

Ionic Strength

Since small ions in solution can shield charges on polyelectrolytes, an increase of the solution's ionic strength results in a decrease of effective net charges and, consequently, the extent of complex coacervation (Michaels 1965; Mortada et al. 1987).

Solution Viscosity

Solution viscosity strongly affects the efficiency of coacervation and the size of the formed coacervates. For most polyelectrolytes, coacervation occurs when the solution viscosity is above a threshold level. Lower solution viscosity will lead to less efficient coacervation and weak bead structures. A higher solution viscosity is also preferred for higher efficiency of drug incorporation, although at higher concentrations a homogeneous distribution of the added counterions is more difficult, frequently resulting in the formation of agglomerates. Since solution viscosity depends on the molecular weight and charge of polyelectrolytes, the ideal solution concentration for a specific system may differ from each material and processing procedure. For example, alginate (solution) concentration in the coacervation process is usually less than 2.2 percent, and chitosan is employed at concentrations of 0.25 percent (w/v) or less (Bodmeier and Paertakul 1989; Mumper et al. 1994; Aslani and Kennedy 1996; Rajaonarivony et al. 1993; Berthold et al. 1996).

Solution pH

The pH of a polyelectrolyte solution affects its solubility and net charge, and the stability of the formed coacervate. These in turn affect the release rate of drugs incorporated into the coacervate. Since the mechanism of complexation is primarily based on ionic cross-linking, the coacervation formation kinetics also depends on the solution pH (Mumper et al. 1994; Shao and Leong 1995).

Molecular Weight

The molecular weight of polyelectrolytes is another parameter to be considered when preparing microspheres by the coacervation method.

Figure 9.1 shows the effect of chitosan's molecular weight on the formation of its coacervate with sodium sulfate (Berthold et al. 1996). Chitosan (0.25 percent, w/v) with different molecular weights was added to sodium sulfate solution while stirring. The formation of microspheres was examined by the turbidity of the suspension at 500 nm. The required amount of counterions

Figure 9.1. *Turbidity monitoring during the precipitation step employed to produce chitosan microspheres.*

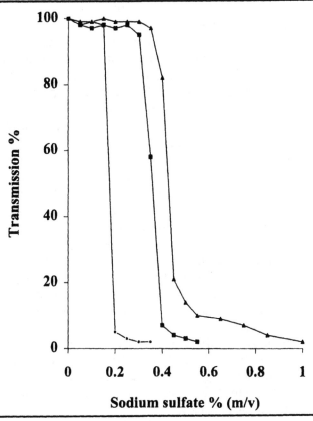

(●) low molecular weight chitosan; (■) medium molecular weight chitosan; (▲) high molecular weight chitosan (mean ± SD; n = 3). Adapted from Berthold et al. (1996).

for coacervation increased with the increase in chitosan's molecular weight. The solubility of chitosan is dependent on the number of positive charges on its surface. High molecular weight chitosan will require a relatively lower number of protonated groups per mass of polymer on its surface, assuming that the unprotonated groups could be concealed in the interior of the rolled up or folded macromolecule.

Spray Drying

Nano/microspheres from natural polymers can also be prepared by spray drying. There are several commercially available systems that are either open or closed cycles (Figure 9.2). In the open cycle spray dryer, the drying gas is heated air that is vented from the atmosphere and exhausted to a fume hood. In the closed cycle, the drying gas is usually nitrogen with a low oxygen concentration, which is heated to and maintained at a temperature that is adequate for use with an organic solvent. The organic vapors are condensed to a liquid state, allowing the gas to be recovered and reused.

In general, the polymer solution is fed into a nozzle, where it is atomized with a stream of hot and dry gas and injected into the main chamber at a high speed. The large surface area of the spray and the intimate contact with the hot gas result in a rapid evaporation of the solvent and solidification of the polymers, yielding microspheres. The size of the droplets and, subsequently, the size of the microspheres are determined by solution viscosity, flow rate, the structure of the nozzle, and pressure. The drying particles are collected from the airsteam in collection vessels connected to the chamber for coarse particles and the cyclon separator for fine particles. It should be noted that the temperature difference between the top and bottom of the chamber can be as large as 20 to 200°C. Injected substances are exposed to these higher temperatures only briefly, which is suitable for heat-labile, low melting point, and low glass transition temperature substances.

The advantages of spray drying include precise control of the particle size, a robust and reproducible process, and ready adaptation to continuous operation at small and large scales (Kissel and Konebeg 1996; Sacchetti and van Oort 1996).

For the encapsulation of proteins, such as antigens, growth factors, or hormones, the application of spray-drying techniques requires the lyophilization of proteins prior to dispersal into the polymer solutions. As the lyophilization process induces aggregation and denaturation of labile proteins, it becomes the major concern in this technique. To prepare biochemically stable, spray-dried polymer/protein products, the addition of water-replacing agents, such as sucrose or polysorbate 20, with the protein solution before lyophilization was proposed (Sarciaux

Figure 9.2. *Preparation of nano/microspheres by the spray-drying method: (A) open-cycle layout; (B) closed-cycle layout.*

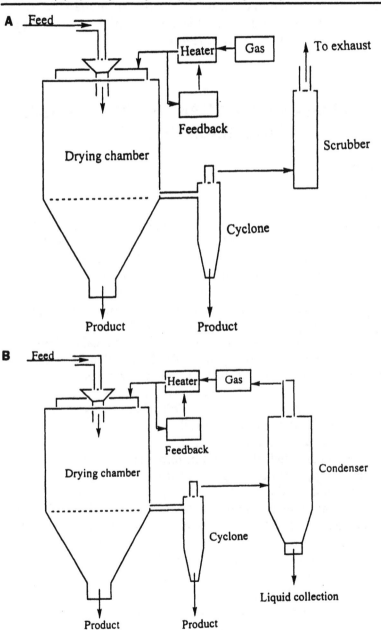

and Hageman 1997; Mumenthaler et al. 1994). For example, the addition of polysorbate 20 or divalent metal zinc ion into the liquid feed of recombinant human growth hormone (rhGH) can supress the formation of soluble and insoluble protein aggregates, respectively (Maa et al. 1998).

EXAMPLES OF NATURAL POLYMER–BASED NANO/MICROSPHERES

Proteins

Albumin

Albumin is one of the most commonly utilized proteins. It has been used for decades for the modification of drugs with small molecular weights to enhance their stability and is employed in diagnostic nuclear medicine for the evaluation of organs and the circulatory system. As a drug delivery carrier, it has been the subject of extensive attention. Albumin nano/microspheres have been prepared on a large scale by chemical or self-cross-linking technology that resulted in stable chemical and physical properties with high reproducibility. Detailed information on the preparation, characterization, and delivery of drugs using albumin-based preformed nano/microspheres can be found elsewhere (Gombotz and Pettit 1995; Bogdansky et al. 1990; Arshady 1990; Poznansky and Juliano. 1984; Davis et al. 1984; Morimoto and Fujimoto 1985).

The major application of albumin nano/microspheres is in chemotherapy, although studies of the encapsulation of insulin (Goosen et al. 1985), urokinase (Bhargave and Ando 1992), and antibacterials (Egbaria and Friedman 1990; Shah et al. 1987; Sugibayashi et al. 1979) have been reported. It is estimated that there have been more than 100 drugs incorporated into albumin nano/microspheres with various features, such as the size and density of the particles, core content, and the extent and nature of cross-linking (Bogdansky et al. 1990). The typical drug release profile from albumin nano/microspheres is generally biphasic, consisting of an initial burst followed by slower, first-order kinetics. Due to the highly porous structure of the albumin matrix produced by suspension cross-linking, the incorporated drugs are usually released from the nano/microspheres within a few days or even hours. According to Gupta et al. (1986a, b, c, d), who studied the in vitro doxorubin hydrochloride (Adriamycin®) release from self-cross-linked albumin microspheres, the initial burst is the combination of drug desorption from the surface and diffusion through pores, while the second phase appears to be related to the hydration of the microsphere's matrix. Since the hydration of the albumin-based matrix occurs in a short period, less than an hour in most cases, the diffusion of drug through the swollen microspheres is considered the rate-controlling step. Experiments done by Sawaya et al. (1987) showed that the ionic strength of the release medium can affect the release kinetics of Adriamycin® from albumin microspheres. Since Adriamycin® is ionically bound to proteins, an increase in ionic strength of the medium enhances drug diffusion and dissolution.

Albumin nano/microsphere stability studies have been reported by several groups (Davis et al. 1984; Morimoto and Fujimoto 1985). In the presence of proteolytic enzymes, such as papain and trypsin, significant microsphere weight loss and surface morphological changes were observed. In vivo studies of albumin microspheres loaded with Adriamycin® that were administered intravenously showed that the release is degradation dependent (Willmott et al. 1985; Willmott and Harrison 1988). The results also suggested that biodegradation of the albumin matrix can be controlled by varying the microsphere preparation process, particularly the degree of cross-linking.

Several factors have been demonstrated to affect drug release from albumin nano/microspheres: drug distribution in the microspheres, drug molecular weight, drug hydrophilicity, drug concentration, drug interaction with the matrix, type and amount of cross-linking agent, cross-linking density, solution pH and polarity, and the presence of enzymes.

Collagen

Collagen is a well-studied and abundant structural protein that is biocompatible and nontoxic in most tissues. Collagen has been utilized for decades in surgical medicine, tissue reconstruction, and as a drug delivery vehicle. A series of bioactive substances have been incorporated into collagen-based delivery systems, such as insulin, growth hormone, bone morphogenic proteins, fibroblast growth factor, antibiotics, heparin, and contraceptives (Silver and Garg 1997). In most cases, collagen-based delivery systems are fabricated in the form of injectable gels or implantable devices, such as films, sponges, monoliths, and microspheres.

Microspheres made from collagen and chondroitin sulfate by complex coacervation were described earlier in this chapter. Since the microspheres can be prepared at low temperatures and in an aqueous solution, the process is suitable for encapsulating labile bioactive agents. The degradation rate of collagen microspheres and, consequently, the release of incorporated substances can be controlled by the cross-linking density, using glutaraldehyde as the cross-linking agent. The degradation rate can also be adjusted by the bacterial collagenase level (Shao and Leong 1995).

Despite the well-documented advantages of collagen as a biomaterial, several drawbacks limit its application: poor dimensional stability (i.e., highly swellable in vivo), low elasticity, and high variability in drug release kinetics due to variable sources of the collagen and its processing. The possible occurrence of immune response to collagen-based materials at the implant site must also be considered (Bogdansky 1990).

Fibrinogen

Fibrinogen, a blood component with a molecular weight of about 350,000 daltons has been used in several clinical applications, such as a surgical sealant and as a temporary tissue replacement. The antigenicity of fibrin is a major concern in biomedical applications. It has been reported that fibrin can induce an immune response in the host tissue after subcutaneous or intravenous administration. However, the immune response to fibrin products can be minimized by heat treatment or chemical cross-linking.

Fibrin microspheres have been used as anticancer drug carriers and evaluated in vitro and in vivo (Miyazaki et al. 1986a). Adriamycin® and 5-fluorouracil encapsulated fibrin microspheres were prepared by suspension technology followed by thermal cross-linking procedures. In vitro studies demonstrated a slow release period of up to 5 and 7 days for 5-fluorouracil and Adriamycin®, respectively. These microspheres were injected intraperitoneally into mice with Ehrlich ascites carcinoma. Those receiving Adriamycin®-releasing microspheres had a longer survival time than those treated with an equivalent dose in solution, while no significant differences were observed between those treated with solution 5-fluorouracil and encapsulated 5-fluorouracil. The authors suggested that the difference could be due to the longer release period of Adriamycin® from microspheres in vivo, as was observed in vitro. Recently, Ho et al. (1994, 1995) prepared a series of fibrin-based microspheres of different particle sizes using an oil-in-oil emulsion method. A number of macromolecules with varying molecular weights, such as insulin, lysozyme, carbonic anhydrase, and ovalbumin were incorporated into the microspheres. The authors evaluated the release profiles in vitro of macromolecules from the fibrin beads with various size distributions and found that the release characteristics are primarily dependent on the molecular size of the macromolecules, the medium used to prepare the fibrin microspheres, and the resulting size distribution. The interaction between incorporated substances and matrix materials also affected the release kinetics (Ho et al. 1994, 1995).

Gelatin

Gelatin is produced by controlled hydrolysis of collagenous tissue, during which collagen fibrils are converted to water-soluble molecules. Gelatin has been used to fabricate films for wound dressing and as a hemostatic agent in surgery. Gelatin-based drug delivery systems, capsules, and nano/microspheres have been extensively examined, and a number of comprehensive reviews have been published (Keenan 1997; Gombotz and Pettit 1995; Domb 1994; Bogdansky 1990; and Oppenheim 1980, 1988).

A series of gelatin-based matrices has been developed by Ikada and his colleagues (Yamada et al. 1995; Fujisato et al. 1996; Tabata and Ikada 1989, 1995; Muniruzzaman et al. 1997). For example, microspheres with interferon (IFN) were formed by sonication of a gelating solution containing IFN in a toluene/ chloroform phase followed by a glutaraldehyde cross-linking procedure. The in vitro degradation profile of the microspheres was found to be inversely proportional to the cross-linking density.

Golumbek et al. (1993) studied gelatin-albumin microspheres that were incorporated with interferon and granulocyte macrophage colony stimulating factor (GM-CSF). In vivo studies showed that the microspheres were degradable and could stimulate an antitumor response.

Since gelatin microspheres can be prepared by thermal cross-linking at a lower temperature than needed for albumin microsphere cross-linking, there is a lower probability of degradation of the encapsulated drugs. Gelatin microspheres were also found to be superior to albumin microspheres in in vivo studies. Swelling occurred in knee joints of rabbits injected repeatedly with albumin microspheres, while no response was observed for similar injections with gelatin microspheres (Marty and Oppenheim 1977; Kennedy 1983).

Polysaccharides

Alginate

Alginate is isolated from seaweed and algae and derived from bacterial sources. The alginate family are linear block copolymers composed of D-mannopyranosyluronate (M) in $\beta(1\text{-}4)$ linkage, and L-gulopyranosyluronate (G) in $\alpha(1\text{-}4)$ linkage. Acetylation may occur at the C-2 and/or C-3 positions in bacterial alginate. The molecular weights of commercial alginates are usually higher than 150 kDa.

In alginate chains, the mannuronic acid residues and glucuronic acid residues are arranged in homopolymeric (GG, MM) and heteropolymelic (MG) sequences in various proportions and distributions. The properties of alginate are determined by the M/G ratio and the ratios and segment lengths of the MM, GG, and MG blocks, as well as the presence of noncarbohydrate substituents (Morris et al. 1982; Soon-Shiong et al. 1994). The ability of hydrogel formation between alginate and multivalent cations is the feature that distinguishes alginate from other polysaccharides. Intermolecular entanglement of alginate solutions can be observed at concentrations as low as 0.01 percent (w/v), while gelation of alginate solutions occurs at higher concentrations (> 0.1 percent, w/v) in the presence of calcium ions. Specific intermolecular cooperative interactions between calcium and G blocks of alginate lead to the well-known proposal of "egg-box" junction zones.

Since the nature of "egg-box" creation is an ionic bonding process, the gel will dissemble in the presence of substances with high affinity for calcium, such as phosphates and citrates. A high concentration of monovalent ions, such as ammonia and sodium, can also disassociate the "egg-box" structure due to ion exchange.

Alginate beads can be easily prepared by coacervation; the size of the beads can range from nanometers to millimeters, primarily dependent on alginate concentration and the preparation method. Rajaonarivony et al. (1993) proposed a simple preparation process. The particles are formed in a sodium alginate solution by the addition of calcium chloride and polylysine. The initial concentrations of sodium alginate and calcium chloride are lower than those required for gel formation and correspond to the formation of a pregel state. The size of the particles formed (250–850 nm) is greatly dependent on the order of addition of calcium and polylysine to the sodium alginate solution. This phenomenon, according to the authors, can be attributed to the difference in the nature of the interaction between calcium and alginate and between polylysine and alginate. Calcium ions induced a parallel packing of the oligopolyguluronic sequences to give egg-box structures, leading to the formation of very compact domains. On the contrary, polylysine reacted with the mannuronic residues of alginate, resulting in weakly organized domains. Evaluation of the drug-loading capacity was done with doxorubicin as a drug model. The results indicate that these alginate nanoparticles have a high drug loading capacity, > 50 mg of doxorubicin per 100 mg of alginate.

Live cells (Lim and Sun 1980) and a number of bioactive proteins, such as interleukin-2 (IL-2) (Liu et al. 1997), transforming growth factor-β1 (Mumper et al. 1994), basic fibroblast growth factor, tumor necrosis factor receptor (Wee and Gombotz 1994), the angiogenic molecule urogastrone (Downs et al. 1992), and epidermal growth factor (EGF), have been incorporated into alginate-based nano/microspheres in mild conditions, room temperature and physiological pH, without the use of an organic solvent.

Liu et al. (1997) found the alginate/chitosan microspheres for the delivery of IL-2 superior to microspheres made from other polymers. Some of these systems, such as gelatin, poly-oxyethylene-polyoxypropylene block copolymer, or pluronic gel, delivered IL-2 for up to 48 h (Donohue and Rosenberg 1983; Morikawa et al. 1987). Poly(DL-lactide-coglycolide) micro-spheres displayed a relatively long-term sustained release of IL-2, but only 1 to 3 percent of the loaded IL-2 could be recovered in the release medium (Hora et al. 1990; Atkins et al. 1994). The porous alginate/chitosan microspheres were prepared by the gelatinization of two polysaccharides, a polyanionic sodium alginate, and a polycationic chitosan followed by lyophilization. IL-2 was incorporated into the alginate microspheres by diffusion from an external aqueous solution of IL-2. Sustained release of IL-2 from the alginate/chitosan microspheres occurred over 5 days, and almost 100 percent of the active IL-2 was recovered in the release medium, fetal calf serum. The activity of released IL-2 was investigated by determining the induction of cytotoxic T lymphocytes (CTLs) when incubated with tumor cells and lymphocytes. It was found that the IL-2 remained active in the alginate/chitosan microspheres since the release of IL-2 triggered induction of CTLs. In addition, IL-2 released in a sustained manner triggered induction of CTLs more efficiently than free IL-2. The tumor-killing activity of CTLs was the same whether induced by sustained release IL-2 or by the addition of free IL-2.

Cellulose

Cellulose was discovered about 150 years ago and has been used in the pharmaceutical industry for several decades. Cellulose is a linear high molecular weight polymer composed of D-glucose in $\beta(1-4)$ linkage. In controlled drug delivery applications, cellulose has been most commonly used in its ester derivatives. Methylcellulose has been effectively used for site-specific delivery of transforming growth factor-$\beta 1$ (TGF-$\beta 1$) for skin wound healing (Beck et al. 1990, 1991a) and for bone defects (Beck et al. 1991b). Hydroxyethylcellulose incorporated with acidic fibroblast growth factor (aFGF) has been applied for wound healing (Matuszewska et al. 1994). Cellulose has been also used as a coating material in preparing granules or capsules in pharmaceutical industry.

Chitin and Chitosan

Chitin is composed of D-N-acetyl glucosamine in $\beta(1\text{-}4)$ linkage and constitutes one of the most abundant glycans after cellulose. The partially or fully deacetylated derivative, chitosan [2-amino-2-deoxy-D-glucose in $\beta(1\text{-}4)$ linkages], is a cationic polymer, which has recently received increasing attention (Kost and Goldbart 1997; Lehr et al. 1992; Berscht et al. 1994; Illum et al. 1994; and Artursson et al. 1994).

Natural chitin and chitosan often exist as copolymers of glucosamine and acetylated glucosamines. All hydroxyl groups in chitin are hydrogen bonded, resulting in the formation of a screw axis in the macromolecular chains and strong bonding between the chitin molecules. Therefore, chitin does not swell or dissolve in water. In contrast to chitin, chitosan can be dissolved in aqueous solutions of most organic and inorganic acids to form salts.

In the medical and pharmaceutical area, chitosan has been used for the preparation of drug carriers for sustained release (Hou et al. 1985; Kawashima et al. 1985; Miyazaki et al. 1988; Sawayanagi et al. 1982; Shiraishi et al. 1993; Berthold et al. 1996) and site-specific release (Gallo and Hassan 1988; Hassan et al. 1992), as well as for mucoadhesive preparations (Lehr et al. 1992; Lueben et al. 1994; Illum et al. 1994; Imai et al. 1991).

Chitosan microspheres can be prepared by suspension followed by chemical cross-linking (Hassan et al. 1992; Akbuga and Durmaz 1994; Thanoo et al. 1992), suspension method combined with solvent evaporation (Gallo et al. 1988), spray drying (Genta et al. 1994), or coacervation (Berthold et al. 1996). Coacervation of chitosan with its counterions such as alginate, chondroitin sulfate, or gelatin was described earlier in this chapter. Chitosan microspheres obtained by coprecipitation of chitosan and an inorganic salt, sodium sulfate, were also reported (Berthold et al. 1996). The amount of sodium sulfate required for the formation of microspheres increased with the increase in the molecular weight of chitosan. An anti-inflammatory drug, prednisolone sodium phosphate (PSP), was incorporated (up to 30 percent) into the microspheres by diffusion. The linearity of the flux and the potential bioadhesiveness of chitosan/PSP microspheres are important properties required for the treatment of chronic gastrointestinal inflammation.

Chitin and its derivatives have also been applied as drug delivery matrices for bone defects, as internal fixation devices for bone fractures, and as immunological activation adjuvants (Maeda et al. 1995).

Dextran

Dextrans are a family of D-glucose in $\alpha(1\text{-}6)$ linkage with branches in varying proportions of $\alpha(1\text{-}4)$, $\alpha(1\text{-}2)$, and $\alpha(1\text{-}3)$ linkage types. Crude dextran preparations are polymers of very high molecular weight, up to 50 million (weight-average). The largest portion of the dextran family consists of $\alpha(1\text{-}6)$ linked backbones with branches at the C-2, C-3, and C-4 positions. The next largest portion consists of an additional $\alpha(1\text{-}3)$ linkage distributed randomly in the backbones and linked branches. The third portion of the dextran family has $\alpha(1\text{-}3)$ linked backbones and $\alpha(1\text{-}6)$ linked branches.

Dextran of relatively low molecular weight (50 to 100 kDa) can be used as a therapeutic agent in restoring blood volume for mass casualties (Gronwall 1957). The most widely used dextran derivatives are obtained by reacting an alkaline solution of dextran with epichlorohydrin to yield cross-linked chains. The products are gels in the form of microspheres that have been known as molecular sieves, became commercially available, and are widely used in purification and bioseparation processes. Dextran and its derivatives have been used as controlled-release vehicles for drugs and bioactive proteins. In drug-linked dextrans, the polysaccharides act as a shield, protecting the active moiety against chemical or biological degradation, and also as a carrier for transferring the active drug via the blood to the organs, without decreasing its bioavailability.

Dextran-based microspheres can be prepared from its derivatives (i.e., acrylated dextran). For instance, methacrylateddextran microspheres with various average particle sizes ranging from 2.5 to 20 µm were obtained in an aqueous solution of polyethylene glycol (PEG) by emulsion polymerization (Stenekes et al. 1998). In one study, polyacryldextran microspheres containing different proteins were prepared by copolymerizing bisacrylamide with acryldextran. The heat stability of carbonic anhydrase encapsulated in the microspheres increased, the enzymatic activity of the immobilized proteins was retained, and polymer degradation was increased in the presence of dextranase (Edman et al. 1980). Battersby et al. (1996) studied dextran-carrying active aldehyde groups. Upon the reaction with periodate, rhGH was linked to dextran via a reversible imine conjugation. The complex formed between rhGH and oxidized dextran had biological activity and was efficacious in hypophysectomized rats based on the observation of significant weight gain in a rat bioassay; the release of rhGH was inversely proportional to the degree of oxidation of the dextran. Dextran-based nano/microspheres therefore demonstrate great potential as injectable drug carriers.

Hyaluronate

Hyaluronic acid (also hyaluronan) is a linear polymer composed of D-glucuronic acid and N-acetylglucosamine in $\beta(1-3)$ and $\beta(1-4)$ linkages. It was first isolated from the vitreous body by Meyer and Palmer in 1934. Shortly after that, it was identified in connective tissue; synovial fluid; and the aqueous humor of the eye, skin, and blood; it is also present on the surface of many cells and in extracellular matrices. It is prepared by bacterial fermentation or purification of tissues. The molecular weight of hyaluronate ranges from several thousand to 9 million Da depending on the source (Laurent 1970), thus showing considerable physiological variation (Table 9.2).

Hyaluronate products are typically aqueous gels that have a limited half-life and are rapidly cleared upon administration. Upon chemical modification or intermolecular cross-linking, a series of hyaluronate derivatives with physicochemical properties that are different from those of original hyaluronate, but retain the desired biological properties, were developed (Balazs et al. 1995; Mctaggart and Halbert 1993; Cortivo et al. 1991;

Table 9.2. *Molecular Weights of Hyaluronic Acid from Different Tissues**

Tissue	Molecular Weight	Technique
Human umbilical cord	3.4×10^6	Light scattering
Bovine viterous body	7.7×10^4 to 1.7×10^6	Light scattering, sedimentation and diffusion
Bovine synovial fluid	14×10^6	Light scattering
Human synovial fluid		
Normal	6×10^6	Light scattering
Rheumatoid	$(2.7 \text{ to } 4.5) \times 10^6$	Light scattering
Rooster comb	1.2×10^6	Ultracentrifugation
Streptococcal cultures	$(0.115 \text{ to } 0.93) \times 10^6$	Sedimentation and viscosity

*Modified from Laurent (1970).

Benedetti et al. 1990; Hunt et al. 1990). Esterified hyaluronate derivatives are the most successful examples that provide a more stable form while maintaining their rheological and biological properties.

Due to its excellent biocompatibility, hyaluronate has been proposed for use in variety of medical applications, including wound healing, adhesion prevention in postoperative surgery, artificial organs, and drug delivery. As described at the begining of this chapter, a number of hyaluronate preparations have been approved for medical applications.

Hyaluronates have also been developed as topical, injectable, and implantable vehicles for the controlled and localized delivery of biologically active agents (Larsen and Balazs 1991; Radomsky et al. 1997, 1998a, b; Meyer et al. 1995; Nobuhiko et al. 1993; Prisell et al. 1992; Drobnik 1991). Microspheres from hyaluronic acid esters were prepared by an emulsion and solvent evaporation technique. Nerve growth factor (NGF) (Ghezzo et al. 1992) and a series of peptides (Papini et al. 1993) were loaded into microspheres by a diffusion method. The release profile of these peptides was found to vary with the type of hyaluronate esters, suggesting that the polymer matrix can be tailored to a desired release rate for a specific peptide.

Starch

Amylose and amylopectin are two main constituents of starch. The former is a linear polymer with a molecular weight of 100 to 500 kDa, consisting of D-glucose in an α(1-4) linkage. In contrast, the later is a highly branched, high molecular weight (> 1,000 kDa) polydisperse glucopyranosyl polymer, composed of a amylose-type α(1-4) linked D-glucose backbone and attached clusters of α(1-6) linked branches with an average length of 20–30 residues.

Starch gels can be formed simply by heating a starch solution (conc. > 6 weight percent) to the gelatinization temperature followed by cooling to ambient conditions. The gelatinization temperature of starch occurs between 60 and 70°C depending on the source. When the gelatinization point is reached, starch granules irreversibly swell several times their original size, and chains entangle in the solution. Starch gels were considered as a composite of a polymer-deficient phase and a polymer-rich phase. The polymer-rich phase may give rise to a gel network above certain critical concentrations (Miller et al. 1985). In the polymer-rich regions of the gel network, amylose crystallization may occur by way of intermolecular associations of branched chain fragments (Ring et al. 1987).

The initial applications of starch were associated with the food, paper, and texile industries. Recently, however, modification of starch has been reported to meet the growing needs for biodegradable polymers as matrices for controlled drug delivery systems (Kost and Shefer 1990; Hag et al. 1990; Vandenbossche et al. 1992; Visavarungro; et al. 1990). Kost and Shefer (1990) developed methods of network formation (cross-linking by calcium or epichlorohydrin) to entrap drug molecules in the starch matrix for controlled drug delivery applications. Starch network formation in both calcium and epichlorohydrin procedures involves two steps: gelatinization and cross-linking. In the first stage, sodium hydroxide solution was added to a suspension of water and starch with the drug. The solution became very viscous after 1 h of continuous stirring at 600 rpm. A solution of calcium chloride or epichlorohydrin, depending on the type of network formed, was added by continuous slow mixing and resulted in coagulation of the mixture. The particles formed were washed with water and air dried at room temperature. An in vitro release study of entrapped bioactive substances demonstrated that the release rate of high molecular weight substances is closely related to the matrix degradation rate, while the release of small molecules was mainly controlled by diffusion. Starch microspheres were also used in clinical trials for occlusion of arteries and delivery of cytotoxic drugs.

In another study (Degling et al. 1993), recombinant mouse IFN-α was covalently coupled to polyacryl starch microspheres using carbonyldiimidazole chemistry. The bound IFN-α at low dose was found to activate cultured macrophages for nitrite production and significantly reduced the load of *Leishmania donovani* in infected mice. In contrast, low doses of INFα in the free form had no effect. Primaquine and trimethoprim, were also covalently conjugated to polyacryl starch microspheres via an oligopeptide spacer using similar chemstry. Upon the enzymatic degradation of microspheres in a specific tissue site, drug is released in the target organelle (Laakso et al. 1987).

CONCLUSIONS

Natural polymer-based nano/microspheres have become increasing interesting vehicles for the controlled release of bioactive substances. Many of the polymeric nano/microspheres described in this chapter are rapidly degraded and eliminated in a few weeks or months from the body after implantation and, therefore, are particularly suitable for short-term drug delivery applications. Most natural polymer-based microspheres are highly porous and swell in physiological solutions. As Bogdansky (1990) points out, an effective way to control the initial burst limitation is a challenge that must be solved before these drug carriers can gain widespread clinical use. More efforts toward in vivo study of the drug release kinetics and the local tissue response to the injected microspheres are also needed.

REFERENCES

Akbuga, J., and G. Durmaz. 1994. Preparation and evaluation of cross-linked chitosan microspheres containing furosemide. *Int. J. Pharm.* 111:217–222.

Arshady, R. 1990. Albumin microspheres and microcapsulates: Methodology of manufacturing techniques. *J. Control. Rel.* 14:111–131.

Arshady, R., and A. Ledwith. 1983. Suspension polymerization and its application to the preparation of polymer supports. *Reactive Polymer* 1:159–174.

Aslani, P., and R. A. Kennedy. 1996. Study on diffusion in alginate gels. I. Effect of cross-linking with calcium or zinc ions on diffusion of acetaminophen. *J. Control. Rel.* 42:75–82.

Atkins, T. W., R. L. McCallion, and B. J. Tighe. 1994. The incorporation and release of glucose oxidase and interleukin-2 from a bead formed macroporous hydrophilic polymer matrix. *J. Biomater. Sci. Polymer Edn.* 6:651–659.

Artursson, P., T. Lindmark, S. S. Davis, and L. Illum. 1994. Effect of chitosan on the permeability of monolayers of intestinal epithelial cells (CACO-2). *Pharm. Res.* 11:13 58–1361.

Balazs, E. A., E. Leshchinger, N. E. Larsen, and P. Band. 1995. Hyaluronan biomaterials: Medical applications. In *Encyclopedic handbook of biomaterials and bioengineering,* edited by D. L. Wise, D. J. Trantolo, D. E. Altobeli, M. J. Yashemski, J. D. Gresser, and E. R. Schwartz. New York: Marcel Dekker, Inc., pp. 1693–1715.

Battersby, J., R. Clark, W. Hancock, E. Puchulu-Campanella, N. Haggarty, D. Poll, and D. Harding. 1996. Sustained release of recombinant human growth hormone from dextran via hydrolysis of an imine bone. *J. Control. Rel.* 42:143–156.

Beck, S. L., T. L. Chen, P. Mikalauski, and A. J. Amman. 1990. Recombinant human transforming growth factor-beta 1 enhances healing and strength of granulation skin wounds. *Growth Factors* 3:267–275.

Beck, S. L., L. Deguzman, W. P. Lee, Y. Xu, L. A. McFatridge, and E. P. Amento. 1991. TGF-β1 accelerates wound healing: Reversal of steroid-impaired healing in rats and rabbits. *Growth Factors* 5:295–305.

Beck, L. S., L. Deguzman, W. P. Lee, Y. Xu, L. A. McFatridge, N. A. Gillet, and E. P. Amento. 1991. TGF-β1 induces bone closure in skull defects. *J. Bone Miner. Res.* 6:1257–1265.

Benedetti, L. M., E. M. Topp, and V. J. Stella. 1990. Microspheres of hyaluronic acid esters–fabrication methods and in vitro hydrocortisone release. *J. Control. Rel.* 13:33–41.

Berscht, P. C., B. Nies, A. Liebendorfer, and J. Kreuter. 1994. Incorporation of basic fibroblast growth factor into methylpyrrolidinone chitosan fleeces and determination of the in vitro release characteristics. *Biomaterials* 15:593–600.

Berthold, A., K. Cremer, and J. Kreuter. 1996. Preparation and characterrization of chitosan microspheres as drug carriers for prednisolone sodium phosphate as model for anti-inflammatory drugs. *J. Control. Rel.* 39:17–25.

Bhargave, K., and H. Y. Ando. 1992. Immobilization of active urokinase on albumin microspheres: Use of a chemical dehydrant and process monitoring. *Pharm. Res.* 9:776–781.

Bodmeier, R., and O. Paertakul. 1989. Spherical agglomerates of water insoluble drugs. *J. Pharm. Sci.* 78:964–967.

Bogdansky, S. 1990. Natural polymers as drug delivery systems. In *Biodegradable polymers as drug delivery systems,* edited by M. Chasin and R. Langer. New York: Marcel Dekker, pp. 231–259.

Chasin, M., and R. Langer, eds. 1990. *Biodegredable polymers as drug delivery systems.* New York: Marcel Dekker.

Cortivo, R., P. Brun, A. Rastrelli, and G. Abatangelo. 1991. In vitro studies on biocompatibility of hyaluronatic acid esters. *Biomaterials* 2:727–732.

Davis, S. S., L. Illum, J. C. Mcvie, and E. Tomlinson, eds. 1984. *Microspheres and drug therapy: Pharmaceutical, immunological, and medical aspects.* Amsterdam: Elsevier.

Degling, L., P. Stjarnkvist, and I. Sjoholm. 1993. Interferon-alpha in starch microparticles: Nitric oxide-generating activity in vitro and antileishmanial effect in mice. *Pharm. Res.* 10:783–790.

Dianty, A. L., K. H. Goulding, P. K. Robinson, I. Simpkins, and M. D. Trevan. 1986. Stability of alginate-immobilized algal cells. *Biotech. Bioeng.* 28:210–216.

Domb, A. J. 1994. Implantable biodegradable polymers for site-specific drug delivery. In *Polymeric site-specific pharmacotherapy,* edited by A. J. Domb. New York: John Wiley & Sons Ltd., pp. 1–22.

Donohue, J. H., and S. A. Rosenberg. 1983. The fate of interleukin-2 after in vivo administration. *J. Immunol.* 130:2203–2208.

Downs, E. C., N. E. Robertson, T. L. Riss, and M. L. Plunkett. 1992. Calcium alginate beads as a slow-release system for delivering angiogenic molecules in vivo and in vitro. *J. Cell Physiol.* 152:422–429.

Drobnik, J. 1991. Hyaluronan in drug delivery. *Adv. Drug Del. Rev.* 7:295–308.

Edman, P., B. Ekman, and I. Sjoholm. 1980. Immobilization of proteins in microspheres of biodegradable polyacryldextran. *J. Pharm. Sci.* 69:838–842.

Egbaria, K., and M. Friedman. 1990. Sustained release albumin microspheres containing antibacterial drugs: Effects of preparation conditions on kinetics of drug release. *J. Control. Rel.* 14:79–94.

Fujisato, T., T. Sajiki, Q. Liu, and Y. Ikada. 1996. Effect of basic fibroblast growth factor on cartilage regeneration in chondrocyte-seeded collagen sponge scaffold. *Biomaterials* 17:155–162.

Gallo, J. M., and E. E. Hassan. 1988. Receptor-mediated magnetic carriers: Basis for targeting. *Pharm. Res.* 5:300–304.

Genta, I., F. Pavanetto, B. Conti, P. Giunchedi, and U. Conte. 1994. Spray-drying for the preparation of chitosan microspheres. *Proc. Int. Symp. Control. Rel. Bioact. Mater.* 231:616–617.

Ghezzo, E., L. M. Benedetti, M. Rochira, F. Biviano, and L. Callegaro. 1992. Hyaluronic acid derivative microspheres as NGF delivery devices: Preparation methods and in vitro release characterization. *Int. J. Pharm.* 87:21–29.

Golumbek, P. T., R. Azhari, E. M. Jaffee, H. I. Levitsky, A. Lazenby, K. W. Leong, and D. Pardoll. 1993. Controlled release, biodegradable cytokine depots: A new approach in cancer vaccine design. *Cancer Res.* 53:5841–5844.

Gombotz, W. R., and D. K. Pettit. 1995. Biodegradable polymers for protein and peptide drug delivery. *Bioconjugate Chem.* 6:332–350.

Goosen, M. F. A., Y. F. Leung, G. M. O'Shea, S. Chou, and A. M. Sun. 1985. Optimization of microencapsulation parameters: Semipermeable microcapsules as an artificial pancreas. *Biotechnol. Bioeng.* 27:146–150.

Gronwall, A. 1957. *Dextran and its use in colloidal infusion solutions.* New York: Academic Press.

Gupta, P. K., C. T. Huang, F. C. Lam, and D. G. Perrier. 1986a. Albumin microspheres. III. Synthesis and characterization of microspheres containing adriamycin and magnetite. *Int. J. Pharm.* 43:167–177.

Gupta, P. K., C. T. Huang, and D. G. Perrier. 1986. Albumiin microspheres. I. Release characteristics of adriamycin. *Int. J. Pharm.* 33:137–146.

Gupta, P. K., C. T. Huang, and D. G. Perrier. 1986. Albumin microspheres. II. Effect of stabilization temperature on the release of adriamycin. *Int. J. Pharm.* 33:147–153.

Gupta, P. K., C. T. Huang, and D. G. Perrier. 1986. Quantitation of the release of doxorubicin from colloidal dosage forms using dynamic dialysis. *J. Pharm. Sci.* 76:141–145.

Hag, E., H. Teder, G. Roos, P. I. Christensson, and U. Stenram. 1990. Enhanced effect of adriamycin on rat liver adenocarcinoma after hepatic artery injection with degradable starch microspheres. *Sel. Cancer Therap.* 6:23–24.

Hassan, E. E., R. C. Parish, and J. M. Gallo. 1992. Optimized formulation of magnetic chitosan microspheres containing the anticancer agent, oxantrazole. *Pharm. Res.* 11:1358–1361.

Ho, H. O., C. C. Hsiao, C. Y. Chen, T. D. Sokoloski, and M. T. Sheu. 1994. Fibrin-based drug delivery systems. II: The preparation and characterization of microbeads. *Drug Dev. Indus. Pharm.* 20:535–546.

Ho, H. O., C. C. Hsiao, T. D. Sokoloski, C. Y. Chen, and M. T. Sheu. 1995. Fibrin-based drug delivery systems. III: The evaluation of the release of macromolecules from microbeads. *J. Control. Rel.* 34:65–70.

Hora, M. S., R. K. Rana, J. H. Nunberg, T. R. Tice, R. M. Gilley, and M. E. Hudson. 1990. Controlled release of interleukin-2 from biodegradable microspheres. *Biotechnology* 8:755–758.

Hou, W. M., S. Miyazaki, M. Takada, and T. Komai. 1985. Sustained release of indomethacin from chitosan granules. *Chem. Pharm. Bull.* 33:3986–3992.

Hunt, J. A., H. N. Joshi, V. J. Stella, and E. M. Topp. 1990. Diffusion and drug release in polymer films prepared from ester derivatives of hyaluronic acid. *J. Control. Rel.* 12:159–169.

Illum, L., N. F. Farraj, and S. S. Davis. 1994. Chitosans as a novel nasal delivery system for peptide drug. *Pharm. Res.* 11:1186–1189.

Imai, T., S. Shiraishi, H. Saito, and M. Otagiri. 1991. Interaction of indomethacin with low molecular weight chitosan, and improvements of some pharmaceutical properties of indomethacin by low molecular weight chitosan. *Int. J. Pharm.* 67:11–20.

Kafarov, V. V., and B. M. Babanov. 1959. Phase contact area of immiscible liquids during agitation by mechanical stirrers. *J. Appl. Chem. USSR* 32:810–814.

Kawashima, Y., T. Handa, A. Kasai, H. Takenaka, S. Y. Lin, and Y. Ando. 1985. Novel method for the preparation of controlled-release theophyolline granules coated with a polyelectrolyte complex of sodium phosphate-chitosan. *J. Pharm. Sci.* 74:264–268.

Keenan, T. R. 1997. Gelatin. In *Handbook of bioerodible polymers*, edited by A. J. Domb, J. Kost, and D. M. Wiseman. Amsterdam: Harwood Academic Publishers, pp. 307–317.

Kissel, T., and R. Konebeg. 1996. Injectable biodegradable microspheres for vaccine delivery. In *Microparticulate systems for the delivery of proteins and vaccines*, edited by S. Cohen and H. Bernstein. New York: Marcel Dekker, pp. 51–83.

Kost, J. and R. Goldbart. 1997. Natural and modified polysaccharides. In *Handbook of bioerodible polymers*, edited by A. J. Domb, J. Kost, and D. M. Wiseman. Amsterdam: Harwood Academic Publishers, pp. 275–289.

Kost, J., and S. Shefer. 1990. Chemically-modified polysaccharides for enzymatically-controlled oral drug delivery. *Biomaterials* 11:695–698.

Larsen, N. E., and E. A. Balazs. 1991. Drug delivery systems using hyaluronan and its derivatives. *Adv. Drug Del. Rev.* 7:279–293.

Laakso, T., P. Stjarnkvist, and I. Sjoholm. 1987. Biodegradable microspheres. VI: Lysosomal release of covalently bound antiparasitic drugs from starch microspheres. *J. Pharm. Sci.* 76:134–140.

Laurent, T. C. 1970. Structure of hyaluronic acid. In *Chemistry and molecular biology of the intercellular matrix*, edited by E. A. Balazs. New York: Academic Press, pp. 703–732.

Lehr, C. M., J. A. Bouwstra, E. H. Schacht, and H. E. Junginger. 1992. In vitro evaluation of mucoadhesive properties of chitosan and some other natural polymers. *Int. J. Pharm.* 78:43–48.

Lim, F. and A. M. Sun. 1980. Microencapsulated islets as a bioartificial endocrine pancreas. *Science* 210:908.

Liu, L. S., S. Q. Liu, S. Ng, M. Froix, O. Tatao, and J. Heller. 1997. Controlled release of interleukin-2 for tumour immunotherapy using alginate/chitosan porous microspheres. *J. Control. Rel.* 43:65–74.

Lueben, H. L., C. M. Lehr, C. O. Rentel, A. B. J. Noach, A. G. Boer, J. C. Verhoef, and H. E. Junginger. 1994. Bioadhesive polymers for the peroral delivery of peptide drugs. *J. Control. Rel.* 29:329–338.

Maa, Y. F., P. A. T. Nguyen, and S. Hsu. 1998. Spray-drying of air-liquid interface sensitive recombinant human growth hormone. *J. Phar. Sci.* 87:152–159.

Maeda, M., Y. Inoue, H. Iwase, and K. Kifune. 1995. Biomedical properties and applications of chitin and its derivatives. In *Encyclopedic handbook of biomaterials and bioengineering*, edited by D. L. Wise, D. J. Trantolo, D. E. Altobeli, M. J. Yashemski, J. D. Gresser, and E. R. Schwartz. New York: Marcel Dekker, pp. 1585–1598.

Marty, J. J. and R. C. Oppenheim. 1977. Colloidal systems for drug delivery. *Australian J. Pharm. Sci.* 6:65–76.

Matuszewska, B., M. Keogan, D. M. Fisher, K. A. Soper, C. Hoe, A. C. Huber, and J. V. Bondi. 1994. Acidic fibroblast growth factor: Evaluation of topical formulations in a diabetic mouse wound healing model. *Pharm. Res.* 11:65–71.

McTaggart, L. E., and G. W. Halbert. 1993. Assessment of polysaccharide gels as drug delivery vehicles. *Intern. J. Pharm.* 100:199–206.

Meyer, K., and J. W. Palmer. 1934. The polysaccharide of the vitreous humor. *J. Biol. Chem.* 107:629–634.

Meyer, J., L. Whitcomb, M. Treuheit, D. Collins. 1995. Sustained in vivo activity of recombinant granulocyte colony stimulating factor (rHG-CSF) incorporated in hyaluronan. *J. Control. Rel.* 35:67–72.

Michaels, A. S. 1965. Polyelectrolyte complexes. *Indust. Eng. Chemis.* 57:32–40.

Miller, M. J., V. J. Morris, and S. G. Ring. 1985. Gelation of amylose. *Carbohydr. Res.* 135:257–269.

Miyazaki, S., N. Hashiguchi, M. Suguyama, M. Takada, and Y. Motimoto. 1986. Fibrinogen microspheres as novel drug delivery systems for antitumor drugs. *Chem. Pharm. Bull. (Tokyo)* 34:1370–1377.

Miyazaki, S., N. Hashiguchi, W. M. Hou, C. Yokouchi, and M. Takada. 1986b. Preparation and evaluation in vitro and in vivo of fibrinogen microspheres containing adriamycin. *Chem. Pharm. Bull. (Tokyo)* 34:3384–3392.

Miyazaki, S., H. Yamaguchi, C. Yokouchi, M. Takada, and W. M. Hou. 1988. Sustained-release and intragastric-floating granules of indomethacin using chitosan in rabbits. *Chem. Pharm. Bull.* 36:4033–4038.

Morikawa, K., F. Okada, M. Kosokawa, and H. Kobayashi. 1987. Enhancement of therapeutic effects of recombinant interleukin-2 on a transplantable rat fibrosarcoma by use of a sustained release vehicle, pluronic gel. *Cancer Res.* 47:37–41.

Morimoto, Y., and S. Fujimoto. 1985. Albumin microspheres as drug carriers. *CRC Crit. Rev. Therapeut. Drug Carrier Sys.* 2:19–63.

Morris, E. R., D. A. Rees, and G. Young. 1982. Chiroptical characterisation of polysaccharide secondary structures in the presence of interfering chromophores: Chain conformation of inter-junction sequences in calcium alginate gels. *Carbohydr. Res.* 108:181–195.

Mortada, S. A., A. M. Egaky, A. M. Motawi, and K. A. Khodery. 1987. Preparation of microcapsules from complex coacervation of Gantrez-gelatin. *J. Microencapsulation* 4:23–27.

Mumenthaler, M., C. C. Hsu, and R. Pearlman. 1994. Feasibility study on spray-drying protein pharmaceuticals: Recombinant human growth hormone and tissue-type plasminogen activator. *Pharm. Res.* 11:12–20.

Mumper, R. J., A. S. Hoffman, P. A. Puolakkainen, L. S. Bouchard, and W. R. Gombotz. 1994. Calcium-alginate beads for the oral delivery of transforming growth factor-β1 (TGF-β1): Stabilization of TGF-β1b by the addition of polyacrylic acid within acid-treated beads. *J. Control. Rel.* 30:241–251.

Muniruzzaman, M., Y. Tabada, and Y. Ikada. 1997. Protein interaction with gelatin hydrogels for tissue engineering. *Mat. Sci. Forum* 250:89–96.

Nobuhiko, Y., J. Nihira, T. Okano, and Y. Sakurai. 1993. Regulated release of drug microspheres from inflammation responsive degradable matrices of cross-linked hyaluronic acid. *J. Control. Rel.* 25:133–143.

Oppenheim, R. C. 1980. Microencapsulation by solvent evaporation and organic phase separation. In *Drug delivery systems: Characteristics and biomedical applications,* edited by R. L. Juliano. New York: Oxford University Press, pp. 135–160.

Oppenheim, R. C. 1988. Solid submicron drug delivery systems. In *Controlled release systems: Fabrication technology,* vol. 2, edited by D. Hsied. Boca Raton, Fla., USA; CRC Press, pp. 221–274.

Papini, D., V. J. Stella, and E. M. Topp. 1993. Diffusion of macromolecules in membranes of hyaluronic acid esters. *J. Control. Rel.* 27:47–57.

Poznansky, M. Z., and R. L. Juliano. 1984. Biological approaches to the controlled delivery of drugs: A critical review. Pharm. Rev. 36:277–336.

Prisell, P. T., O. Camber, J. Hiselius, and G. Norstedt. 1992. Evaluation of hyaluronan as a vehicle for peptide growth factors. *Int. J. Pharm.* 85:51–56.

Radomsky, M. L., A. Merck, M. Gonsalves, G. Anudokem, and J. Poser. 1997. Basic fibroblast growth factor in a hyaluronic acid gel stimulates intramembranous bone formation. *Trans. Ortho. Res. Soc.* 22:510.

Radomsky, M. L., T. B. Aufdemorte, L. D. Swain, W. C. Fox, R. C. Spiro, and J. W. Poser. 1999. A novel formulation of FGF-2 in a hyaluronan gel accelerates fracture healing in non-human primates. *J. Orthop. Res.* in press.

Radomsky, M. L., A. Y. Thompson, R. C. Spiro, and J. W. Poser. 1998b. Potential role of fibroblast growth factor in enhancement of fracture healing. *Clin. Orthop. Relat. Res.* 355S:283–293.

Rajaonarivony, M., C. Vauthier, G. Couarraze, F. Puisieux, and P. Couvreur. 1993. Development of new drug carrier made from alginate. *J. Pharm. Sci.* 82:912–917.

Ring, S. G., P. Colonna, K. J. I'Anson, M. T. Kalichevsky, M. J. Miles, V. J. Morris, and P. D. Orford. 1987. The gelation and crystallisation of amylopectin. *Carbohydr. Res.* 162:277–293.

Sacchetti, M., and M. M. van Oort. 1996. Spray-drying and supercritical fluid particle generation techniques. In *Spray drying inhalation aerosols,* edited by A. J. Hickey. New York: Marcel Dekker, pp. 337–384.

Sarciaux, J. M. E., and M. Hageman. 1997. Effects of bovine somatotropin (rbSt) concentration at different moisture levels on the physical stability of sucrose in freeze-dried rbSt/sucrose mixture. *J. Pharm. Sci.* 86:365–371.

Sawaya, A., J-P. Benoit, and S. Benita. 1987. Binding mechanism of doxorubicin in ion-exchange albumin microcapsules. *J. Pharm. Sci.* 76:475–480.

Sawayanagi, Y., N. Nambu, and T. Nagai. 1982. Use of chitosan for sustained-release preparations of water-soluble drugs. *Chem. Pharm. Bull.* 30:4213–4215.

Scully, D. B. 1976. Scale up in suspension polymerization. *J. Appl. Polym. Sci.* 20:2299–2303.

Shah, M. V., M. D. Degennaro, and H. Suryakasuma. 1987. An evaluation of albumin microcapsules prepared using a multiple emulsion technique. *J. Microencapsul.* 4:223–238.

Shao, W., and K. W. Leong. 1995. Microcapsules obtained from complex coacervation of collagen and chondroitin sulfate. *J. Biomater. Sci. Polymer Edn.* 7:389–399.

Shiraishi, S., T. Imai, and M. Otagiri. 1993. Controlled release of indomethacin by chitosan-polyelectrolyte complex: Optimization and in vivo/in vitro evaluation. *J. Control. Rel.* 25:217–225.

Silver, F. H., and A. K. Garg. 1997. Collagen: characterization, processing and medical application. In *Handbook of bioerodible polymers,* edited by A. J. Domb, J. Kost, and D. M. Wiseman. Amsterdam: Harwood Academic Publishers, pp. 325–346.

Soon-Shiong, P., R. Heintz, N. Merideth, X. Yao Qiang, Z. Yao, T. Zheng, M. Murphy, M. Moloney, M. Schmehl, M. Harris, R. Mendez, and P. Sandford. 1994. Insulin independance in a type I diabetic patient after encapsulated iselet transplantation. *Lancet* 21:343.

Stenekes, R. J. H., O. Franssen, E. M. G. Bommel, D. J. A. Crommelin, and W. Hennink. 1998. The preparation of dextran microspheres in all-aqueous system: Effect of the formulation parameters on particle characteristics. *Pharm. Res.* 15:557–561.

Sugibayashi, K., Y. Morimoto, T. Nadai, Y. Kato, A. Hasegawa, and T. Arita. 1979. Drug carrier properties of albumin microspheres entrapped 5-fluorouracil. *Chem. Pharm. Bull.* 27:204–209.

Tabata, Y., and Y. Ikada. 1989. Synthesis of gelatin microspheres containing interferon. *Pharm. Res.* 6:422–427.

Tabata, Y., and Y. Ikada. 1995. Potentiated in vivo biological activity of basic fibroblast growth factor by incorporation into polymer hydrogel microsphere. In *Proc. 4th Japan Intl. SAMPE Symp.,* pp. 577–582.

Thanoo, B. C., M. C. Sunny, and A. Jayakrishnan. 1992. Cross-linked chitosan microspheres: Preparation and evaluation as a matrix for the controlled release of pharmaceuticals. *J. Pharm. Pharmacol.* 44:283–286.

Vandenbossche, G., R. Leffebvre, G. De wilde, and J. P. Remon. 1992. Performance of a modified starch hydrophilic matrix for the sustained release of theophylline in healthy volunteers. *J. Pharm. Sci.* 81:245–248.

Visavarungroj, N., J. Herman, and J. P. Remon. 1990. Crosslinked starch as sustained release agent. *Drug Dev. Ind. Pharm.* 16:1091–1108.

Wee, S., and W. R. Gombotz. 1994. Controlled release of recombinant human tumor necrosis factor from alginate beads. *Proceed. Intern. Symp. Control. Rel. Bioact. Mater.* 21:730–731.

Willmott, N., and P. J. Harrison. 1988. Characterisation of freeze-dried albumin microspheres containing the anti-cancer drug adriamycin. *Int. J. Pharm.* 43:161–166.

Willmott, N., C. Cummings, and A. T. Florence. 1985. In vitro release of adriamycin from drug-loaded albumin and haemoglobin microspheres. *J. Microencapsul.* 2:293–304.

Yamada, K., Y. Tabada, K. Yamamoto, S. Miyamoto, I. Nagata, H. Kikuchi, and Y. Ikada. 1997. Potential efficacy of basic fibroblast growth factor incorporated in biodegradable hydrogels for skull bone regeneration. *J. Neurosurgery* 86:871–875.

Yapel, A. 1979. Albumin medicament carrier system. U.S. Patent 4,147,767.

10

In Situ Gelling Systems

Arthur J. Tipton
Southern BioSystems, Inc.
Birmingham, Alabama

Richard L. Dunn
Atrix Laboratories, Inc.
Fort Collins, Colorado

OVERVIEW AND RATIONALE FOR IN SITU GELLING SYSTEMS

The use of injectable implants to sustain the delivery of drugs is well known. Both biodegradable and nonbiodegradable implant versions have been marketed since the 1980s. Perhaps the most familiar implant examples are Zoladex®, the biodegradable device for prostate cancer, and Norplant®, the nonbiodegradable silicone device for contraception. Small, injectable, preformed microspheres have also served as the basis for long-term drug delivery, with the most well-known example being Lupron Depot®, the leuprolide acetate microsphere product based on poly(lactic acid) (PLA) and poly(lactide-co-glycolide) (PLGA) for the treatment of prostate cancer, endometriosis, and precocious puberty. An inherent drawback of these preformed delivery systems is administration. Cylindrical rods, such as those used in Zoladex® and Norplant®, are placed subcutaneously using a relatively large-bore needle (10–16 gauge). The microspheres are injected using smaller-bore needles, but the need for an intramuscular (IM) injection and dispersion in an aqueous vehicle are drawbacks.

Several organizations have recognized the drawbacks associated with preformed depot delivery systems and have begun research into systems that can be easily injected and, after injection, undergo a change to a depot system. This chapter's descriptive title, "In Situ Gelling Systems," is an attempt to capture the broad application of these technologies. We use the term *gelling* broadly here to describe any process by which a liquid system undergoes a change to a depot upon injection.

Although in situ gelling systems vary greatly in chemical composition and the method of gelling, they all share two attributes:

1. They are relatively low-viscosity fluids prior to injection.

2. They undergo a rapid change in physical form to a depot upon injection.

These types of technologies offer several advantages. Because the systems are easily injected, they are well suited to local placement and delivery. Because they change form upon injection, many of them assume amorphous shapes and can flow into and fill voids. Although the postinjection characteristics of the different technologies vary greatly, all of the technologies offer some potential to function as medical devices. For example, a material that is injected as a liquid but then forms a solid can block arterial blood flow or provide a matrix for tissue ingrowth.

A formulation in a vial or syringe encounters a different environment after subcutaneous (SC) or IM injection. Changes include pH, osmolarity, temperature, and water concentration. In addition to these environmental changes, there is also the possibility for the practitioner to initiate a chemical reaction, such as acrylate cross-linking. All of these methods have been explored in some detail. In addition, some of these systems have undergone clinical trials and are available on the market. While a tremendous number of possible drug delivery solutions are available to trained chemists and formulators, the challenge is to work within somewhat narrow windows that are acceptable from the basis of safety and toxicity. For example, biodegradable polyesters can be heated to a molten state, but the temperature at which ease of flow is obtained is generally unacceptable biologically. The multitude of technologies under development is a testimony to the talents of the formulators pursuing this new method of drug delivery.

For description here, we broadly divide the range of in situ gelling technologies into two subclasses: those that undergo a *physical change*—for example, a polymer in a nonaqueous solvent precipitating upon injection—and those that undergo a *chemical change*—for example, a prepolymer cross-linked after injection. Systems that undergo no change—that is, those injected as a gel, such as collagen and fibrin—are not covered in this chapter.

TYPES OF SYSTEMS

In situ gelling can occur as the result of either a physical or a chemical change. Four systems that undergo a physical change to form the gel are described below: (1) exceeding a lower critical solution temperature (LCST), (2) change of liquid crystalline phase, (3) polymer precipitation, and (4) a sucrose ester precipitation system. Next we describe two systems that are formulated to transform upon injection by a chemical cross-linking reaction: (1) biodegradable-acrylate terminated polymers and (2) polyethylene glycol (PEG)–PLA–acrylates that are photopolymerized. We use these six systems as examples illustrative of the methods being used for in situ gelling systems and to define some of the important trends in drug delivery today. Several other technologies that rely on physical or chemical changes, or in some cases both, are also being researched. We cover several of these in a separate section.

Lower Critical Solution Temperature

Systems in the LCST group are among the most well-studied technologies for gelling upon injection. While a variety of subtle changes and several polymer compositions have been described, these systems all display a change characterized by the transition from a sol to a gel as the temperature is raised. As most commonly practiced, LCST systems rely on water-based solutions of polymer undergoing a phase separation when the temperature is increased. By controlling the properties of the polymer and the solvent, scientists can cause this transition to occur when heating the systems from room temperature to body temperature. Specifically, from a chemical standpoint, the low temperatures permit enough hydrogen bonding to keep the polymer in solution. As the temperature is raised, the amount of hydrogen bonding decreases to a point that phase separation occurs. In most chemical systems, the entropy term dominates as the temperature is raised, leading to greater solubility; in LCST systems, however, the entropy term is relatively low (a characteristic generally associated with high molecular weight materials), and an enthalpy-effect change dominates as the temperature is raised. In most systems used for controlled release, this enthalpy effect is due to hydrogen bonding with water. Loss of hydrogen bonding upon heating decreases water solubility. Within somewhat narrow chemical and physical parameters, this transition can occur as the temperature is raised from room temperature to body temperature.

The polymers most commonly used in LCST systems are polyethylene oxide–polypropylene oxide (PEO–PPO) block copolymers. The example most described in the literature is Poloxamer® 407 (Pluronic® F 127), manufactured by BASF (Johnston and Miller 1985; Yoshida 1994; Gilbert et al. 1987). This system has been described for the release of human growth hormone (Katakam et al. 1997), interleukin-2 (Johnston et al. 1992), urease (Fults and Johnston 1990), pilocarpine (Brown et al. 1976), and melanotan-1 (Bhardwaj and Blanchard 1995). The duration of drug release is uniformly less than one week. The concentration of polymer in water is generally in the range of 25 to 40 percent, resulting in solutions with injectable solution viscosities. Additives such as methycellulose and hydroxypropylcellulose have been incorporated to slow release (Bhardwaj 1995), but these additions raise solution viscosity and make injection more difficult.

In addition to PEO–PPO diblocks, several other LCST chemistries have been described, including systems based on polymers prepared with N-isopropylacrylamide (NIPA) as the thermoresponsive component (Okano et al. 1990, 1995; Bae et al. 1987, 1990; Yoshida et al. 1991, 1992; Grainger and Yu 1994; Hoffman et al. 1986). Several copolymers have been prepared, including N-hydroxysuccininmidyl acrylate (Shah et al. 1995) and acrylic acid (Baudys et al. 1997; Vernon et al. 1996). The LCST of NIPA (32°C) is raised by incorporating a hydrophilic monomer and lowered by incorporating a hydrophobic one (Shah et al. 1995; Feil et al. 1993). Release for these systems generally occurs over a period of less than 1 week.

The copolymerization described for both the PEO–PPO systems and NIPA with several comonomers has been useful in expanding the LCST range or incorporating other functionality into the polymer. This copolymerization is also possible for graft copolymers. The use of an adhesive polyacrylic acid (PAA) backbone has been described (Hoffman et al. 1995) in cases where the adhesive properties of the PAA would have been diminished by mixing or classical copolymerization. PEO–PPO has been grafted to the backbone (Chen et al. 1995a; Hoffman et al. 1997b). The NIPA was grafted to the PAA backbone, retaining both adhesive properties and thermal responsiveness. As expected, these types of systems also show a pH effect, with the LCST decreasing as pH is decreased. More recently, the use of chitosan as the polymer backbone has been described (Chen and Hoffman 1995a, b) with PEO–PPO grafts. For a Pluronic® L-122 graft chitosan polymer (Pluronic® L-122 is on average PEO_{13}–PPO_{67}–PEO_{13}) using 750,000 MW chitosan, a 2–3 percent aqueous solution had a viscosity of 5,000 cP at 20°C, which increased to > 50,000 cP at 37°C (Hoffman 1997a). The release of protein has been reported in vitro and is in the range of 1 week. A 1 percent aqueous solution of a PAA-g-PEO-b-PPO that underwent a 50 cP to 6,000 cP viscosity increase upon warming from ambient to 37°C has demonstrated more than a 12-hour release of estradiol (Ron et al. 1997).

Because the systems described above contain only C-C or C-O moieties in the backbone, they are not biodegradable. Biological elimination, if it occurs at all, is through excretion. The advantage of a biodegradable linkage is the ease of elimination from a degradative mechanism, followed by complete aqueous solubility. Diblocks of PEO–PLA and triblocks of PEO–PLA–PEO have been described (Lee et al. *1996;* Jeong et al. *1996),* along with a range of LCST systems dependent on molecular weight, polymer concentration, and composition. The use of polycaprolactone–PEO copolymer has also been described.

Although LCST in situ gelling systems have been well studied, the technology has yet to form the basis for an injectable marketed product. The examples of delivered drugs encompass many of the drug classes, as described above. These systems contain a high concentration of water. As such, water diffusion in and out of the implant is rapid, leading to relatively rapid drug release. Examples of release are seen in Figures 10.1–3.

Figure 10.1. *Cumulative in vitro release of recombinant human growth hormone (rhGH) from poloxamer gel as analyzed by reverse phase high performance liquid chromatography (RP-HPLC).*

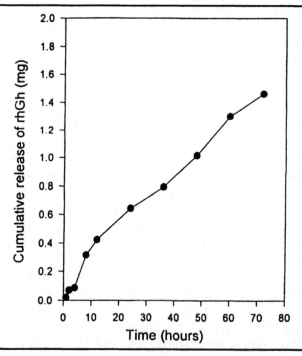

The gel used 4.8 µg of rhGH per mg of poloxamer 407. Reprinted from Katakam et al. (1997) with permission of the Controlled Release Society, Inc.

Figure 10.2. *Cumulative in vitro release of timolol maleate released from various polymer matrices versus time.*

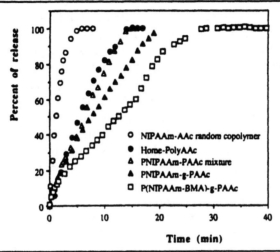

Figure 10.3. *In vitro insulin release profiles.*

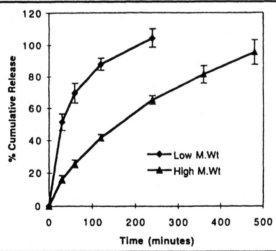

Cumultive in vitro insulin release profiles at pH 7.4 and 37°C for the low and high molecular weight polymers with NiPAAm/BMA/AA mole percent ratio 85/5/10. Reprinted from Baudys et al. (1997) with permission of the Controlled Release Society, Inc.

Liquid Crystalline

Liquid crystals are a well-known technology used for a variety of applications, including flat-screen displays and ultrahigh strength polymers. Their use in in situ gelling systems has been based on a variety of molecules—fatty acids of glycerol—that exist in one phase at high concentrations in water and transform into a different, more highly viscous form upon injection and dilution with more water. The most well-described molecule possessing these properties is glycerol monooleate (GMO) (Wyatt and Dorschel 1992; Engström 1990; Larsson 1989; Engström et al. 1992). GMO has been the subject of several publications and is the basis for one marketed product. As with the LCST systems, liquid crystalline systems maintain a high water content and generally offer a relatively rapid drug release of less than 1 week. GMO, for example, is injected as a lamellar phase that absorbs water to an equilibrium content of 35 percent at 37°C into the cubic phase (Himmelstein and Kumar 1995). The systems are stable to hydrolysis but are degraded in vivo by lipases (Appel et al. 1994).

Additives have been incorporated into liquid crystalline systems to ease injectability and to affect drug delivery. Because the cubic phase is formed upon hydration, the lamellar phase is an intermediate stage, but the GMO does not have to be applied as an aqueous mixture. To lower the viscosity of the GMO for injectability, other additives, such as sesame oil (Norling et al. 1992), can be used in formulating anhydrous products.

The systems also offer the advantages of some bioadhesion and a stabilization effect on incorporated proteins. For a pure GMO–based system, the equilibrium content of water in the lamellar phase is 10–20 percent, which increases to 35 percent after contact with additional water. As such, more than one-third of the cubic phase system is water. This high water content gives rapid drug release—on the order of less than 5 days.

The GMO–based system has led to an approved product: Elyzol® Dental Gel marketed by Dumex in several European countries. This product delivers metronidazole into a periodontal pocket for local management of subgingival flora. The product is administered via a syringe and changes to a semisolid in the periodontal pocket, where it adheres to the mucosa and slowly releases the drug, accompanied by degradation (Norling et al. 1992). Figure 10.4 shows in vitro release of metronidazole benzoate as a function of drug concentration (Norling et al. 1992).

Polymer Precipitation

Biodegradable polymers have been used extensively over the past 20 years as medical and drug delivery devices (Baker 1987; Chasin and Langer 1990; Hollinger 1995). Most of the biodegradable polymers exist as solids at room temperature; they therefore must be placed in the body as solid implants. As we

Figure 10.4. *Cumulative in vitro release of metronidazole dental gel using USP apparatus.*

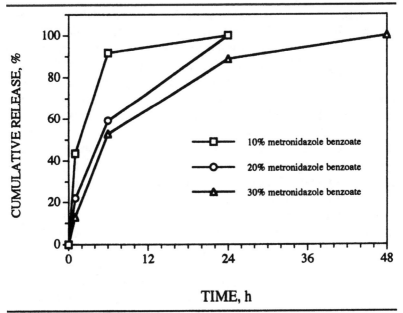

Adapted from Table 3 in Norling et al. (1992) with permission of Munksgaard International Publishers.

noted earlier, these systems have to be implanted with large trochars or during surgical procedures. Only microparticles formed from these biodegradable polymers can be injected using normal syringes and needles. Recently, a liquid polymer system, ATRIGEL® (Atrix Laboratories, Fort Collins, Colo.), has been developed that combines the advantages of a solid implant with the ease of administration of microparticles (Dunn et al. 1990, 1991). In this liquid polymer system, water-insoluble biodegradable polymers are dissolved in water-soluble biocompatible solvents to form solutions that can be injected into the body using normal syringes and needles. Upon contact with body fluids, the water-soluble solvent diffuses away from the injection site, and water permeates into the liquid polymer matrix. Because the polymer is insoluble in water, it precipitates or coagulates to form a solid implant. When a drug is incorporated in the liquid polymer solution, it is entrapped within the solid polymer matrix and is slowly released by diffusion, dissolution, or polymer erosion (Shah et al. 1993; Lambert and Peck 1995; Shively et al. 1995).

Almost all of the water-insoluble biodegradable polymers commonly used in the medical device and drug delivery fields can be used in the liquid polymer system, as long as they have sufficient solubility in the selected solvents. These include the polylactides, copolymers of lactide with glycolide or caprolactone, polyanhydrides, polyorthoesters, polyurethanes, and polycarbonates. The most frequently used biodegradable polymers in the liquid polymer system are PLA, PLGA, and the copolymers of DL-lactide with caprolactone because of their amorphous structure, known degradation rates, commercial availability, and acceptance by the U.S. Food and Drug Administration (FDA).

A number of biocompatible solvents can be used with the in situ forming polymeric depot system. These range from highly water-soluble solvents such as N-methyl-2-pyrrolidone (NMP), 2-pyrrolidone, ethyl lactate, dimethylsulfoxide (DMSO), solketal, and glycerol formal to the less water-soluble propylene carbonate, ethyl acetate, and triacetin. Although all these solvents have some toxicity, the amounts used in the polymeric implant system are well below the established adverse effects levels. The major concern has been the potential for local irritation at the site of injection, but a number of animal studies have shown that the polymer implant appears to modulate the release of the solvent over a period of 24 to 48 h, thereby reducing any local irritation expected from simple injection of the solvent alone.

When the biodegradable polymers are dissolved in the solvents at concentrations of 20 to 80 percent by weight, they form flowable compositions that range in consistency from liquids to gels to putties. The viscosity of the polymer solution depends on the molecular weight and concentration of the polymer in the solvent. Low molecular weight polymers, or solutions with relatively low concentrations of polymer (around 20 to 40 percent by weight), provide formulations that can be easily injected using standard syringes and needles or even aerosolized for spray applications. Higher molecular weight polymers coupled with higher polymer concentrations provide gels or putty-like materials that can be used to fill tissue defects and provide some structural support once they have fully solidified. Nonpolymeric materials can also be added to the polymer formulations to reduce the solution viscosity, increase the porosity of the implant after solidification, or to modify the release of a drug from the formed implant. A further modification to the system to control the initial drug burst and to lower the solution viscosity is to incorporate the drug at high concentrations into particles of polymers that have low solubility in the solvents and to add these particles to the polymer solution immediately before injection. These polymer particles control the initial release of drug just long enough for the polymer solution to form an implant that regulates the release of the drug.

The in situ forming polymer system can be used as both a medical device and a drug delivery system. One of the first products developed with this system and approved by the FDA is a barrier membrane for the regeneration of periodontal tissue (Polson et al. 1994, 1995a, b; Bogle et al. 1997). This product, the ATRISORB®GTR barrier membrane (manufactured by Atrix Laboratories, Inc., and sold by Block Drug Company, Jersey City, N.J.), is a solution containing 37 percent by weight PLA dissolved in NMP. The polymer solution is sterilized by exposure to gamma irradiation and supplied with a barrier-forming kit to periodontists, who prepare a semisolid film of the polymer material at chairside. They then cut the film to the desired dimensions and place it over the periodontal defect, where the final coagulation of the material occurs to produce a rigid barrier membrane. This membrane allows new bone and periodontal tissue to form while preventing epithelium growth into the defect. Recently, the same polymer solution has been used to form barrier membranes in situ by first using demineralized bone matrix to fill the periodontal defect and then applying the polymer solution directly over the bone matrix to form the membrane. This new product, ATRISORB® Free Flow, was approved by the FDA in September 1998 and is being manufactured and sold by the same two companies.

A second medical device application for the in situ polymer system has been the prevention of surgical adhesions (Fujita et al. 1994). In this application, the concentration of polymer in the solutions has been reduced to about 20 percent by weight. At this concentration, the polymer solutions can be aerosolized and sprayed directly onto tissues to form a thin, adherent, and flexible film to prevent the formation of adhesions following an operation. Although several polymers and solvents have been investigated, the best candidate appears to be a copolymer of poly(lactide-co-caprolactone) (PLC) dissolved in caprolactone (Dunn 1997).

In addition to its use as a medical device, the in situ polymer system has been investigated as a drug delivery platform. One of its first applications as a drug delivery system is the local delivery of an antibiotic to treat periodontal disease (Polson et al. 1993, 1996, 1997). The antibiotic doxycycline is incorporated into a solution of PLA in NMP and placed into a diseased periodontal pocket containing high levels of bacteria. A syringe attached to a 23-gauge cannula is used to place the polymer solution into the pocket. Upon contact with the gingival crevicular fluid, the polymer coagulates to form a solid implant that conforms to the irregular shape of the pocket. Over a period of about 10 days, the drug is released from the implant at levels well above the minimum inhibitory concentration of the antibiotic for the periodontal pathogens. Within this time, the polymer starts to degrade and fragment to the point where it is lost or flushed out of the pocket. This product (ATRIDOX®) has been tested in more than 800 patients and found to be both safe and effective. In September 1998 it was approved by the FDA for treating periodontal disease; it is currently manufactured by Atrix Laboratories, Inc., and sold by Block Drug Company. A similar product containing doxycycline, which was approved by the FDA in November 1997 for treating periodontal disease in companion animals, is marketed by Heska Corporation, Fort Collins, Colo.

In addition to these periodontal treatment products, a wide range of other drugs has been evaluated for in vitro and in vivo release with the in situ forming polymer system. These include antibiotics such as sanguinarine, tetracycline, and quinolone (Dunn et al. 1995); nonsteroidal anti-inflammatories such as naproxen, diclofenac, and flurbiprofen (Tipton et al. 1991, 1992); anesthetics such as bupivacaine (Sherman et al. 1994); narcotic antagonists such as naltrexone (Coonts 1993); hormones such as estradiol and trenbolone acetate (Dunn et al. 1995); antitumor agents such as doxorubicin, vinblastine, and cisplatin (Duysen et al. 1993a; Dunn et al. 1996, 1997b); proteins such as fibroblast growth factor, platelet-derived growth factor, insulin-like growth factor, and bone morphogenic proteins (Yewey et al. 1997; Duysen et al. 1993b; Chandrashekar et al. 1997); and LHRH peptides such as nafarelin, ganirelix, and leuprolide acetate (Radomsky et al. 1993; Dunn et al. 1997a). Several of these, including the delivery of bupivacaine, bone morphogenic protein, and leuprolide acetate, have advanced to the stage where they have demonstrated efficacy in animal models. Figures 10.5–8 show the release of several drugs from the polymer system.

Because the in situ polymer system forms a solid implant inside the body, it can also be used as a combination medical device and drug delivery system. One example is the addition of doxycycline to the ATRISORB®GTR barrier membrane described earlier. In this application, the barrier membrane functions as a medical device to guide periodontal tissue regeneration. A slow release of doxycycline incorporated into the barrier membrane serves to decrease the levels of periodontal pathogens at the surgical site and thereby reduces the risk of infections that adversely affect the outcome of the regeneration procedure (Botz et al. 1998). In studies with dogs, this product has shown increased periodontal tissue regeneration, and pivotal clinical studies are in progress. Another combination product incorporates drugs such as tissue plasminogen activator (t-PA) into an aerosol polymer formulation used to prevent surgical

Figure 10.5. *Cumulative in vitro release of doxycycline hyclate from the ATRIGEL® system.*

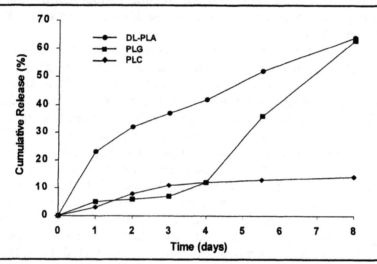

Adapted from Dunn et al. (1991) with permission of the Controlled Release Society, Inc.

Figure 10.6. *In vivo release of platelet-derived growth factor (PDGF) from the ATRIGEL® system.*

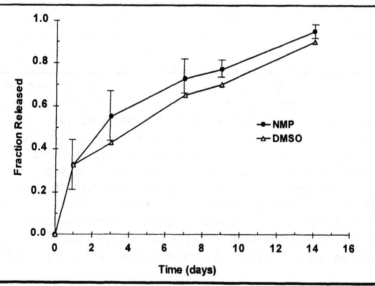

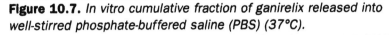

Figure 10.7. *In vitro cumulative fraction of ganirelix released into well-stirred phosphate-buffered saline (PBS) (37°C).*

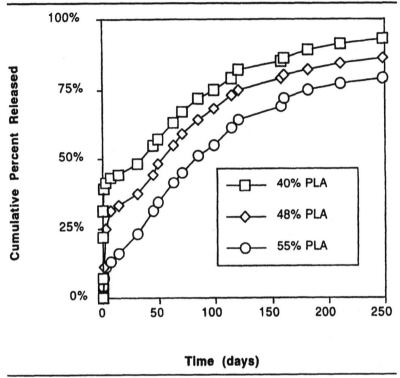

The formulations contained 15 percent ganirelix. Reprinted from Radomsky et al. (1993) with permission of the Controlled Release Society, Inc.

adhesions. The polymer film acts as a mechanical barrier, while the release of t-PA breaks down the blood clot that normally leads to scar tissue (Dunn 1997).

Sucrose Ester Precipitation

Solvent-based polymer systems have been studied by several groups and have recently been approved as a periodontal system for the delivery of doxycycline. However, such systems have two known disadvantages: They require relatively high polymer

Figure 10.8. *Cumulative in vitro release of naproxen from selected* ATRIGEL® *formulations.*

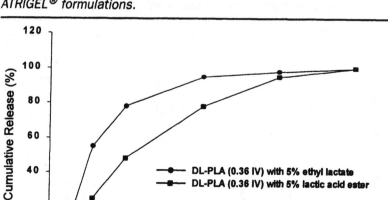

Adapted from Tipton et al. (1991) with permission of the Controlled Release Society, Inc.

loadings to minimize the initial drug burst that leads to high viscosity, and they depend on a somewhat small range of solvents that both dissolve the commonly used polymers and are biologically well tolerated. Using a small molecule as the basis for the system would offer improvements in both areas. First, the lack of polymer entanglement dramatically diminishes solution viscosity. Second, with a much more favorable entropy of mixing, a lower molecular weight material is soluble in a wider range of solvents than a high molecular weight polymer.

One example of a small molecule useful for drug delivery is a sucrose derivative, sucrose acetate isobutyrate (SAIB), which has been described in the literature (Smith and Tipton 1996; Tipton and Holl 1998). SAIB is a mixed ester of sucrose, fully esterified at a ratio of 6 isobutyrate groups to 2 acetate groups. The resulting ester is completely noncrystalline, with a viscosity of greater than 100,000 cP, in the range of noncrystalline polymers

(Eastman 1989). Unlike polymers, however, the solution viscosity of SAIB at excipient concentrations as high as 85–90 percent is on the order of 50–200 cP, similar to the viscosity of vegetable oil. A wide range of solvents is possible for this molecule, from hydrophobic hydrocarbons to the hydrophilic ethanol (Eastman 1989). While SAIB is one reported example of a small molecule with the properties of amorphous, high viscosities, it is expected that other examples of mixed and partial esters of other sugars would also have this property, as would other randomly substituted molecules.

A range of solvents has been described for SAIB. Some that may have biological interest include ethanol, ethyl lactate, propylene carbonate, NMP, 2-pyrrolidone, and DMSO. The viscosity undergoes a change of more than 3 orders of magnitude with only 15 percent solvent. The drugs that have been described with this system include the antimicrobials and anti-inflammatories chlorhexidine, naproxen, and flurbiprofen (Smith and Tipton 1996); the luteinizing hormone releasing hormone (LHRH) analog deslorelin (Burns et al. 1997); the anti-cancer agents 5-fluorouracil, doxorubicin, and paclitaxel (Taxol®) (Sullivan et al. 1997); and heparin (Espinal et al. 1995). Figures 10.9 and 10.10 show the in vitro release for some of these small molecules. The published in vivo studies have included heparin delivered from vascular grafts in a canine model and deslorelin delivered subcutaneously in an equine model to regulate estrus (Burns et al. 1997). These studies show in vivo delivery for 24 to 72 h. Recent data show release of progesterone for a period of up to 4 weeks, as seen in Figure 10.11 (Sullivan et al. 1998).

As with the polymer-based solution systems, a number of formulation variables can be used to affect the delivery of actives. These variables include drug concentration, solvent type, and solvent concentration. Additives have perhaps the most profound effect, changing release rates by as much as two orders of magnitude.

Biodegradable Acrylate-Terminated Polymers

Another approach to a liquid injectable polymer system that forms a solid implant within the body has been the development of biodegradable polymers with acrylic end groups that can be

Figure 10.9. *Cumulative in vitro release of 5 percent chlorhexidine from SAIB in ethanol.*

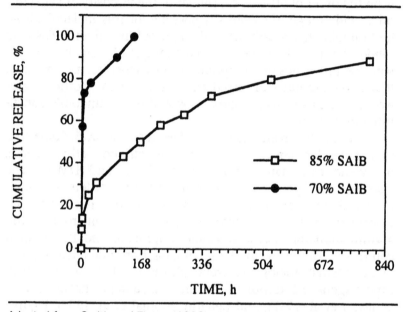

Adapted from Smith and Tipton (1996) with permission of the American Association of Pharmaceutical Scientists.

polymerized in situ to form a cross-linked solid polymer (Dunn et al. 1990). Although almost any biodegradable polymer can be used, the limiting criterion is that the polymers with the acrylic end groups must be liquids. To produce this injectable liquid system, low molecular weight oligomers of the biodegradable polymers are prepared with terminal hydroxyl groups. As an example, the synthesis of hydroxy-terminated PLC is achieved by using a bifunctional initiator such as ethylene glycol to copolymerize a mixture of DL-lactide and ε-caprolactone. A trifunctional initiator molecule, such as trimethylolpropane, produces a polymer chain with three hydroxyl end groups. The hydroxy-terminated prepolymers are converted to acrylic ester-capped prepolymers via a reaction with acryloyl chloride under

Figure 10.10. *Cumulative in vitro release of taxol from 85:15 SAIB:ethanol formulations showing the affect of drug loading.*

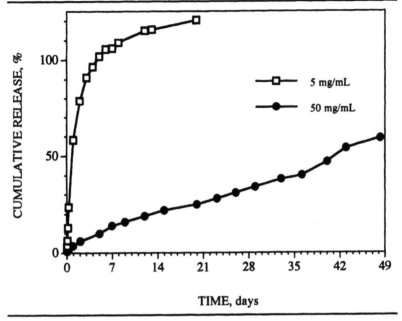

Reprinted from Sullivan et al. (1997) with permission of the American Association of Pharmaceutical Scientists.

Schotten-Baumann–like conditions. The liquid acrylic ester-capped prepolymers are purified and then mixed with a cross-linking agent, such as benzoyl peroxide, immediately before injection into the body.

In one report describing this in situ curing polymer system, polycaprolactone triol was reacted with acryloyl chloride to form a trifunctional acrylic ester-terminated prepolymer (Moore et al. 1995). The liquid prepolymer was then divided into two portions. Benzoyl peroxide, the catalyst at a level of 0.5 percent by weight, was added to one of the portions, and the initiator, N,N-dimethyl-p-toluidine at a level of 0.25 percent, was added to the other portion. The two liquid solutions were then placed in syringes and mixed just prior to injection by coupling the two

Figure 10.11. *In vivo release of progesterone from SAIB:ethanol formulations.*

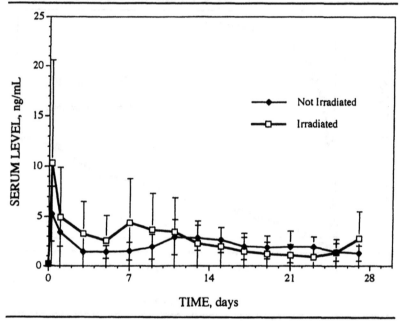

Reprinted from Sullivan et al. (1998) with permission of the American Association of Pharmaceutical Scientists.

syringes together and passing the contents back and forth for a number of cycles. The syringes were decoupled, a 22-gauge syringe needle attached, and the product injected into the subcutaneous region of rats. In vitro studies showed that this material began to cure in less than 5 min and was fully cured in 30 min. Implants retrieved from the animals at 30 and 60 min confirmed that the polymer system also cured in vivo at approximately the same rate, with no evidence of tissue irritation or toxicity. Formulations of the same polymer system containing 5 percent by weight flurbiprofen showed a low initial burst of drug and extended release out to 7 days (Figure 10.12).

Figure 10.12. *In vitro flurbiprofen release from acrylate-terminated polycaprolactone.*

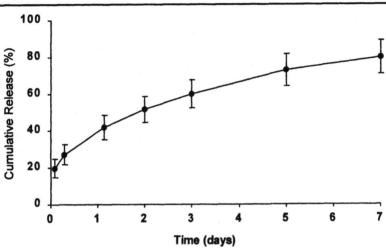

Reprinted from Moore et al. (1995) with permission of the Society for Biomaterials.)

Acrylate-Terminated PEG–PLA

The acrylate-terminated PLGs described above are water-insoluble polymers that are cross-linked in vivo. A highly hydrophilic polymer has also been described with small amounts of penultimate lactide linked to acrylate groups. The synthesis of these PEG triblocked polymers with lactide linkages to acrylic groups has been well described from a synthetic route (Sawhney et al. 1993). They combine the properties of the biologically safe PEG with the cross-linking capability of acrylates and degradation via the lactide polymer unit. Their use as medical devices has been well described and includes investigation for efficacy in blocking surgical adhesions, in preventing restenosis, and as a lung sealant (Hill-West et al. 1994a, b; Sawhney et al. 1996). From an application standpoint, the important parameter is that these materials are aqueous-based solutions, which can be applied as a solution and converted to the cross-linked gel only after exposure to a specific wavelength of light.

The researchers evaluating these systems for medical devices have also realized their utility for drug delivery. Not surprisingly, one of the active areas of research is in local delivery, working in conjunction with a medical device function. For example, heparin has been incorporated into such a system that is photopolymerized on the endoluminal surface of an artery, obtaining high concentrations of heparin locally in a rat model for more than 48 h, as shown in Figure 10.13 (Philbrook et al. 1995).

The delivery of other agents has also been described, including a variety of proteins, insulin, immunoglobulin G (IgG), and platelet-derived growth factor (West and Hubbell 1995), where the molecular weight of the active was shown to have a strong effect on delivery rate, as shown in Figure 10.14.

In this system, which may typically be composed of 8,000 MW PEG and be applied from a 10 percent aqueous solution (Philbrook et al. 1995), release is first order and rapid, with a half-life of a few hours to 2 days. It is expected that more-concentrated solutions and higher lactide/PEG ratios would decrease the drug release rate. It is also possible to use degradable

Figure 10.13. *Release profile of heparin from bulk acrylate-terminated PEG–PLA hydrogel.*

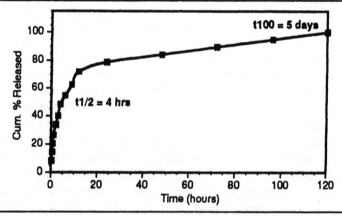

Reprinted from Philbrook et al. (1995) with permission of the Controlled Release Society, Inc.

Figure 10.14. *Cumulative in vitro release of proteins from bulk acrylate-terminated PEG–PLA hydrogel.*

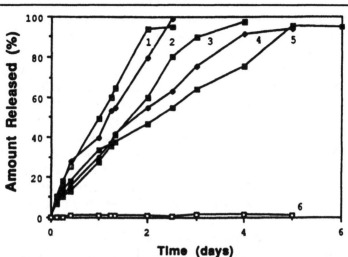

Curve 1, insulin, MW 6,000. Curve 2, lysozyme, MW 14,300. Curve 3, LDH, MW 36,500. Curve 4, ovalbumin, MW 45,000. Curve 5, BSA, MW 66,000. Curve 6. IgG, MW 150,000. Reprinted from West and Hubbell (1995) with permission of the Controlled Release Society, Inc.

linkages other than lactide. The use of peptide sequences has been described, where degradation occurs in the presence of a protease such as collagenase (West and Hubbell 1996).

Other Systems

In the previous sections, we have covered a great deal of the development on in situ gelling systems. A number of other systems have also been described in the literature. In addition to the above six types of systems, which undergo either a physical or a chemical gelling phenomenon, single systems that take advantage of both types of change have also been described. Using a Pluronic® for the center block of acrylate-terminated PLA can produce a system that both experiences an LCST phenomenon and can be cross-linked via acrylate end groups (Pathak 1995).

This combination offers a wide range of synthetic possibilities for affecting degradation and the drug delivery rate.

The use of PLGs as block copolymers with PEO–PPO for LCST was described above. It is also possible to synthesize these materials so that they do not exhibit LCST behavior but exhibit the more expected gelling behavior at higher temperatures, driven by the hydrophobic interaction of the PLA block. Through proper synthetic control, materials can be prepared that are sols at about 45°C and convert to gels upon cooling to body temperature (Jeong 1997). These systems have shown in vitro release on the order of one to two weeks.

Systems based on response to a specific ion have also been described. The ions most commonly used are H^+ and Ca^{+2} (Kaushik et al. 1995; Kumar and Himmelstein 1994; Cohen 1997). These systems generally are based on a carboxylic acid functionality in the polymer. For H^+ responsive systems, the polymer is synthesized and formulated to be the ionized form prior to injection. Upon injection and exposure to physiological pH, the acid is protonated, decreasing the aqueous solubility of the polymer. With proper synthetic choices, this change can affect overall solubility. For Ca^{+2} systems, the chemistry is generally the same; however, ionic cross-linking similar to that seen in ionomer chemistry is possible. A system based on both pH and the use of an organic solvent similar to these systems has also been described (Haglund and Himmelstein 1994), as has a combination system that uses both pH effect and cross-linking (Barman et al. 1997).

One of the more recent reactive systems is related to glucose. Here, polymeric systems are prepared with glucose-containing polyvinyl pyrrolidone and concanavalin A (Obaidat and Park 1995, 1996). Concanavalin A serves as a binding site for glucose, and, being tetrafunctional, it can also serve as a site for cross-linking. Because free glucose competes for the binding sites, the transition is reversible, forming a gel at low free-glucose levels and a sol at high glucose levels. This system has potential as the basis for a self-regulating glucose system.

The use of liquid crystalline–based systems in which the excipient is GMO was described above. It is also possible for the drug itself to form a liquid crystalline phase, which will form a gel upon injection and slowly redissolve. The structure of linear peptides is the most common geometry for these types of systems (Wong et al. 1994a, b) and can be affected to some extent by pH (Wan et al. 1994).

The application of bone cements has also provided a system that has been the basis for drug delivery. Bone cements are typically poly(methyl methacrylate) based and are composed of a prepolymer, a cross-linker, and an initiator. Before use, this prepolymer blend is mixed with monomer and formed into a paste. The paste undergoes cross-linking to form a rigid thermoset. A drug mixed into the paste or a component of the paste is entrapped and slowly released from the thermoset (Salvati et al. 1986). A variety of other cements based on calcium phosphate to deliver drugs have also been described (Otsuka et al. 1996, 1997).

DRUG CANDIDATES FOR IN SITU GELLING SYSTEMS

As can be seen from many of the examples above, a wide variety of drug types have been evaluated for release from in situ gelling systems. It is worthwhile to evaluate what types of drugs might be the best matches for in situ gelling systems. In general, any parenteral administered drug is a likely candidate, but there are some general guidelines to selecting optimal drugs. Also, some drug candidates will be a better matches with certain in situ gelling technologies.

The usual requirements for delivery systems hold for in situ gelling systems: Duration and efficacious levels must match the payload, the drug must withstand the biological environment for the duration of release, the drug must have acceptable shelf-life stability in the delivery system or be able to be mixed into the system at the time of administration, and the drug must be able to withstand the formulation process.

Some of the technologies are aqueous based, such as LCST or liquid crystalline (Johnston and Miller 1985; Yoshida 1994; Brown et al. 1976; Engstöm 1990; Norling et al. 1992), and water-soluble drugs are likely candidates. Likewise, the technologies based on biodegradable polymers or SAIB in organic solution are often good matches for hydrophobic drugs. As both of these systems often use solvents with high solubility parameters (Dunn et al. 1990; Polson et al. 1997; Smith and Tipton 1996; Tipton and Holl 1998), such as N-methyl pyrrolidone or ethanol, a range of water-soluble drugs is often soluble in these systems as well. While drug solubility in the system is generally desired from a stability, sterility, and clinical administration standpoint, a drug that is insoluble in the system will generally show slower release (Smith and Tipton 1996) due to the additional barrier of drug dissolution.

With the caveats of drug stability and solubility, virtually any small molecule drug can be used with any of the technologies. As noted above, high equilibrium water contents in a system generally corresponds to rapid release. So systems based on PEG or GMO will usually display a duration of drug release for at best 1 week. Longer duration release can be obtained with more hydrophobic systems, such as those based on biodegradable polymers or SAIB. As with any drug delivery technology, the technology must match the clinical requirement.

Like their usefulness in delivering small-molecule drugs, most of the systems also have shown promise for the delivery of peptides. Examples for most of the major types of systems discussed in this chapter have been described in the literature (Radomsky et al. 1993; Burns et al. 1997; Bhardwaj and Blanchard 1995).

The known difficulties of protein delivery have also challenged researchers evaluating in situ gelling systems. There have been a number of reports of success in delivery at a research scale, but no system has yet made it into clinical trials. The cases based on water-based systems (Johnston et al. 1992; Katakam et al. 1997; West and Hubbell 1995) have been the most studied and have generally shown protein release for a short period of 3 days or so. Several reports have also been made for protein delivery from polymer precipitation systems (Yewey et al. 1997; Duysen et al. 1993b; Chandrashekar et al. 1997) and have shown longer term delivery both in vitro and in vivo.

TECHNOLOGY LIMITATIONS

As with all technologies, there are some limitations to applications of in situ gelling drug delivery systems. The most notable one is simply that "something happens" upon injection. Because the systems are undergoing some physical or chemical change to a gel-like state, there are a number of considerations. Of broad concern for all of the technologies is the kinetics of the transition. Specifically, during the time that the gel formation is occurring, the issue of what happens to the drug must be addressed. If the gelling process occurs too slowly, an unacceptable initial burst of drug will be obtained. Also, from a chemical stability standpoint, the drug must be able to withstand the gelling process. For example, with acrylate-based systems, the gelling may involve free-radical reaction, and the drug must not participate in this reaction.

There are cases where somewhat narrow formulation choices are available. For LCST and liquid crystalline systems, excipient and drug must be formulated with water, and the appropriate transition from an injectable liquid to a gel relies on specific ratios. For example, high drug concentrations may interfere with the hydrogen bonding needed in most LCST systems or the cubic phase structure needed in most liquid crystalline systems. For systems in organic solvents, formulators are challenged with keeping solvent concentrations low to minimize drug burst; very low concentrations generally cannot be used without facing difficult injectablity due to high viscosity.

A well-appreciated, but little-studied concern is shape. Because the systems are applied as liquids, they can flow during the application period and for at least part of the time during the gelling transition. As opposed to preformed implants with a known surface area, these systems may have different surface area for a set injection size due to shape differences. There does not seem to be a published study on this effect, but it would be expected that more surface area would lead to faster release. While it is true that the shape is not controlled, and that this loss of control may lead to variability, the reader is directed to Figures 10.3, 10.6, 10.8, 10.11, and 10.12 to see relatively tight in vitro and in vivo release curves. Another point to consider in terms of burst is that while surface area is not controlled, there is generally less burst than technologies based on microspheres that have a much larger cumulative surface area.

As noted previously for water-based systems, such as LCST, liquid crystalline, or acrylate-terminated PEG–PLA, water is a majority component of the system. Consequently, diffusion is relatively rapid, and drug release is generally limited to a period of days.

For many of these technologies, there are additional regulatory issues that must be addressed. Some of them rely on new compounds or combinations of compounds that do not have a proven history of regulatory acceptance. Others are using known compounds in new ways or at different levels. For example, two of the technologies rely on the use of organic solvents, and while a great deal of safety information has been published, there is often initial reluctance from the pharmaceutical industry to these compounds. Others rely on chemical reactions such as acrylate chemistry that may need to be further studied to guarantee there is no toxicity.

FUTURE TRENDS WITH IN SITU GELLING SYSTEMS

Many researchers and clinicians alike refer to oral delivery systems, particularly for protein delivery, as the "Holy Grail" of drug delivery. The reason is simple: Most patients prefer to take their drug regimes orally. For many drugs and conditions, however, oral administration is not possible, and parenteral delivery will continue to be the preferred route of administration. And for those drugs that must be delivered over a week to a month to a year, a delivery system that eliminates the need for daily or more-frequent injections is desired.

As we stated at the beginning of this chapter, a number of systems have been proven in the marketplace, and these combine one month up to five years of delivery from a single placement. These systems include Zoladex®, the Norplant® implant, Lupron Depot®, and Decapeptyl®. (See Chapter 3 for a comprehensive list.) What is it about in situ gelling systems that offers advantages in place of these market-proven alternatives?

First, in situ gelling systems have, and will continue to have, a record of approval. The Elyzol® Dental Gel for periodontal delivery of metronidazole (Dumex) and the ATRIGEL® system (Atrix Laboratories, Inc., Fort Collins, Colo.) system (with no drug) for periodontal guided tissue regeneration (ATRISORB®) are on the market now. The ATRIGEL® system with doxycycline (ATRIDOX®) for periodontal treatment was recently approved as well. Although the concepts are relatively new, the advantages to the patient and clinican have allowed for relatively quick market introduction.

It is expected that in situ gelling systems will also offer advantages in local delivery. Not only are the most clinically advanced systems and the marketed ones focused on local delivery, but local delivery is also an area where in situ–formed implants may offer an advantage over preformed implants. Simply keeping a solid implant or microspheres in a local environment may be problematic. The in situ gelling systems offer some advantage here, even if it is primarily due to physical interlocking. In addition, the fact that some in situ gelling systems can be placed using a cannula or minimally invasive surgical techniques allows for pinpoint placement that is not possible with microspheres or implants. Another advantage is the payload of drug that can be delivered. Compared to microspheres, in situ gelling systems offer the potential for drug loads that are larger by roughly an order of magnitude.

The ease of injection that in situ gelling systems afford is a tremendous advantage over preformed implants. In the midst of revolutionary changes in health care, and with all involved wanting to minimize hospital and professional in-house care, in situ gelling systems offer greater potential for self-administration. It is difficult to imagine a patient self-administering an SC implant that requires a 16-gauge or larger needle. It is also difficult to imagine self-administration of an IM injection even through a needle as fine as 23 gauge for microspheres. An in situ gelling system, however, can be injected using needles as fine as 27 gauge. SC administration of such a system can be accomplished with an in-house treatment by a medical professional or by the patient. While a three-month delivery system may be possible with an implant or microspheres, the regime will require four visits to the physician's office or hospital over one year. This treatment may not offer as great an advantage as a one- to two-month in situ gelling system that can be administered at home.

As with all delivery systems, it is a certainty that in situ gelling systems will not work for all drugs or conditions. However, the availability of these exciting new systems adds a valuable alternative to the formulator's and pharmacologist's arsenal for novel delivery regimes.

REFERENCES

Appel, L., K. Engle, J. Jensen, L. Rajewski, and G. Zentner. 1994. An in vitro model to mimic in vivo subcutaneous monoolein degradation. *Pharm. Res.* 11 (10):217.

Bae, Y. H., T. Okano, and S. W. Kim. 1987. Thermo-sensitive polymers as on-off switches for drug delivery. *Makromol. Chem. Rapid Commun.* 8:481.

Bae, Y. H., T. Okano, and S. W. Kim. 1990. Temperature dependence of swelling of cross-linked poly(N,N'-alkyl substituted acrylamides) in water. *J. Polym. Sci. B: Polym. Phys.* 28:923–927.

Baker, R. 1987. *Controlled release of biologically active agents.* New York: John Wiley and Sons.

Barman, S. P., C. P. Pathak, and A. J. Coury. 1997. Molecular chain structure effects on release properties of hydrogels modulated by external stimuli. *Proc. Intern. Symp. Control. Rel. Bioact. Mater.* 24:889–890.

Baudys, M., A. Serres, C. Ramkissoon, and S. W. Kim. 1997. Temperature and pH-sensitive polymers for polypeptide drug delivery. *J. Control. Rel.* 48:304–313.

Bhardwaj, R., and J. Blanchard. 1995. Poloxamer 407 based controlled release delivery for melanotan-1. *Pharm. Res.* 12 (9):225.

Bogle, G., S. Garret, N. H. Stoller, D. D. Swanbom, J. C. Fulfs, P. W. Rodgers, S. Whitman, R. L. Dunn, G. L. Southard, and A. M. Polson. 1997. Periodontal regeneration in naturally occurring class II furcation defects in beagle dogs after guided tissue regeneration with bioabsorbable barriers. *J. Periodontol.* 68:536–544.

Botz, M. L., K. C. Godowski, R. L. Dunn, and G. L. Southard. 1998. Antimicrobial effects of a bioabsorbable GTR barrier containing doxycycline. *J. Dent. Res.* 77:140.

Brown, H. S., G. Meltzer, R. C. Merril, M. Fisher, C. Ferre, and U. A. Place. 1976. Visual effects of pilocarpine in glaucoma. Comparative study of administration of eye drops or by ocular therapeutic systems. *Arch. Ophthalmol.* 94:1716–1719.

Burns, P., D. Thompson Jr., F. Donadue, L. Kincald, B. Leise, J. Gibson, R. Swaim, and A. Tipton. 1997. Pharmacodynamic evaluation of the SABER® delivery system for the controlled release of the GnRH analogue deslorelin acetate for advancing ovulation in cyclic mares. *Proc. Intern. Symp. Control. Rel. Bioact. Mater.* 24:737–738.

Chandrashekar, B., S. Jeffers, K. Holland, S. Garret, R. L. Dunn, and J. Benedict. 1997. Sustained release and osteoinductive effect of bone protein: Evaluation of an in situ forming bioabsorbable polymeric implant as a delivery vehicle. *Proc. Portland Bone Symp.* pp. 583–589.

Chasin, M., and R. Langer. 1990. *Biodegradable polymers as drug delivery systems.* New York: Marcel Dekker, Inc.

Chen, G., and A. S. Hoffman. 1995a. Temperature-induced phase transition behaviours of random vs. graft copolymer of n-isopropylactylamide and acrylic acid. *Macromol. Rapid Commun.* 16:175.

Chen, G., and A. S. Hoffman. 1995b. Graft copolymers that exhibit temperature-induced phase transitions over a wide range of pH. *Nature* 373:49–50.

Chen, G., A. S. Hoffman, and E. S. Ron. 1995a. Novel hydrogels of a temperature-sensitive Pluronic® grafted to a bioadhesive polyacrylic acid backbone for vaginal drug delivery. *Proc. Intern. Symp. Control. Rel. Bioact. Mater.* 22:167–168.

Cohen, S., E. Lobel, A. Trevgoda, and Y. Peled. 1997. A novel in situ-forming ophthalmic drug delivery system from alginate undergoing gelation in the eye. *J. Control. Rel.* 44:201–208.

Coonts, B. A., B. K. Lowe, R. L. Norton, A. J. Tipton, G. L. Yewey, and R. L. Dunn. 1993. Plasma concentrations of naltrexone base following subcutaneous and intramuscular injections of ATRIGEL® formulations in dogs. *Pharm. Res.* 10:S192.

Dunn, R. L. 1997. An in-situ formed biodegradable polymer barrier for prevention of post-operative surgical adhesives. In *Post-surgical adhesions: Linking business and science to further the development of adhesion prevention devices.* NMHCC Conference, 26–27 June, Baltimore, MD.

Dunn, R. L., A. J. Tipton, and E. M. Menardi. 1991. A biodegradable in-situ forming drug delivery system. *Proceed. Intern. Symp. Control. Rel. Bioact. Mater.* 18:465–466.

Dunn, R. L., J. P. English, D. R. Cowsar, and D. Vanderbilt. 1990. Biodegradable in-situ forming implants and methods of producing the same. U.S. Patent 4,938,763.

Dunn, R., G. Hardee, A. Polson, A. Bennett, S. Martin, R. Wardley, W. Moseley, N. Krinick, T. Foster, K. Frank, and S. Cox. 1995. In-situ forming biodegradable implants for controlled release veterinary applications. *Proceed. Intern. Symp. Control. Rel. Bioact. Mater.* 22:91–92.

Dunn, R. L., G. L. Yewey, S. M. Fujita, K. R. Josephs, S. L. Whitman, G. L. Southard, W. S. Dernell, R. C. Straw, S. J. Withrow, and B. E. Powers. 1996. Sustained release of cisplatin in dogs from an injectable implant delivery system. *J. Bioactive Compat. Polymers* 67:1176–1184.

Dunn, R. L., B. Chandrashekar, D. Anna, S. Garrett, and L. Southard. 1997a. An in-situ forming delivery system for peptides and proteins. U.S./Japan Symposium on Drug Delivery, 14–19 December, in Hawaii.

Dunn, R. L., S. A. Jeffers, K. M. Holland, K. L. Tow, C. M. Balliu, J. P. Mitchell, and B. A. Coonts. 1997b. An in-situ forming biodegradable implant system for intratumoral delivery of cisplatin. *Proceed. Intern. Symp. Control. Rel. Bioact. Mater.* 24:163–164.

Duysen, E. G., S. L. Whitman, N. L. Krinick, S. M. Fujita, and G. L. Yewey. 1993a. An injectable biodegradable delivery system for antineoplastic agents. *Pharm. Res.* 10:S88.

Duysen, E. G., G. L. Yewey, G. L. Southard, R. L. Dunn, and W. Huffer. 1993b. Release of bioactive growth factors from the ATRIGEL® drug delivery system in tibia defects and dermal wound models. *Pharm. Res.* 10:S83.

Eastman. 1989. Technical Bulletin No. ZM-90. Kingsport, Tenn., USA: Eastman Chemical Company.

Engström, S. 1990. Cubic and other lipid-water phases as drug delivery systems. *Lipid Technol.* 2:42–49.

Engström, S., K. Larsson, and B. Lindman. 1992. Method of preparing controlled-release preparations for biologically active materials and resulting compositions. U.S. Patent 5,151,272.

Espinal, E., J. Bulgrin, M. Chapman, S. Schmidt, and A. J. Tipton. 1995. Controlled release of heparin from vascular grafts over a 55 day period. Paper presented at American College of Surgeons Meeting, 13–16 March, in Akron, Ohio.

Feil, H., Y. H. Bae, J. Feijen, and S. W. Kim. 1993. Mutual influence of pH and temperature on the swelling of ionisable and thermosensitive hydrogels. *Macromolecules.* 26:2496–2502.

Fujita, S. M., J. L. Southard, G. L. Yewey, and R. L. Dunn. 1994. Prevention of surgical adhesions using aerosoled biodegradable polyesters. *Trans. Soc. Biomater.* 17:384.

Fults, K. A., and T. P. Johnston. 1990. Sustained-release of urease from a poloxamer gel matrix. *J. Paren. Sci. Technol.* 44:58–63.

Gilbert, J. C., J. L. Richardson, M. C. Davies, and K. J. Paliin. 1987. The effect of solutes and polymer on the gelation properties of Pluronic F-127 solution for controlled drug delivery. *J. Control. Rel.* 5:113–119.

Grainger, D. W., and H. Yu. 1994. Novel thermo-responsive amphiphilic poly N-isopropylacrylamide-co-sodium acrylate-co-n-N-alkylacrylamide networks. *J. Control. Rel.* 28:319–320.

Haglund, B. O., and K. J. Himmelstein. 1994. An in situ gelling delivery system for parenteral injection. *Pharm. Res.* 11(10):267.

Hill-West, J. L., M. C. Sanghamitra, M. L. Slepian, and J. A. Hubbell. 1994a. Inhibition of thrombosis and intimal thickening by in situ photopolymerization of thin hydrogel barriers. *Proc. Natl. Acad. Sci. USA* 91:5967–5972.

Hill-West, J. L., S. M. Chowdhury, A. S. Sawhney, C. P. Pathak, R. C. Dunn, and J. A. Hubbell. 1994b. *Obst. & Gynec.* 83:59.

Himmelstein, K. J., and S. Kumar. 1995. An injectable enzymatic bioreactor. *Proc. Intern. Symp. Control. Rel. Bioact. Mater.* 22:212–213.

Hoffman, A. S., A. Afrassiabi, and L. L. Dong. 1986. Thermally reversible hydrogels: Delivery and selective removal of substances from aqueous solutions. *J. Control. Rel.* 4:213–220.

Hoffman, A. S., G. Chen, S. Kaang, and D. T. Priebe. 1995. New bioadhesive polymer compositions for prolonged drug release in the eye. *Proc. Intern. Symp. Control. Rel. Bioact. Mater.* 22:159–160.

Hoffman, A. S., X. Wu, Z. Ding, M. Schiller, and E. Ron. 1997a. Novel drug carriers based on physical gels of Pluronic® polyethers grafted to chitosan backbones: Synthesis and solution properties. *Proc. Intern. Symp. Control. Rel. Bioact. Mater.* 24:563–564.

Hoffman, A. S., G. Chen, X. Wu, Z. Ding, B. Kabra, K. Randeri, M. Schiller, E. Ron, N. Peppas, and C. Brazel. 1997b. Graft copolymers of PEO-PPO-PEO triblock polyethers on bioadhesive polymer backbones: Synthesis and properties. *ACS Polym. Preprints* 38 (1):524–525.

Hollinger, J. O. 1995. *Biomedical applications of synthetic biodegradable polymers.* Boca Raton, Fla., USA: CRC Press.

Jeong, B., Y. H. Bae, D. S. Lee, and S. W. Kim. 1996. Biodegradable thermoresponsive hydrogel. *Pharm. Res.* 13 (9):333.

Jeong, B., Y. H. Bae, D. S. Lee, and S. W. Kim. 1997. Biodegradable block copolymers as injectable drug-delivery systems. *Nature* 388:860–862.

Johnston, T. P., and S. C. Miller. 1985. Toxicological evaluation of poloxamer vehicles for intramuscular use. *J. Paren. Sci. Technol.* 39:83–88.

Johnston, T. P., M. A. Punjabi, and C. J. Froelich. 1992. Sustained delivery of interleukin-2 from a poloxamer 407 gel matrix following intraperitoneal injection in mice. *Pharm. Res.* 9:425.

Katakam, M., W. R. Ravis, and A. K. Banga. 1997. Controlled release of human growth hormone in rats following parenteral administration of poloxamer gels. *J. Control. Rel.* 49:21–26.

Kaushik, S., C. Toris, and K. Himmelstein. 1995. Evaluation of an in situ gelling system for ophthalmic use—a rheological and in vivo study. *Pharm. Res.* 12 (9):224.

Kumar, S., and K. J. Himmelstein. 1994. Comparison of different nonionic cellulose derivatives for use in in situ gel forming drug delivery systems. *Pharm. Res.* 11 (10):291.

Lambert, W. J. and K. D. Peck. 1995. Development of an in-situ forming biodegradable poly-lactide-co-glycolide system for the controlled release of proteins. *J. Control. Rel.* 33:189–195.

Larsson, K. 1989. Cubic lipid-water phases: Structures and biomembrane aspects. *J. Phys. Chem.* 93:7302.

Lee, D. S., B. Jeong, Y. H. Bae, and S. W. Kim. 1996. New thermoreversible and biodegradable block copolymer hydrogels. *Proc. Intern. Symp. Control. Rel. Bioact. Mater.* 23:228–229.

Moore, L. A., R. L. Norton, S. L. Whitman, and R. L. Dunn. 1995. An injectable biodegradable drug delivery system based on acrylic terminated poly(ε-caprolactone). *Trans. Soc. Biomater.* 18:186.

Norling, T., P. Lading, S. Engström, K. Larsson, N. Krog, and S. S. Nissen. 1992. Formulation of a drug delivery system based on a mixture of monoglycerides and triglycerides for use in the treatment of periodontal disease. *J. Clin. Periodontol.* 19:687–692.

Obaidat, A. A., and K. Park. 1995. Gel-sol phase-reversible hydrogel sensitive to glucose. *Pharm. Res.* 12 (9):225.

Obaidat, A. A., and K. Park. 1996. Characterization of the phase transition of glucose-sensitive hydrogels. *Proc. Intern. Symp. Control. Rel. Bioact. Mater.* 23:214–215.

Okano, T., Y. H. Bae, H. Jacobs, and S. W. Kim. 1990. Thermally on-off switching polymers for drug permeation and release. *J. Control. Rel.* 11:255–261.

Okano, T., A. Kikuchi, Y. Sakurai, Y. Takei, and N. Ogata. 1995. Temperature-responsive poly(N-isopropylacrylamide) as a modulator for alteration of hydrophilic/hydrophobic surface properties to control activation/inactivation of platelets. *J. Control. Rel.* 36:125–133.

Otsuka, M., Y. Nakahigashi, Y. Matsuda, J. L. Fox, W. I. Higuchi, and Y. Sigiyama. 1997. A novel skeletal drug delivery system using self-setting calcium phosphate cement. VIII: The relationship between in vitro and in vivo drug release from indomethacin-containing cement. *J. Control. Rel.* 43:115–122.

Otsuka, M., K. Yoneka, Y. Matsuda, J. L. Fox, W. I. Higuchi, and Y. Sugiyama. 1996. Osteoporosis-responsive estradiol release from a self-setting apatitic bone cement. *Proc. Intern. Symp. Control. Rel. Bioact. Mater.* 23:186–187.

Pathak, C. P., S. P. Barman, A. J. Coury, A. S. Sawhney, and J. A. Hubbell. 1995. Biodegradable thermoresponsive hydrogel and macromonomers. *Proc. Intern. Symp. Control. Rel. Bioact. Mater.*22:85–86.

Philbrook, M., E. Weselcouch, L. Roth, M. Lovich, M. Gallant, E. Edelman, and R. Leavit. 1995. Local sustained delivery of heparin via in situ photopolymerized biodegradable hydrogels. *Proc. Intern. Symp. Control. Rel. Bioact. Mater.* 22:19–20.

Polson, A. M., R. L. Dunn, J. C. Fulfs, K. C. Godowski, A. P. Polson, G. L. Southard, and G. L. Yewey. 1993. Periodontal pocket treatment with subgingival doxycycline from a biodegradable system. *J. Dental. Res.* 72:360.

Polson, A. M., G. Southard, R. Dunn, and A. Polson. 1994. Healing patterns associated with an ATRISORB® barrier in guided tissue regeneration. *Compend. Contin. Educ. Dent.* 14:1162–1172.

Polson, A. M., G. Southard, R. Dunn, A. P. Polson, J. Billen, and L. Laster. 1995a. Initial study of guided tissue regeneration in class III furcation results after use of a biodegradable barrier. *Int. J. Periodont. Rest. Dent.* 15:45–55.

Polson, A. M., G. Southard, R. Dunn, and A. P. Polson, G. Yewey, D. Swanbom, J. Fulfs, and P. Rodgers. 1995b. Periodontal healing after guided tissue regeneration with ATRISORB® barriers in beagle dogs. *Int. J. Periodont. Rest. Dent.* 15:575–589.

Polson, A., G. Southard, R. Dunn, G. Yewey, K. Godowski, A. Polson, J. Fulfs, and L. Laster. 1996. Periodontal pocket treatment in beagle dogs using subgingival doxycycline from a biodegradable system. I. Initial clinical responses. *J. Periodontol.* 67:1176–1184.

Polson, A., S. Garrett, N., Stoller, C. Bandt, P. Haner, W. Killoy, G. Southard, S. Duke, G. Bogle, C. Drisko, and L. Friesen. 1997. Multicenter comparative evaluation of subgingivally delivered sanguinarine and doxycycline in the treatment of periodontitis. II. Clinical results. *J. Periodontol.* 68:119–126.

Radomsky, M. L., G. Brouwer, F. J. Floy, D. L. Loury, F. Chu, A. J. Tipton, and L. M. Sanders. 1993. The controlled release of ganirelix from the ATRIGEL® injectable system. *Proceed. Intern. Symp. Control. Rel. Bioact. Mater.* 20:458–459.

Ron, E. S., L. Bromberg, S. Luczak, M. Kearney, D. R. Deaver, and M. Schiller. 1997. Smart Hydrogel®: A novel mucosal delivery system. *Proc. Intern. Symp. Control. Rel. Bioact. Mater.* 24:407–408.

Salvati, E. A., J. J. Callaghen, B. D. Brause, R. F. Klein, and R. D. Small. 1986. Reimplantation in infection, elution of gentamicin from cement and beads. *Clin. Orthop.* 207:83–88.

Sawhney, A. S., C. P. Pathak, and J. A. Hubbell. 1993. Bioerodible hydrogels based on photopolymerised poly(ethylene glycol)-co-poly(alpha-hydroxy acid) diacrylate macromers. *Macromolecules* 26:581–590.

Sawhney, A. S., M. D. Lyman, F. Yao, M. A. Levine, and P. K. Jarett. 1996. A novel in situ formed hydrogel for use as a surgical sealant or barrier. *Proc. Intern. Symp. Control. Rel. Bioact. Mater.* 23:236–237.

Shah, N. H., A. S. Railbar, F. C. Chen, R. Tarantino, S. Kumar, M. Murgani, D. Palmer, M. H. Infeld, and A. W. Malick. 1993. A biodegradable injectable implant for delivering micro and macromolecules using poly(lactic-co-glycolic) acid (PLGA) copolymers. *J. Control. Rel.* 27:139–147.

Shah, S. S., J. Wertheim, and C. G. Pitt. 1995. Polymer-drug conjugates: Dependence of the rate of hydrolysis on the LCST. *Proc. Intern. Symp. Control. Rel. Bioact. Mater.* 22:34–35.

Sherman, J. M., S. M. Fujita, A. T. Bennett, J. L. Southard, S. L. Whitman, R. L., Dunn, and G. L. Yewey. 1994. Localized delivery of bupivacaine HCl from ATRIGEL® formulations for the management of post-operative pain. *Pharm. Res.* 11:S318.

Shively, M. L., B. A. Coonts, W. D. Renner, J. L. Southard, and A. T. Bennett. 1995. Physicochemical characterization of a polymeric injectable implant delivery system. *J. Control. Rel.* 33:237–243.

Smith, D. A., and A. J. Tipton. 1996. A novel parenteral delivery system. *Pharm. Res.* 13 (9):300.

Sullivan, S. A., R. M. Gilley, J. G. Gibson, and A. J. Tipton. 1997. Delivery of Taxol and other antineoplastics agents from a novel system based on sucrose acetate isobutyrate. *Pharm. Res.* 14 (11):291.

Sullivan, S. A., P. J. Burns, J. W. Gibson, and A. J. Tipton. 1998. Comparison of in vitro and in vivo release of progesterone and estradiol from the SABER® delivery system. *Proc. Intern. Symp. Control. Rel. Bioact. Mater.* 25:918–919.

Tipton, A. J., S. M. Fujita, K. R. Frank, and R. L. Dunn. 1991. A biodegradable injectable delivery system for nonsteroidal anti-inflammatory drugs. *Pharm. Res.* 8:S196.

Tipton, A. J., S. M. Fujita, K. R. Frank, and R. L. Dunn. 1992. Release of naproxen from a biodegradable injectable delivery system. *Proc. Intern. Symp. Control. Rel. Bioact. Mater.* 19:314.

Tipton, A. J., and R. J. Holl. 1998. High viscosity liquid controlled delivery system. U.S. Patent 5,747,058.

Vernon, B., S. W. Kim, and Y. H. Bae. 1996. In vitro insulin release of rat islets entrapped in thermally reversible polymer gel. *Proc. Intern. Symp. Control. Rel. Bioact. Mater.* 23:216–217.

Wan, J., K. Trimble, D. Lidgate, B. Floy-Laidlow, and L. Sanders. 1994. Characterization of factors that control ganirelix release from its liquid crystal gel. *Pharm. Res.* 11 (10):291.

West, J. L., and J. A. Hubbell. 1995. Localized intravascular protein delivery from photopolymerized hydrogels. *Proc. Intern. Symp. Control. Rel. Bioact. Mater.* 22:17–18.

West, J. L., and J. A. Hubbell. 1996. Proteolytically degradable hydrogels. *Proc. Intern. Symp. Control. Rel. Bioact. Mater.* 23:224–225.

Wong, G. K., T. Schultz, and G. V. Buskirk. 1994a. Measurement of surface activities and performulation studies on a gel forming peptide. *Pharm. Res.* 11 (10):226.

Wong, G. K., D. Guazzo, T. Schultz, M. Bornstein, and G. V. Buskirk. 1994b. Formulation and sterilization of a sustained release delivery system for a gel forming peptide. *Pharm. Res.* 11 (10):312.

Wyatt, D. M., and D. Dorschel. 1992. A cubic-phase delivery system composed of glyceryl monooleate and water for sustained release of water-soluble drugs. *Pharm. Technol.* 16:116–130.

Yewey, G. L., E. G. Duysen, S. M. Cox, and R. L. Dunn. 1997. Delivery of proteins from a controlled release injectable implant. In *Protein delivery: Physical systems,* edited by L. M. Sanders and T. Hendren. New York: Plenum Press, pp. 93–117.

Yoshida, R. 1994. Temperature-responsive polymers and drug delivery. In *Advances in polymeric systems for drug delivery,* edited by T. Okaro, N. Yui, M. Yokoyama, and R. Yoshida. Yverdon, Gordon and Breach Science Switzerland.

Yoshida, R., K. Sakai, T. Okano, and Y. Sakurai. 1992. Surface-modulated skin layers of thermal responsive hydrogels as on-off switches. II: Drug permeation. *J. Biomater. Sci. Polym. Edn.* 3:243–249.

Yoshida, R., K. Sakai, T. Okano, Y. Sakurai, Y. H. Bae, and S. W. Kim. 1991. Surface-modulated skin layers of thermal responsive hydrogels as on-off switches. I: Drug release. *J. Biomater. Sci. Polym. Edn.* 3:155–162.

11

Guidelines for Selecting Sustained-Release Technology

Praveen Tyle
Pharmacia & Upjohn
Kalamazoo, Michigan

Ge Chen
Aronex Pharmaceuticals
The Woodlands, Texas

Leo Pavliv
Cato Research Ltd.
Durham, North Carolina

There are a number of reasons for the immense scientific and commercial interest in developing new delivery systems for sustained-release injectable products:

- To extend the product life cycle of successful, first generation, immediate-release drugs by developing value-added sustained-release products.

- New systems are required to effectively deliver novel, genetically engineered pharmaceuticals, i.e., peptides, proteins, and plasmid DNA (deoxyribonucleic acid), to their sites of action without incurring significant immunogenicity or biological inactivation. This will be essential as gene therapy becomes a reality.

- The treatment of certain disease states, such as enzyme deficiencies and cancer, can be significantly improved by drug targeting and sustained-release delivery.

- The therapeutic efficacy and safety of drugs administered by conventional methods can be improved by more precise spatial and temporal placement within the body by utilizing sustained-release injectable technology, thereby reducing both the size and number of doses.

Two prerequisites come to mind when conceptualizing the ideal sustained-release drug delivery system. The first prerequisite is that it should deliver drug at a rate dictated by the needs of the body for the intended application over the required period of treatment. This may necessitate delivery at a constant rate for drugs that have a clear relationship between steady state plasma levels and the resultant therapeutic response. For other drugs, a pulsatile delivery may be optimal, resulting in a series of peaks and troughs that could maximize the therapeutic effect while minimizing toxicity. Extending this concept further, pulsatile delivery can be synchronized to follow the body's natural circadian rhythms, i.e., chronotherapy. Therapeutic areas that can benefit from chronotherapy include inflammatory, respiratory, oncological, cardiovascular, and gastrointestinal diseases (Mangione 1997). The second prerequisite is that it should deliver the active moiety primarily to the site of action, eliminating or minimizing adverse reactions. Conventional injectable formulations are unable to control either the rate of release or the delivery to the site of action. Advanced sustained-release delivery systems, on the other hand, can overcome these problems. They can be designed to deliver a drug more selectively to a specific site of action. This results in a product that is easy to administer, requires less frequent dosing, and exhibits decreased variability in systemic drug concentrations.

Drug delivery strategies can be categorized as physical, chemical, biological, or mechanical (Robinson and Mauger 1991; Weiner 1989). The physical approach relies on specific physical properties of both the selected excipients and the drug (e.g., solubility, permeability, and viscosity) to control the location, duration, and rate of drug release. Biological approaches link drug delivery with biological characteristics, such as the specific affinity of a receptor site or other biological target for a biological carrier. Examples of biological approaches include the use of monoclonal antibodies and lipids. For example, liposomes were selected as a delivery system for amphotericin B (AMB) on the premise that their small size, surface charge, and lipophilic nature would alter the kinetics, tissue distribution, and therapeutic to toxic ratio of the drug. This premise was substantiated when it was demonstrated that AMB selectively partitions from liposomes into fungus rather than mammalian cells, which increases the therapeutic activity while decreasing the toxicity associated with the free drug (Juliano et al. 1990).

This chapter describes the factors influencing the selection of sustained-release technology for injectable products. Decision factors such as physicochemical properties of the drug, delivery issues, biological considerations, and formulation factors are discussed in detail, and several case studies are presented.

CHARACTERISTICS OF AN IDEAL VEHICLE

The ideal vehicle for a sustained-release injectable would be biocompatible, easy to administer, and not require removal. It would be capable of high drug loading with a variety of active agents. The vehicle would deliver drugs to the site of action at the desired rate.

UNDESIRABLE DRUG CHARACTERISTICS FOR SUSTAINED–RELEASE DELIVERY

Undesirable drug characteristics for sustained release delivery are essentially the same as conventional delivery. They include narrow therapeutic index, insolubility, high dose, and either a very short half-life or a very long half-life. These factors are discussed below.

DECISION FACTORS

Physicochemical Properties of Drugs

Drug release from the dosage form will depend largely on the physicochemical properties of the drug and must be appropriate for the therapeutic indications supported by the product label. These physicochemical properties can at times prohibit or restrict the incorporation of the drug into a specific sustained-release dosage form. Also, the physicochemical properties of the drug, e.g., solubility, can dramatically influence the biological properties of a drug, such as pharmacokinetics or toxicity. Thus, both drug properties and routes of administration will influence the design of controlled-and sustained-release systems (Li and Robinson 1989). This discussion focuses on the physicochemical properties of drugs and their impact on the absorption, distribution, metabolism, and excretion (ADME) characteristics of a drug when selecting a specific delivery system for a drug.

Aqueous Solubility

Typically, drugs with a very low solubility in aqueous solutions or in pharmaceutically acceptable lipophilic excipients are poor candidates for injectable delivery. However, utilizing micronized drug and encapsulation and stabilization techniques has expanded the number and types of drugs that may be delivered as sustained-release injectables.

The dissolution rate of a drug must be considered in selecting polymer coatings for a sustained-release drug delivery system. High molecular weight drugs may have reasonably good aqueous solubility but very slow dissolution rates. The solubility of drugs in various carriers, such as liposomes, erythrocytes, and other microparticles, will also affect the loading efficiency of the drugs.

Half-Life

The biological half-life, hence the duration of action of a drug, is very important in selecting a drug for sustained-release delivery. Factors influencing the biological half-life of a drug include its elimination, metabolism, and distribution pattern. Drugs with short half-lives require frequent dosing in order to minimize fluctuations in blood levels (Dittert 1974; Niebertgall et al. 1974). Hence, sustained-release dosage forms would appear very desirable for such drugs. A generalized lower limit for the biological half-life of sustained-release products cannot be defined because the applications are drug specific (Li and Robinson 1989).

Particle Size

For intravenous parenteral sustained-release dosage forms, a very small particle size is important to avoid blocking small capillaries. Particle size is also an important factor for nonintravenous parenteral sustained-release dosage forms. Particle size affects the dissolution rate and, hence, the rate of release. For example, a controlled particle size distribution of microcrystalline medroxyprogesterone acetate delivers an effective contraceptive dose for three months (Kaunitz 1994). Also, the release rate of levonorgestrel microencapsulated in a biodegradable polymer [poly(lactide-co-glycolide), PLGA] is affected by the particle size of the microspheres (Beck et al. 1985), which must be compatible with the desired size of the syringe needle.

Biological Factors Influencing the Selection of Sustained-Release Systems

The design of a sustained-release product should be based on a comprehensive picture of drug disposition. This entails a thorough examination of the ADME characteristics of a drug following appropriate dosing. The following factors will affect the dose and rate of drug release, hence product formulation.

Duration of Action

The amount of drug required for a particular therapeutic indication in a given delivery system is determined by the drug potency, the release characteristics of the delivery system, and the amount of any rate-limiting excipient required. Optimizing these factors results in maximum duration of drug release for that particular dosage form. The more potent a drug and the smaller the amount of any rate-limiting excipient required, the smaller the total mass needed per day and the longer the duration of drug release by the formulation. Usually, the potency of the drug and the physicochemical limitations of the sustained-release technology limit the desired duration of action. The duration of action can be selected to fit in with a patient's visits for a particular treatment (Tice and Tabibi 1992).

Distribution

The distribution of drugs into tissues can be important in the overall drug elimination kinetics, since it not only lowers the concentration of circulating drug but it can also be rate limiting in its equilibrium with blood and extracellular fluid. In general, the bound portion of a drug can be considered inactive and unable to cross membranes. Drugs that have a high degree of binding also typically exhibit a longer half-life.

The apparent volume of distribution of a drug is frequently used to describe the magnitude of distribution. Conceptually, this pharmacokinetic parameter can be viewed as a proportionality constant relating plasma or serum concentration of drug to the total amount of drug in the body. Since rate processes are driven by concentration, it is this quantity in which we are interested. When designing a sustained-release product, one would like to have as much information on drug disposition as possible. However, in reality, decisions are usually based on only a few pharmacokinetic parameters, one of which is the apparent volume of distribution. The apparent volume of distribution influences the concentration and amount of drug circulating in the blood or target tissues.

Metabolism

The metabolism of a drug can either inactivate an active drug or convert an inactive drug to an active metabolite, as in the case of a prodrug. Metabolic alteration of a drug can occur in a variety of tissues, some of which are richer in enzymes than others. For example, the organ most responsible for metabolism is the liver, thus, the greatest metabolic conversion occurs after a drug has been absorbed into the general circulation. Clearly, for optimal bioavailability, the route of drug administration may be dictated by the drug's metabolic pattern.

Delivery Issues

The external environment into which the drug is released affects the rate of release of the active agent, although to a lesser degree than conventional formulations. The release profile is predominantly controlled by the design of the system and may provide a variety of profiles. For pulsatile delivery, the release profile occurs in multiple, discrete, and controlled pulses of time following a single injection and releases drug in a temporal manner.

Pulsatile release provides advantages in vaccine delivery. A pulsatile release of Staphylococcal enterotoxin B (SEB) using PLGA polymers was achieved. This system was designed to provide distinct "pulses" of antigen release after injection of a mixture of vaccine-containing microspheres of various sizes and degradation times, providing discrete primary and booster doses following a single injection. A mixture of two microsphere size distributions, 1–10 μm and 20–125 μm, containing SEB encapsulated in a 50:50 copolymer of PLGA induced both a primary and a secondary anti-SEB response following single-dose administration (Lewis 1990).

FORMULATION CONSIDERATIONS

Liposomes

Liposomes have been utilized to deliver a wide variety of drugs (Table 11.1). The criteria for selecting a liposome composition are high drug encapsulation efficiency, preservation of the full biological activity of the drug after liposome encapsulation, stability in the dosage form and in plasma, a favorable tissue distribution pattern, decreased toxicity, and an increased therapeutic index.

Table 11.1. *Liposome Products on the Market and in Clinical Trials*

Product	Developer	Marketer	Status	Route of Administration	Indication
Abelcet® (Amphotericin B lipid complex)	The Liposome Company (TLC)	The Liposome Company (TLC)	Approved	Intravenous	Invasive fungal infection
AmBisome® (Amphotericin B)	NeXstar Pharmaceuticals	NeXstar Pharmaceuticals	Approved	Intravenous	Empirical therapy for presumed fungal infection
Doxil® (Doxorubicin)	Sequus (as Doxil®)	Sequus (as Doxil® in U.S.) Schering-Plough (as Caelyx® outside U.S.)	Approved	Intravenous	Ovarian cancer
Amphotericin B (cholesterol sulfate complex)	Sequus	Sequus (as Amphotec® in U.S. Various as Amphocil®)	Approved	Intravenous	Invasive aspergillosis
DaunoXome® (Daunorubicin citrate)	NeXstar Pharmaceuticals	NeXstar Pharmaceuticals	Approved	Intravenous	First-line cytotoxic therapy for advanced HIV–associated Kaposi's sarcoma
Nyotran® (Nystatin)	Aronex Pharmaceuticals	Grupo Ferrer Internacional SA in Spain and Portugal	Phase III	Intravenous	Empirical therapy of presumed fungal infections

Table 11.1 continued on next page.

Table 11.1 continued from previous page.

Product	Developer	Marketer	Status	Route of Administration	Indication
Atragen® (All-trans retinoic acid)	Aronex Pharmaceuticals/ Genzyme Corp.	Genzyme Corp.	Phase III	Intravenous	Promyelocytic leukemia
Mikasome® (liposomal amikacin)	NeXstar Pharmaceuticals	NeXstar Pharmaceuticals	Phase II	Intravenous	Bacterial infection
SPI-77® (Cis-platin PEGylated liposome)	Sequus	Sequus	Phase I	Intravenous	Advanced cancer
DepoCyt®	SkyePharma plc/ Chiron Corporation	Chiron Corporation	Phase IV	Intracerebral spinal fluid	Neoplastic Meningitics
DepoMorphine™	SkyePharma plc	SkyePharma plc	Phase II	Epidural	Postoperative pain management
DepoAmitracin™	SkyePharma plc	SkyePharma plc	Phase II	Subcutaneous/ Intralesionial	Local infections
DepoBupivacaine™	SkyePharma plc	SkyePharma plc	Phase I	Local Subcutaneous	Pain management

Lipid Selection

The first step for the preparation of well-defined liposomes is to use well-characterized lipids. The selection of the bilayer components is an important issue both for toxicity reasons and for shelf-life optimization. Commercial sources of lipids vary widely in purity depending on the manufacturer and the manufacturing process. These differences can even vary from lot to lot (Ponpipom and Bugianesi 1980). Thus, lipids should be examined before use by either thin layer chromatography (TLC) or high performance liquid chromatography (HPLC) in at least two solvent systems to confirm purity. The lipid component of these products must meet stringent pharmaceutical requirements in order to obtain regulatory approval for large-scale human testing and marketing. These include suitable purity, safety, and microbial/endotoxin limits, as well as adequate stability (Amselem et al. 1993).

Although many lipid compositions can be employed for liposomal delivery systems, from a pharmaceutical point of view, stability and cost are the important factors. Thus, acidic lipids such as phosphatidylserine (PS), cardiolipin, and phosphatidic acid (PA) are undesirable due to high costs and the labile nature of these compounds. Similarly, very unsaturated lipids, such as soy phosphatidylcholine (PC) or naturally occurring PS, phosphatidylethanolamine (PE), and cardiolipin, should be avoided because of the considerable oxidation problems encountered. Therefore, given similar loading and retention characteristics, liposomal formulations composed of egg PC or hydrogenated varieties of egg PC are more desirable for the pharmaceutical product.

Selection of Manufacturing Methods

There is now a wide range of techniques for generating multilamellar and unilamellar liposomes. From a pharmaceutical point of view, optimal liposome products should avoid or minimize the use of organic solvents and detergents, exhibit high encapsulation efficiency, yield well-defined and reproducible liposomes, and be amenable to scale-up. The three most important factors are encapsulation efficiency, drug retention properties, and the drug/lipid ratios. In addition to lipid selection issues, the long-term stability of the liposomal formulation should be addressed. The formulator should be aware that the stability of conventional liposome-drug formulations achieved by standard laboratory methods are unlikely to meet pharmaceutical demands for a product shelf life of two or more years at room temperature. Thus, a suitable storage temperature needs to be defined in each case (Betageri et al. 1993).

Drug Retention

Assuming that drug retention is the rate-limiting step, most drug-liposome formulations do not exhibit sufficiently low leakage rates to allow a shelf life of two or more years. However, if the encapsulation efficiencies remain sufficiently high (e.g., 90 percent or more), nonentrapped drug need not be removed. In this case, no net leakage of drug would occur on extended storage due to the absence of a transmembrane drug concentration gradient.

Drug/Lipid Ratio

The optimum drug/lipid ratio of a liposomal formulation will likely be dictated by the biological efficacy and toxicity of the preparation. For a pharmaceutical product, high drug/lipid ratios are the most economical but not always possible due to formulation or stability factors.

In summary, optimal liposome pharmaceuticals will exhibit drug encapsulation efficiencies in excess of 90 percent, employ inexpensive and relatively saturated lipids, such as egg PC and cholesterol, and provide the highest possible drug/lipid ratio, while maintaining efficacy of the product.

Polymeric Delivery Systems

Polymeric delivery systems have been used extensively in cancer chemotherapy, contraception, and protein/peptide vaccine delivery. Some clinical reports of polymeric chemotherapy using microparticles are listed in Table 11.2. There are several major types of polymers used for sustained-release systems. These systems require no subsequent surgical removal of the product once the therapeutic agent is depleted. The selection of the matrix for a particular therapeutic application must take into account such factors as drug potency, physicochemical and biological properties of the active agent, local injection site environment, requirements for degradation time, the drug release rate, and cost.

Polymeric delivery systems provide four unique advantages:

1. Polymers have diverse physicochemical properties that facilitate the design of systems for specific drugs and drug delivery patterns.

2. Polymers can be readily incorporated or manufactured into a variety of useful systems.

3. The number of available biocompatible and bioerodible polymers is increasing.

4. A mechanism for drug release kinetics can be determined and predicted, at least qualitatively, for particular drug polymer systems.

Polymer Type

The type of polymer used in the formulation can significantly affect the release rates of certain proteins and peptides (Figure 11.1). For example, the in vitro release profile of zona pellucida protein antigen is dependent on the ratio of the more hydrophobic poly (lactic acid) (PLA) polymers to PLGA polymers in the delivery system. In addition to hydrophobicity, other polymer properties important in drug delivery include reactivity to water, which influences swelling, solubility, permeability, and viscosity; mechanical and physical properties, such as plasticity, elasticity, viscosity, gel strength, and density; chemical structure, which determines the ability to react, complex, and ionize; and biological properties, including the extent to which they are biodegradable, bioerodible, bioadhesive, and biocompatible. For details of specific polymer types, see Chapter 8.

Table 11.2. *Clinical Reports of Polymeric Chemotherapy Using Microparticles*

System Used	Function of Device	Drug Used/ Encapsulated	Indication	Number of Patients	Route of Administration	Reference
Starch microspheres	Chemotherapy	Mitomycin C	Incurable metastatic liver cancer	24	Intravenous	Wollner et al. (1986)
Starch microspheres	Chemotherapy	5-Fluorouracil	Gastric cancer	18	Intravenous	Kitamura et al. (1989)
Starch microspheres	Chemotherapy	5-Fluorouridine	Hepatic metastasis of colorectal carcinoma	14	Intravenous	Thom et al. (1989)
PLGA microspheres	Hormonal therapy	Goserelin acetate	Prostate cancer	38	Intravenous	Dijkman et al. (1990)
PLGA microspheres	Hormonal therapy	Leuprolide acetate	Prostate cancer	53	Intravenous	Sharifi and Soloway (1990)
Albumin microspheres	Embolization	None	Colorectal liver cancer	7	Intravenous	Goldberg et al. (1991)
Gelatin microspheres	Embolization	None	Inoperable hepatic metastasis	9	Intravenous	Chen et al. (1992)
Starch microspheres	Chemo-occlusion	Mitomycin C	Irressectable liver metastasis	39	Intravenous	Civalleri et al. (1994)

Table 11.2 continued on next page.

Table 11.2 continued from previous page.

System Used	Function of Device	Drug Used/ Encapsulated	Indication	Number of Patients	Route of Administation	Reference
Vinyl polymer	Interstitial chemotherapy	Leuprolide acetate	Prostate cancer	5	Intravenous	Imai et al. (1986)
PMMA	Interstitial chemotherapy	Mitomycin, Adriamycin®, ACNU, 5-fluorouracil	Malignant brain tumors	55	Intravenous	Kubo et al. (1986)
Polymer needles	Interstitial chemotherapy	Mitomycin C	Nonresectable cancers of pancreas, biliary duct,etc.	220	Intravenous	Hanyu et al. (1988)
PCPP-SA	Interstitial chemotherapy	Carmustine	Recurrent malignant glioma	21	Intravenous	Brem et al. (1991)
PCPP-SA	Interstitial chemotherapy	Carmustine	Recurrent malignant glioma	222	Intravenous	Brem et al. (1995)
PLA	Interstitial chemotherapy	Nimustine chloride	Nonresectable or recurent glioma	11	Intravenous	Kuroda et al. (1994)
Albumin microspheres	Chemotherapy	Adriamycin®	Breast cancer	1	Intravenous	Doughty et al. (1994)

PMMA = Polymethyl methacrylate

PCPP-SA = Poly(carboxyphenoxy propane-co-sebacic acid)

ACNU = 1-(4 amino-2 methyl-5-pyrimidinyl)-methyl-3-(2-chloroethyl)-3-nitrosourea hydrochloride

Figure 11.1. *Effect of polymer type on the release of a zona pellucida protein antigen from ATRIGEL® formulations.*

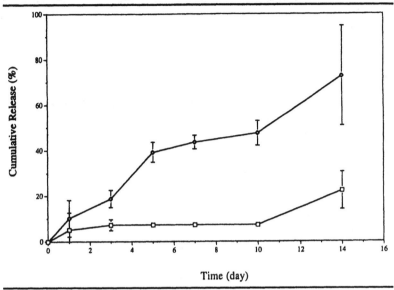

Time (day)

□, 45 percent PLA (inherent viscosity, 0.2); ◊, 45 percent PLGA (inherent viscosity, 0.2). Formulation solvent was N-methyl-2-pyrrolidone (NMP), and the protein antigen load was 0.2 percent. Cumulative release profiles were generated in phosphate-buffered saline (PBS), pH 7.4. Reproduced from Yewey et al. (1997) with permission.

Polymer Concentration

In general, as the concentration of polymer increases in a formulation, the release of the protein is retarded. As shown in Figure 11.2, as the concentration of PLA in the formulation increased from 5 percent to 30 percent, the cumulative release of follicle-stimulating hormone (FSH) decreased accordingly (Yewey et al. 1997).

Polymer Molecular Weight

The molecular weight of the different polymers used in the formulation also plays a role in the release kinetics of proteins. As polymer chain lengths become longer and more entangled, a

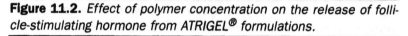

Figure 11.2. *Effect of polymer concentration on the release of follicle-stimulating hormone from ATRIGEL® formulations.*

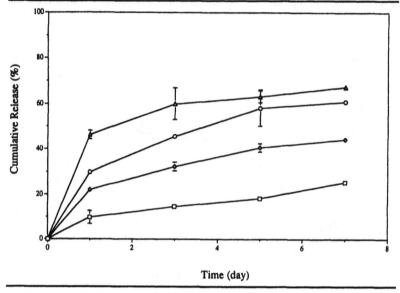

△, 5 percent PLA; ○, 10 percent PLA; ◊, 20 percent PLA; □, 30 percent PLA. Inherent viscosity of the PLA was 0.75, and the solvent used was NMP. Protein load was 1 percent. Cumulative release profiles were generated in PBS, pH 7.4. Reproduced from Yewey et al. (1997) with permission.

characteristics of high molecular weight polymers, proteins are hindered in their ability to be released from the matrix. To illustrate this, two PLA formulations of equal polymer concentration but different molecular weights were mixed with equivalent loads of myoglobin. Figure 11.3 depicts the cumulative release profiles of the two PLA formulations, one of low molecular weight and the other of medium molecular weight. With time, the lower molecular weight polymer released roughly 10 percent more myoglobin than did its high molecular weight counterpart (Yewey et al. 1997).

Figure 11.3. *Effect of polymer molecular weight on the release of myoglobin from ATRIGEL® formulations.*

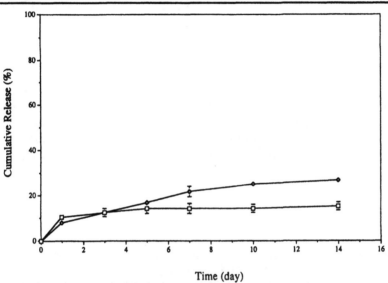

◊, inherent viscosity 0.05; □ inherent viscosity 0.33. The polymer used was 45 percent PLA. Formulation solvent was NMP, and the protein load was 10 percent. Cumulative release profiles were generated in PBS, pH 7.4. Reproduced from Yewey et al. (1997) with permission.

Solvent

Numerous solvents are used with various polymeric delivery systems, including acetaldehyde dimethyl acetal, acetone, dichloromethane, dimethylsulfoxide (DMSO), dioxane, ethyl acetate, ethyl vinyl ether, N-methyl-2-pyrrolidone (NMP), nitromethane, tetrahydrofuran, 1,1,1-trichloroethane, and 1,1,2-trichloroethylene. These solvents differ greatly in their physicochemical characteristics, such as boiling point, vapor pressure, miscibility, interfacial tension, and solubility (Gander et al. 1995). They can also have different effects on the release of certain peptides and proteins. For example, polymers dissolved in NMP often have different coagulation rates than polymers dissolved in DMSO. Also, proteins dissolved in the two solvents behave differently owing to solution or aggregation effects. Figure 11.4 shows the release kinetics of bovine growth hormone-releasing factor from two PLA formulations prepared with NMP or DMSO as the solvent (Yewey et al. 1997).

Protein Load

In protein delivery, it is also possible to control the release of proteins by varying the protein load within the formulation. Figure 11.5 shows that as the protein load is increased, a smaller percentage of the total protein in the formulation is released. This effect may be due to increasing protein-protein interactions

Figure 11.4. *Effects of formulation solvent on the release of a peptide hormone from ATRIGEL® formulations.*

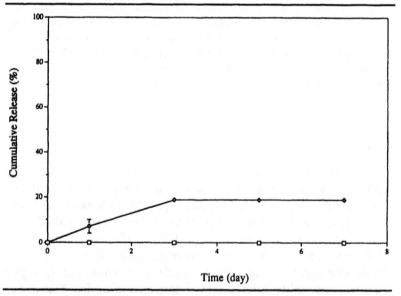

◊, 59 percent DMSO; □, 59 percent NMP. The polymer used was 40 percent PLA (inherent viscosity, 0.22), and the peptide load was 1 percent. Cumulative release profiles were generated in PBS pH 7.4. Reproduced from Yewey et al. (1997) with permission.

Figure 11.5. *Effect of protein load on the release of bovine serum albumin from ATRIGEL® formulations.*

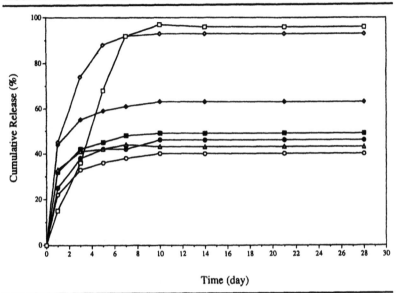

□, 0.01 percent bovine serum albumin (BSA); ◊, 0.1 percent BSA; ○, 1.0 percent BSA; △ 2.5% BAS; ■, 5 percent BSA; ♦, 10 percent BSA; ● 20 percent BSA. Formulations consisted of PLA (inherent viscosity, 0.05), with polyvinyl-pyrrolidone and calcium phosphate as additives. NMP was used as the solvent. Cumulative release profile were generated in PBS, pH 7.4. Reproduced from Yewey et al. (1997) with permission.

within the polymer as the load is increased. However, the contrary has also been shown due to an increased amount of protein released as an initial burst because of its higher loading.

Additives

Sometimes, it is necessary to incorporate an additive such as a surfactant into the formulation in order to prevent aggregation of the protein. Because proteins used at relatively high loads in a polymer formulation are prone to giving a poor overall release, surfactants such as sodium dodecyl sulfate (SDS) are added to formulations. Figure 11.6 shows the release profiles of two PLA formulations

Figure 11.6. *Effect of an additive on the release of ovalbumin from ATRIGEL® formulations.*

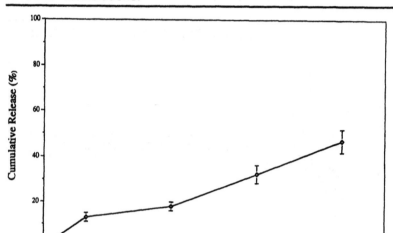

Time (day)

☐, 40 percent PLA (inherent viscosity, 0.36); ◊, 40 percent PLA (inherent viscosity, 0.36) with 5 percent SDS. The formulation solvent was NMP, and the protein load was 2.5 percent. Cumulative release profiles were generated in PBS, pH 7.4. Reproduced from Yewey et al. (1997) with permission.

containing ovalbumin, one with 5 percent SDS incorporated and one without. Approximately 47 percent of the protein load was released from the formulation with SDS, whereas only about 2 percent of the protein was released from the formulation without SDS.

Biocompatibility

The safety and biocompatibility of various drug delivery systems and their components have been extensively tested. PLA is nontoxic, and the hazard potential of NMP at typical use levels is insignificant. Additional preclinical tests to evaluate the tissue irritation potential, implantation effects, and biodegradation have been completed for formulations prepared with PLA, PLGA, and poly(lactide-co-caprolactone) (PLC) polymers dissolved in NMP or DMSO. There were no significant differences between formulations of the different polymer types prepared with NMP or DMSO. Further studies have demonstrated that the polymer formulations can decrease the local tissue response to irritating drugs (Lewis 1990).

CASE STUDIES

Sustained-release technology has been widely used in the delivery of a wide range of pharmaceuticals (Table 11.3). The following case histories illustrate some interesting pharmaceutical applications for sustained-release injectable products.

An Emulsion Formulation of Growth Hormone

A parenteral, biologically active, sustained-release multiple water-in-oil-in water emulsion was developed to increase and maintain increased levels of growth hormones for extended periods of time (Tyle 1989). The therapeutic indication was to increase weight gain, growth rate, milk production, and muscle size; improve feed efficiency; decrease body fat; and increase the lean meat to fat ratio in cattle. The emulsion consists of an internal aqueous phase (W1) of water, buffer, growth hormone, growth factor, and somatomedin or a biologically active fragment or derivatives. The oil phase is comprised of a pharmaceutically acceptable oil (such as light mineral oil or cottonseed oil) or other water immiscible liquid and a thickening or gelling agent consisting of a mixture of nonionic surfactants. The external aqueous phase W2 is comprised of water, buffer, nonionic surfactant, and thickening or gelling agent. The emulsion can be prepared by either a syringe technique or homogenization. The efficacy of the injectable composition is demonstrated by utilizing a hypophysectomized (hypox) rat assay. The hypox rat does not produce its own growth hormone and is sensitive to injected bovine growth hormone. The response measure is growth over a

Table 11.3. *Sustained-Release Injectable Products on the Market and in Clinical Trials*

Product	Developer	Marketer	Administration	Status	Route of Administration	Indication
Caelyx® (Doxorubicin)	Sequus (as Doxil®)	Sequus (as Doxil® in U.S.) Schering-Plough (as Caelyx® outside U.S.)	Liposome	Approved	Intravenous	Ovarian cancer
DepoCyt®	SkyePharma plc	SkyePharma plc/ Chiron, Pharmacia & Upjohn outside U.S.	Injectable, sustained release	Awaiting approval	Intravenous	Neoplastic meningitis
Accusite® (Fluorouracil and epinephrine)	Matrix Pharmaceuticals, Inc.	Matrix Pharmaceuticals, Inc.	Sustained-release injectable gel	Phase III	Intravenous	Basal cell cancer
Aerosomes™	ImaRx Pharmaceutical Corp.	NR	NR	Phase III	Intravenous	Ultrasound contrast agent
Nyotran® (Nystatin)	Aronex Pharmaceuticals	Grupo Ferrer Internacional SA in Spain and Portugal	Liposome	Phase III	Intravenous	Empirical therapy of presumed fungal infections

Table 11.3 continued on next page.

Table 11.3 continued from previous page.

Product	Developer	Marketer	Administration	Status	Route of Administration	Indication
TLC D-99® (Doxorubicin)	The Liposome Co., Inc., and Pfizer, Inc.	Pfizer Inc.	Liposome	Phase III in Europe and U.S.	Intravenous	Metastatic breast cancer
Prolease™ hgh	Alkermes/ Genentech	Genentech	Injectable	Phase III	Intravenous	Growth hormone deficiency
Atragen® (All-trans retinoic acid)	Aronex Pharmaceuticals, Inc., and Genzyme Corp.	Genzyme Corp.	Liposome	Phase III	Intravenous	Acute promyelocytic leukemia
Allovectin 7®	Vical, Inc.	Vical, Inc.	Cationic lipid formulation	Phase II	Intravenous	Metastatic melanoma, head & neck cancer
Leuvectin®	Vical, Inc.	Vical, Inc.	Lipid–DNA complex	Phase I/II	Intravenous	Melanoma, kidney cancer, sarcoma, and prostate cancer
PowderJect® alprostadil	PowderJect Pharmaceuticals, plc	PowderJect Pharmaceuticals, plc	Injectable	Phase I	Intravenous	Male erectile dysfunction

period of time. The results show that those hypox rats gained an average of 9.98 g after administration of 2,400 mg doses of bovine growth hormone in 0.2 mL of w/o/w multiple emulsion. Among all the compositions, one achieved an average weight gain of 15.2 g/animal within 10 days.

AmBisome®

AmBisome® is a liposomal formulation of AMB. Conventional AMB has been generally considered the drug of choice for many types of systemic fungal infections. These infections are a major threat to those whose immune systems are compromised, such as patients undergoing chemotherapy for cancer, bone marrow transplant recipients, and AIDS, (acquired immunodeficiency syndrome) patients. However, AMB is very toxic, especially to the kidneys. For these patients at a high rate of morbidity and mortality, there is a need for a delivery system to overcome the toxicity problem.

AmBisome® for injection is a sterile, nonpyrogenic lyophilized product for intravenous injection. AmBisome® is a true, single bilayer, liposomal drug delivery system consisting of hydrogenated soy PC, cholesterol, NF; distearoylphosphatidylglycerol, alpha tocopherol, USP; sucrose, NF; and disodium succinate hexahydrate as buffer. Following reconstitution with Sterile Water for Injection, USP, the resulting pH of the suspension is between 5.0 and 6.0.

In this formulation, hydrogenated soy PC, a relatively saturated phospholipid, is selected to form a rigid phospholipid bilayer. The use of distearoylphosphatidylglycerol is to incorporate negative charge into the phospholipid vesicle without losing its bilayer rigidity. The rigid phospholipid bilayer enables AMB to remain encapsulated in the liposomal vesicles in the blood circulation before reaching the targeting site. In addition, cholesterol is also incorporated to enhance the stability of liposome vesicles in plasma. This formulation exhibits a low lipid to drug ratio of 3:1, making the product more economical.

To enhance the physical stability of the liposomal formulation, an antioxidant, alpha tocopherol, is used to prevent any oxidation of the phospholipids. The pH of the formulation, between 5.0 and 6.0, is assumed to be the pH at which both phospholipids and AMB exhibit maximal stability. In this formulation, the pH is controlled by disodium succinate hexahydrate. The final formulation is a lyophilized powder, requiring a lyophilization cycle. Sucrose is the cryoprotectant in the formulation to ensure that the liposome vesicles are intact throughout the lyophilization cycle.

AmBisome® has shown equivalent therapeutic success rates to AMB deoxycholate in clinical studies. The liposomal formulation is stable when stored at 2–8°C.

Lupron Depot®

Lupron Depot® contains leuprolide acetate, a synthetic nonapeptide analog of naturally occurring gonadotropin-releasing hormone (GnRH). It is the first superactive agonistic analog of luteinizing hormone releasing hormone (LHRH). This analog has almost 100 times the biological activity of LHRH. However, chronic treatment with LHRH analogs has the practical disadvantage of requiring long-term daily injections, as these analogs are water-soluble nona- or decapeptides with a short biological half-life and are poorly absorbed from the gastrointestinal tract. To overcome these problems, a sustained-release delivery system is warranted.

The sustained-release dosage form of leuprolide acetate is available in several dosages: Lupron Depot® 3.75 mg and 7.5 mg are intended as a monthly intramuscular injection. Lupron Depot®-3 month 11.25 mg is to be given once every 3 months. Lupron Depot® 3.75 mg and 7.5 mg are available in a single and in a prefilled dual-chamber syringe. Both contain sterile lyophilized microspheres, which when mixed with diluent become a suspension.

The single-dose vial of Lupron Depot® contains leuprolide acetate, purified gelatin, PLGA copolymers, and D-mannitol. The accompanying diluent consists of carboxymethylcellulose sodium, D-mannitol, Water for Injection, and glacial acetic acid to control pH. In this formulation, PLGA is used as the wall material for polycore microspheres of the peptide. These polymers have already been utilized as biodegradable surgical sutures and are known to be biocompatible. The peptide in the formulation is released in two phases, an initial diffusional release from near the surface of the swollen microsphere and elution with the erosion of the polymer. High trap efficiency is achieved by adding gelatin to increase the viscosity (Okada and Toguchi 1995). Mannitol and carboxymethyl cellulose are also key compounds to help disperse the semidry microspheres before the lyophilization process. These excipients prevent aggregation of the microspheres during the drying process and also during distribution and storage of the commercial product.

The Lupron Depot® formulation can provide a fairly constant release of the peptide over 1-month or 3-month periods in animals and humans after intramuscular injection. Therefore, this formulation shows reliable efficacy for the treatment of patients with advanced prostate cancer, endometriosis, precocious puberty, and other hormone-dependent diseases through persistent chemical castration without any need for repeated daily injection; thus, it has markedly improved patient compliance. The effective dose of the peptide using the depot formulation is reduced to one-quarter to one-eighth of that for repeated parenteral administration of the solution due to the promotion of down-regulation with continuous hits on the receptors by the sustained peptide concentrations in the target organ.

CONCLUSION

There is great interest in developing sustained-release parenteral products for a variety of therapeutic and economic reasons. Decision factors in selecting delivery systems include the physicochemical properties of the drug, biological factors effecting its metabolism, and specific delivery issues. A number of different sustained-release delivery technologies are available, including liposomes, polymers, and emulsions. Several highly successful products have been developed and are being marketed throughout the world.

REFERENCES

Amselem, S., A. Gabizon, and Y. Barenholz. 1993. A large-scale method for the preparation of sterile and nonpyrogenic liposomal formulations of defined size distributions for clinical use. In *Liposome technology,* edited by G. Gregoriadis. Boca Raton, Fla., USA: CRC Press, pp. 501–626.

Beck, L. R., V. Z. Pope, T. R. Tice, and R. M. Gilley. 1985. Long-acting injectable microsphere formulation for the parenteral administration of levonorgestrel. *Adv. Contracept.* 1 (2):119–129.

Betageri, G., S. A. Jenkins, D. L. Parsons. 1993. *Liposome drug delivery systems.* Lancaster, Penn, USA: Technomic Publishing Co., Inc., pp. 19–20.

Brem, H., S. M. Mahaley Jr., N. A. Vick, K. L. Black, S. C. Schold Jr., P. C. Burger, A. H. Friedman, I. S. Ciric, T. W. Eller, J. W. Cozzens, and J. N. Kenealy. 1991. Interstitial chemotherapy with drug polymer implants for the treatment of recurrent gliomas. *J. Neurosurg.* 74:441–446.

Brem, H., S. Piantadosi, P. C. Burger, M. Walker, R. Selker, N. A. Vick, K. Black, M. Sisti, G. Brem, G. Mohr, P. Muller, R. Morawetz, and S. C. Schold. 1995. Placebo-controlled trials of safety and efficacy of intraoperative controlled delivery by biodegradable polymers of chemotherapy for recurrent gliomas. *Lancet* 345:1008–1012.

Chen, X., Y. Wu, D. Zhong, L. Li, T. Tan, X. Xie, C. Yan, and X. Li. 1992. Hepatic carcinoma treated by hepatic arterial embolization using 131I and chemotherapeutic agent gelatin microspheres report of 9 cases [Chinese]. *J. West China Univ. Med. Sci.* 23 (4):420–423.

Civalleri, D., J. C. Pector, L. Hakansson, J. P. Arnaud, M. Duez, and M. Buyse. 1994. Treatment of patients with irresectable liver metastates from colorectal cancer by chemo-occlusion with degradable starch microspheres. *Brit. J. Surg.* 8 (9):1338–1341.

Dijkman, G. A., P. F. del Moral, J. W. Plasman, J. J. Kums, K. P. Delaere, F. J. Debruyne, F. J. Hutchinson, and B. J. Furr. 1990. A new extra long acting depot preparation of the LHRH analogue Zoladex. First endocrinological and pharmacokinetic data in patients with advanced prostate cancer. *J. Ster. Biochem. & Molec. Biol.* 37 (6):933–936.

Dittert, L. W. 1974. Pharmacokinetic considerations in clinical drug trials. *Drug Intell. Clin. Phar.* 8:222.

Doughty, J. C., J. H. Anderson, N. Willmott, and C. S. Asardle. 1994. Intra-arterial administration of adriamycin-loaded albumin microspheres for locally advanced breast cancer. *Postgrad. Med. J.* 71:47–49.

Gander, B. E. Wehrli, R. Alder, and H. P. Merkle. 1995. Quality improvement of spray-dried, protein-loaded D,L-PLA microspheres by appropriate polymer solvent selection. *J. Microencapsul.* 12 (1):83–97.

Goldberg, J. A., N. S. Willmott, J. H. Anderson, G. McCurrach, R. G. Bessent, and C. S. McArdle 1991. The biodegradation of albumin microspheres used for regional chemotherapy in patients with colorectal liver metastasis. *Nuc. Med. Comm.* 12 (1):57–63.

Hanyu, F., M. Nakamura, T. Takasaki, Y. Kasai, Y. Sato, H. Sato, Y. Sukurai, A. Yamada, T. Okawa, and T. Salabe. 1988. Clinical study of controlled-release preparation of mitomycin C in the treatment of inoperable cancer patients. *Japanese J. Cancer & Chemo.* 15 (11):3087–3093.

Imai, K., H. Yamanaka, H. Yuasa, M. Yoshida, M. Asano, I. Kaetu, I. Yamazaki, and K. Suzuki. 1986. The sustained release of LHRH agonist-polymer composite in patients with prostatic cancer. *Prostate* 8 (4):325–332.

Juliano, R. L., S. Daoud, H. J. Krause, and C. W. M. Grant. 1990. Membrane-to-membrane transfer of lipophilic drugs used against cancer or infectious disease. *Annals N. Y. Acad. Sci.,* pp. 89–103.

Kaunitz, A. M. 1994. Long-acting injectable contraception with depot medroxyprogesterone acetate. *Am. J. Obstet. Gynecol.* 170 (5 Pt 2): 1543–1549.

Kitamura, M., K. Arai, K. Miyashita, and G. Kosaki. 1989. Arterial infusion chemotherapy in patients with gastric cancer in liver metastasis and long-term survival after treatment. *Japanese J. Cancer & Chemo.* 16 (8 part 2):2936–2939.

Kubo, O., H. Himuro, N. Inoue, Y. Tajika, T. Tajika, M. Tohyama, M. Sakairi, M. Yoshida, I. Kaetsu, and K. Kitamura. 1986. Treatment of malignant brain tumors with slowly releasing anticancer drug-polymer composites. *No Shinkei Geka Neuro. Surg.* 14 (10): 1189–1195.

Kuroda, R., F. Akia, H. Iwasaki, J. Nakatani, T. Uchiyama, M. Ioku, and I. Kaetsu. 1994. Interstitial chemotherapy with biodegradable ACNU pellet for glioblastoma. *Stereotactic & Functional Neurosurg.* 63 (1–4):154–159.

Lewis, D. H. 1990. Controlled release of bioactive agents from lactide/glycolide polymers. In *Biodegradable polymers as drug delivery systems,* edited by P. Tyle. New York: Marcel Dekker, pp. 1–41.

Li, V. H-K., and J. R. Robinson. 1989. Influence of drug properties and routes of drug administration on the design of sustained and controlled release systems. In *Controlled drug delivery: Fundamentals and applications,* edited by J. Robinson and V. Lee. New York: Marcel Dekker, pp. 4–61.

Mangione, R. 1997. Chronotherapeutics and pharmaceutical care. *U.S. Pharmacist* (July).

Niebertgall, P. J., T. Sugita, and R. L. Schnaare. 1974. Potential dangers of common drug dosing regimen. *Am. J. Hosp. Pharm.* 31:53.

Okada, H., and H. Toguchi. 1995. Biodegradable microspheres in drug delivery. *Crit. Rev. Therap. Drug Carrier Sys.* 12 (1):1–99.

Ponpipom, M. M., and R. L. Bugianesi. 1980. Isolation of 1,3-distearoyl-glycero-2-phosphocholine (-lecithin) from commercial 1,2-distearoyl-sn-glycero-3-phosphocholine. *J. Lipid Res.* 21:136–139.

Robinson, D. H., and J. W. Mauger. 1991. Drug delivery systems. *AJHP* 48:14–24.

Sharifi, R., and M. Soloway. 1990. The leuprolide study group: Clinical study of leuprolide depot formulation in the treatment of advanced prostate cancer. *J. Uro.* 143:68–71.

Thom, A. K., E. R. Sigurdson, M. Bitar, and J. M. Daly. 1989. Regional hepatic arterial infusion of degradable starch microspheres increases fluorodeoxyurine (FUdr) tumor uptake. *Surgery* 105 (3):383–392.

Tice, T. R., and S. E. Tabibi. 1992. Parenteral drug delivery: Injectables. In *Treatise on controlled drug delivery,* edited by A. Kydonieus. New York: Marcel Dekker, Inc., pp. 315–316.

Tyle, P. 1989. Sustained release growth hormone compositions for parenteral administration and their use. U.S. Patent 4,857,506. Also European Patent 0278103.

Weiner, A., J. B. Cannon, and P. Tyle. 1989. Commercial approaches to the delivery of macromolecular drugs with liposomes. In *Controlled release of drugs: Polymers and aggregate systems,* edited by M. Rosoff. Florida, USA: VCH Publishers, pp. 217–253.

Wollner, I. S., S. C. Walker-Andrews, J. E. Smith, and W. D. Ensminger. 1986. Phase II study on hepatic arterial degradable starch microspheres and mitomycin. *Cancer Drug Delivery* 3 (4):1037–1041.

Yewey, G. L., E. G. Duysen, S. M. Fox, and R. L. Dunn. 1997. Delivery of proteins from a controlled release injectable implant. *Pharma. Biotech.* 1:93–117.

12

Veterinary Sustained-Release Parenteral Products

Jung-Chung Lee
Cellergy Pharmaceuticals Inc.
S. San Francisco, California

Michael Putnam
Fort Dodge Animal Health
Fort Dodge, Iowa

There are several means for designing sustained-release products for food-producing and companion animals. Such products may be designed for oral, ruminal, topical, rectal, or subcutaneous (SC) administration. The selection of the route of administration is critical to the design of a sustained-release device, due to differences in biological environments. A formulation for subcutaneous implantation must be nonirritating and mechanically strong enough to withstand flexion or impact. An oral or ruminal formulation must be designed for resistance to gastric acidity and sensitivity to pH change. A topical formulation, such as a dip, pour-on, tag, or collar, must be functional and strong enough to withstand long periods of weathering and/or animal abuse.

Although many sustained- or controlled-release formulations are designed for intraruminal and topical administration, this chapter will focus on sustained-release parenteral products. Typically, when a formulation is injected into animals, the systemic drug concentration far exceeds the therapeutic concentration initially and gradually declines to the therapeutic level and finally to an ineffective level. Sustained-release formulations must be designed to deal with this undesirable pharmacokinetic situation by replacing the immediate high release pattern with more constant and therapeutically sound concentrations. Thus, the same total drug dose can have a longer-term release action from the injection site.

Sustained-release injectable formulations can be used for both biological and pharmaceutical preparations. The applications can be extended to both companion animals and food-producing animals. Before beginning the design, however, economics, regulations, and practicality need to be seriously considered. In many cases, the delivery system design needs to be targeted to the therapeutic index of the compound selected for the disease state or treatment. A general approach is not feasible for all categories of a compound.

Solutions, suspensions, emulsions, and implants are the most common approaches used for parenteral sustained-release preparations. The duration of action for these products ranges from 12 h for pharmacological products to 200+ days for growth stimulating products. A range of animal health sustained-release parenteral products with their therapeutic categories, formulations, and durations of activity is given in Table 12.1.

SUSTAINED–RELEASE DOSAGE FORM DESIGN CONSIDERATIONS

The parenteral routes of drug administration normally include intravenous (IV), intramuscular (IM), SC, intraperitoneal (IP), and intrathecal (IT). Not all of these routes are useful for the administration of sustained-release dosage forms. The SC and IM

Table 12.1. *The Sustained-Release Parenteral Products Currently Available for Veterinary Pharmaceuticals*

Classification	Company	Product	Generic Name	Use	Dosage	Mechanism
Allergens	Center Laboratories	Allergenic extract —Flea antigen Allergenic Extracts —pollens, dust, molds, inhalants, epidermals, insects, and prescription products	N/A		< 0.5 mL	
	Bioproducts DVM, Inc.	Bioproducts brand allergenic extract— prescription product				
Anabolic Agents	Pharmacia & Upjohn	Winstrol-V injection	Stanozolol	Anabolic steroid	0.5 to 1.0 mL	Suspension
	Fort Dodge Animal Health	Equipoise injection	Boldenone undecylenate	Anabolic steroid	0.5 mg/lb	Oil depot
		Synovex H implant	Testosterone proprionate and estradiol benzoate	Growth promotion	1 implant	Solid implant
		Synovex S/C implant	Progesterone and estradiol benzoate	Growth promotion	1 implant	Solid implant
		Synovex plus implant	Trenbolone acetate and estradiol benzoate	Growth promotion	1 implant	Solid implant

Table 12.1 continued on next page.

Table 12.1 continued from previous page.

Classification	Company	Product	Generic Name	Use	Dosage	Mechanism
	Hoechst-Roussel AgriVet	Finaplix-H implant	Trenbolone acetate	Growth promotion	1 implant	Solid implant
Antibacterials/ Antibiotics	Pharmacia & Upjohn	Naxcel injection	Ceftiofur	Bacterial infections	1 tube	Salt form
		Albadry Plus intramammary	Penicillin G procaine	Subclinical mastitis		Suspension
	Sanofi	Erythro-200 injection	Erythromycin	Bacterial infections	4 mg/lb	Oil solution
	Elanco	Micotil 300 injection	Tilmicosin phosphate	Bovine respiratory disease	10 mg/kg	Partitioning
	Solvay	Crystiben injection	Penicillin G benzothine and penicillin G procaine	Bacterial infections	2 mL/150 lb	Salt Form
	Durvet	Dura-Pen injection	Penicillin G benzothine and penicillin g procaine	Bacterial infections	2 mL/150 lb	Salt Form

Table 12.1 continued on next page.

Table 12.1 continued from previous page.

Classification	Company	Product	Generic Name	Use	Dosage	Mechanism
	Fort Dodge Animal Health	Tomorrow Intramammary	Cephapirin benzathine	Mastitis	1 tube	Oil Suspension/Salt Form
Anti-Inflammatory Agents	Luitpold	Adequan IA/IM	Polysulfated glycosaminoglycan	Proteolytic enzyme inhibitor	500 mg for 4 days	Partitioning
	Fort Dodge	Hyalovet	Hyaluronate sodium	Equine arthritis	2 mL twice a week	Partitioning
	Boehringer Ingelheim	Hyvisc	Hyaluronate sodium	Equine arthritis	2 mL twice a week	Partitioning
Corticosteroids	Pharmacia & Upjohn	Depo-Medrol®	Methylprednisolone acetate	Musculor skeletal conditions	2–120 mg (dogs) 10–20 mg (cats) 200 mg (Horses)	Suspension
	Fort Dodge Animal Health	Vetalog Parenteral	Triamcinolone acetonide	Glucocorticoid	0.05–.2 mg/lb (dogs and cats) 0.01 to 0.02 mg/lb (horses)	Suspension

Table 12.1 continued on next page.

Table 12.1 continued from previous page.

Classification	Company	Product	Generic Name	Use	Dosage	Mechanism
	Schering-Plough	Betsone	Bethamethasone dipropionate and betamethasone	Anti-inflammatory	0.25–0.5 mg/20 lb	Suspension
			Sodium phosphate	Anti-pruritic		
	Pharmacia & Upjohn	Predef 2X	Isoflupredone acetate	Glucocorticoid	10–20 mg (cattle) 5–20 mg (horses) 5 mg–300 lb (swine)	Suspension
Vitamins	DurVet	A-D injection	Vitamin A and Vitamin D3	Supplemental nutritive	0.5 to 4 mL	Oil-based vehicle

routes are the only parenteral routes that are commonly used for the administration of sustained-release preparations. For sustained-release, a depot needs to be formed in order to provide slow and constant release of the active compound from the injection site. Depots can be formed by injecting suspensions, emulsions, oil solutions, or implants (solid or semisolid).

A study of the drug's pharmacokinetics and toxicology is the initial step in sustained-release formulation design. With some compounds, e.g., aminoglycosides, the normal peak and trough plasma concentrations produced by periodic dosing are important in producing a therapeutic treatment without endangering the patient. Beta-lactam antibiotics, on the other hand, achieve best results with a constant therapeutic level. The drug loading needed to maintain therapeutic levels over the desired period must then be considered. It may require several grams of penicillin to achieve therapeutic plasma concentrations for just a few days. If the amount of drug required exceeds the capacity of the proposed sustained-release dosage form, such a formulation will not be feasible. If the drug is rapidly absorbed into muscle or fat and/or is effective at low concentrations, the likelihood of achieving a viable sustained-release formulation is greatly enhanced.

The methods used to achieve the sustained release of active compounds in veterinary parenteral preparations are mostly physical modifications of the formulations. These include the use of polymer-based solutions (to increase viscosity or to form a gel), aqueous solutions of drug complexes, aqueous suspensions, suspensions of the polymer-coated drug (microencapsulation), oil solutions, oil suspensions, emulsions, pellets, and implants. Alternate methods include the preparation of salt forms to decrease the drug's dissolution rate from the depot or the formation of a prodrug to decrease the metabolism rate. Most such preparations are suitable for SC or IM injections. However, the volume of injection, site of injection, animal body movement, and tissue injury need to be carefully considered before designing sustained-release parenteral dosage forms. Also, residues of drugs at the injection or implant site and blemishes in the edible tissues are not acceptable in food-producing animals.

The physical and chemical stability of the formulation is critical for sustained-release dosage forms. At the concentrations used in sustained-release preparations, some physical characteristics may change over time: Solutions may gel, stable compounds may begin to complex or degrade, and emulsions may equilibrate in unfavorable ways. It is critical to evaluate the formulation's performance over time under accelerated conditions using a variety of criteria. Standard concentration testing should be enhanced with viscosity measurements, particle size determinations, appearance, dissolution rate, and phase concentration analysis, as appropriate. Sustained-release formulations are always potentially hazardous due to the large relative amount of drug used. The importance of maintaining the formulation in a state of control over the desired shelf life and release interval cannot be overemphasized.

Absorption Parameters

As previously mentioned, several different factors affect the release of drug from a sustained-release injection site. Sustained-release depots must be rugged and only marginally affected by potential in vivo challenges.

As with any dosage form, a diffusion gradient is established between the depot and the local in vivo environment. This gradient is especially vulnerable to highly perfused tissues and capillary beds. High blood flow through these areas may accelerate drug release by shifting the diffusion gradient. Poor selection of the administration site location (Schou 1961; Sund and Schou 1964) and poor injection technique are two of the most significant reasons for failures of sustained-release dosage forms.

Drug release from the depot is also affected by the compound's physical and chemical characteristics. For suspensions and implants, particle size has a significant effect on the release rate. Granular materials tend to exhibit slower release patterns than more finely ground products. Actives with multiple crystal states often display a variety of release characteristics. Crystalline drugs often demonstrate slower dissolution times than amorphous powders, and polymorphic forms can have unique release patterns associated with each crystal state. These latter materials are often metastable and more associated with formulation difficulties than with successful dosage design.

The lipophilic/hydrophilic nature of the drug must be thoroughly understood during formulation development. When formulated into oil-in-water (o/w) emulsions, the natural tendency of lipophilic materials to remain in the oil phase can be used to slow the absorption rate. Likewise, when placed in an entirely oleaginous vehicle, further reductions in depot release rates can be obtained. Hydrophilic agents can be somewhat slowed by the use of water-in-oil (w/o) emulsions, but their natural solubility in aqueous body fluids makes formulation development of a liquid or semisolid dosage form more challenging. Some chemical modifications can often be performed to alter the lipophilic/ hydrophilic balance, or pH modifications can be used to aid the generation of sustained-release formulations.

As with all formulations, excipients often control the release parameters of the final sustained-release dosage form. For implants, high molecular weight waxes (e.g., polyethylene glycol 8000) or coatings can be used to form a matrix that extends the release time. With aqueous formulations, the viscosity can be increased somewhat, although significant increases are difficult while maintaining syringability. Oil vehicles can be thickened or made even more lipophilic through the application of appropriate excipients. Because of their effects on release rates, care must be exercised during formulation development to characterize all critical excipients thoroughly.

Species Variations

The principal differences in sustained-release products among species are more based on differences in therapeutic intent rather than on physiological constraints. Some species (e.g., felines) can demonstrate a daunting array of idiosyncratic metabolic behaviors. Pharmaceutical companies invest a large amount of time, money, and effort in developing each species claim. Formulations are optimized for a single species and are usually not suitable for cross-species use.

An important variable among formulations designed for different species is injection/implant size. The range of animal sizes in veterinary medicine is very large. Treatments can be desired for elephants, horses, dogs, or birds. For each species, the maximal dosage size will be different. Important consideration must also be given to the type of patient. Formulations for companion animals are often less confined in scope than those used for food animals, due to government regulations concerning tissue residue levels at injection sites. Restraints imposed by consideration of the target species must be identified early in the development program to avoid costly delays in approval.

Handling of Injection Equipment

In order to facilitate the administration of sustained-release parenterals to animals, special syringes or implanters have been designed (Pope 1983). The food-animal industry has generated demand for a variety of devices to make administrations easier and quicker.

Most implant manufacturers have developed unique injection guns for handling their products. Each vendor places implants in a special holder or cartridge. A single implant may have one or more solid pellets or rods within the dosing cavity. These holders are then loaded into the gun. The user actuates a spring-loaded or mechanical device to advance the holder for the next dose, carefully inserts the needle, and pulls a trigger mechanism, while withdrawing the needle to administer the dose. While some minor modernizations have been made, most are fundamentally unchanged from their introduction 30 years ago.

Injection guns are often used for liquid or semisolid dosing in agricultural settings. These devices may be actuated mechanically, air operated, or electrically powered. Usually, the operator places a filled parenteral bottle on the device's inlet using a needle or spike attachment. As animals pass through a restraint chute, they can be individually dosed.

Injection Sites

The most important factors influencing the absorption rates of the active drugs from the SC or IM depot are blood flow and lymph flow at that site. The dissolution rate of drug particles, the lipophilicity of the active drug, the binding of actives to the drug depot, and the shape and surface area of the formulation depot at the injection site can also affect the absorption of actives from the sustained-release formulation.

Normally, the release rate from the injection site is controlled by the release of actives from the dosage form rather than by the absorption rate of the drug into the bloodstream. When a depot is formed after injecting a sustained-release dosage form, the release of actives from the depot depends on the following: the amount of actives in the depot; the surface area available for drug dissolution; the nature of the solvent, gelling agent, and vehicle used; the pH and viscosity of the solution; the polymer used to alter the dissolution properties of the active particles; and the physicochemical properties of the drug.

Many factors may influence the release rate of actives from a sustained-release depot. The injection site, e.g., subcutaneous versus intramuscular, does not play a major role in drug absorption from the depot. However, the injection technique and location can sometimes create a difference in the bioavailability of the active preparation. If the SC implantation of hormone pellets in the ear for cattle is too close to the cartilage bone, the vascular flow through the drug depot is so limited that the hormone will not be depleted at the desired rate. If the implantation is not deep enough, the pellets could fall from the ear implantation site, and no improvements in weight gain or feed efficiency would be seen for the cattle receiving the hormone implants.

For IM injection, the active preparation is injected into the intermuscular area rather than the intramuscular location; consequently, the depot cannot be depleted as predicted because of limited blood flow through the intermuscular area (Marshall and Palmer 1980). Also, injection into various muscle sites can generate different peak levels of the drug due to different blood flow for the muscles. For example, the deltoid muscle has much greater blood flow than the gluteus muscle; therefore, the gluteus injection site will be preferred if sustained-release action of the drug is desired (Evans et al. 1975). IM injection is generally not acceptable in food-producing animals unless no viable alternative exists.

One drawback for the development of sustained-release parenterals is the possibility of local irritation and tissue necrosis at the injection site. In many cases, the irritation and tissue damage is due to the active drug itself (Rasmussen 1980); in other cases, it is caused by the excipients used in the formulation (Spiegel and Noseworthy 1963). Therefore, care in choosing the solvents, vehicle, gelling agent, suspending agent, polymers, and other excipients is very important in reducing the risk of formulating a sustained-release product that can cause pain to the animals when injected by parenteral routes. A high concentration and small volume dosage form is preferred, since this type of dosage form can create a more defined depot in the injection site and can reduce pain and irritation upon injection.

BASIS FOR THE SUSTAINED–RELEASE DOSAGE FORM DESIGN

Most parenteral solutions are designed for rapid onset of the therapeutic action, especially for the IV injection. Although SC and IM aqueous injections have slower onsets of action than IV injections, the absorption of the drug from the injection site is rather fast when compared to implants. Without modification of the formulation, the SC and IM routes require frequent injection to maintain sufficient drug concentration in the blood. Repeated injections are not economical for food-producing animals or convenient enough for use in companion animal medicine. Therefore, physical and/or chemical modification of the injections will be required in designing sustained-release injectable solutions for SC and IM routes.

The most commonly used approach for designing sustained-release parenteral dosage forms is to create a depot at the injection site. A depot is, in fact, a drug reservoir that can release drug molecules at an appropriate rate over a predetermined period of time. Several approaches used for slowing the release of active drugs from the injection site are illustrated in the following sections.

Solutions

For aqueous solutions, the vehicle needs to be modified to create a viscous solution or a gel at the injection site so that the active molecules can be trapped in the depot for a sustained-release action. The viscosity-building or gelling agent used can be water-miscible agents, natural gums, or synthetic gums.

Gel/Viscosity-Building Solution

By adding gelling or viscosity-building agents such as polyvinyl alcohol, methylcellulose, hydroxypropyl methylcellulose, sodium carboxymethylcellulose, or polyvinyl pyrrolidone to the injectable formulation, a bioerodible matrix of the drug molecules in the gel can be formed. The release of drug molecules from the gel matrix in the IM or SC depot will depend on the concentration of the free drug molecule in the depot. Once the free drug molecules are absorbed from the surface of the gel, a concentration gradient is established, and dissociation of the drug molecule from the gel occurs. The sustained-release characteristics of this type of dosage can be controlled somewhat by selecting the appropriate gels or viscosity-building polymers to control the viscosity of the dosage form or by varying the concentration of these materials.

Complex Formation

A water-soluble drug normally cannot be used effectively to design a sustained-release injectable product because the formation of a slow-release depot will not occur. However, by converting the water-soluble drug to its poorly soluble salt (Gower et al. 1973; Nicolas et al. 1973; Barrios et al. 1975) or by adding a complexation agent to form a low water-soluble complex, a sustained-release parenteral dosage form can be formulated. A good example is the 1:1:1 complex formed among acetaminophen, theophylline, and caffeine, which has much lower aqueous solubility than the individual drugs, thus reducing the dissolution rate from the injection site (Chow and Repta 1972). Other complexes such as penicillin G benzathin (Ober et al. 1958) and protamine zinc insulin (Hagedorn et al. 1936) have much lower aqueous solubilities than their parent compounds, and thus prolonged release effect can be achieved.

Oil Solutions

The use of an oil solution to achieve sustained release for veterinary parenteral dosage forms is common, since this is the simplest and the most economical way to produce a product. In most of cases, the drug molecule is released by partitioning through the oil-water interfaces, and the release rate is controlled by the partition coefficient of the drug. However, in some cases, the drug molecule may precipitate in a very short time after administration to form a drug depot, and release will be prolonged by slow dissolution of the drug particles. The latter will require a highly oil-soluble drug (very high partition coefficient) with a fairly high loading of the drug in the formulation. In addition, this type of formulation will require some surfactant or water-soluble organic phases, such as propylene glycol or polyethylene glycol, to ensure a good mixing of the oil with body fluids at the injection site.

Suspensions

Aqueous

Simple aqueous suspensions can be formulated only for those drugs with very low water solubilities. Drug release from the aqueous suspension is governed by the dissolution rate from the injection site. Therefore, the particle size, crystalline form, and diffusion coefficient become very important parameters that need to be taken into consideration when designing an aqueous suspension. Larger particles provide a smaller surface area and, thus, can slow down dissolution when compared to smaller particles used in the suspension (Bates et al. 1969). Also, careful selection of the polymorphic form is important to control the release rate as well as to ensure the consistency of the sustained-release action because crystalline polymorphs can show dramatically different lattice energies and have different dissolution rates. Lastly, by choosing a different suspending agent, the suspension's viscosity can be controlled to some extent so that the diffusion of the drug molecule can be prolonged (Hussain et al. 1991).

When designing an aqueous suspension for water-soluble drugs such as proteins and peptides (Cady and Langer 1992), a carrier system needs to be in place to protect the drug from quick dissolution at the injection site and to ensure the good stability of such drugs during the sustained-release injection. The most commonly used carrier systems are polymeric-coated or filled microspheres or microcapsules (Cady et al. 1989; Hsieh et al. 1983; Carter et al. 1988). The polymers used for the encapsulation must be biocompatible and can either be bioerodible or nonbioerodible (Visscher et al. 1988; Tice and Tabibi 1992). The vehicle used for this type of suspension normally contains osmotic or thickening agents to facilitate or control the release of the active drug (Carter et al. 1988).

Microspheres or microcapsules can be generated using a variety of processes. Often, liposomes are formed in a volatile environment containing the drug. When the volatiles are removed, microspheres (dried liposomes) remain. Other techniques, such as spray drying, perform similar functions. The primary objective for this type of formulation is to deliver the active protected by a carrier that will control the release rate from the injection site.

Two primary aspects control the release rate from microspheres: microsphere diameter and drug loading. As with any geometric sustained-release dosage form, there is an inverse relationship to size versus release rate. Smaller spheres will release drug faster than larger spheres at a given drug loading. There is a direct correlation between drug loading and the release rate. As more drug is contained (on a percentage basis) in each microsphere, the release rate increases. This is caused by the formation of channels as the exterior drug dissolves/diffuses from the shape. Small, high drug load, microcapsules will release faster than large, low drug load, microcapsules. The challenge is to develop a formulation that optimizes both microsphere size and drug loading to achieve the desired release rate without excessive final injection volume.

Oil

Oil-based vehicles offer some advantages to aqueous vehicles. High concentration oil suspensions are a historically sound method for generating a sustained-release injectable. They are inherently more lipophilic and tend to resist aqueous incursions from the body. This aspect protects water-soluble actives, creating a better reservoir effect. Oil viscosity can be increased by including waxes or metal-stearate components during formulation. While harder to filter than aqueous systems, oil suspensions can be easily irradiated as a terminal sterilization technique.

In the oil suspension, the drug particles suspended in the oil phase serve as the reservoir. The drug molecules must be dissolved in the oil phase and then partition out of the oil phase to aqueous body fluids (Rasmussen and Svendsen 1976; Wagner 1961). In many cases, this type of formulation behaves similar to an oil solution with high loading of a low water-soluble active drug. However, in general, the oil suspension formulation demonstrates a longer duration than oil solutions due to the extra dissolution step in the oil phase.

Liposomes

Liposomes are made from phospholipids with the addition of sterols, glycolipids, other synthetic lipids, and organic acids and bases. Two types of liposomes, multilamellar vesicles (MLVs) with diameters of 25 nm to 4000 nm and small unilamellar vesicles (SUVs) with diameters of 300 Å to 500 Å are most commonly used as carriers for small molecules and macromolecules. Liposomes can be used for both lipophilic and hydrophilic drug molecules due to their lipid bilayer and aqueous core structure.

Many studies have been conducted in using liposomes for drug-targeting purposes (McCullough and Juliano 1979; Dingle et al. 1978). Because of the unique physical and chemical properties of the liposomal preparation, it can be injected directly into the bloodstream, thus serving as an intravascular drug depot. With drugs with low therapeutic indices or short half-lives, the liposome IV injection can protect the drug from metabolism immediately after injection and can persist for hours (Juliano and Stamp 1975, 1978; Kimelberg et al. 1978; Rahman et al. 1978).

Also, with a low water solubility drug, the affiliation of drug molecules with the lipid bilayer of liposomes can allow the drug to be injected intravenously and maintained in the circulation. This is an advantage over aqueous suspensions, which can be injected only through SC or IM routes (Kataoka and Kobayashi 1978; Mayhew et al. 1978).

Implants

Many therapeutic agents are rapidly metabolized or cleared from the subject's system, necessitating frequent administration of the drug to maintain a therapeutic concentration (Preston and Rains 1993). There is a need for a sustained-release device capable of administering an active compound at a relatively constant rate, where the rate is high enough to maintain an effective concentration. Preferably, such a device would be inexpensive and easily manufactured.

There are many ways to design a sustained-release device for implantation. Some devices are a matrix type (Dzuik et al. 1968; Nash et al. 1978) and consist of an active compound dispersed in a matrix carrier material. The carrier material may be either porous or nonporous, solid or semisolid, and permeable or impermeable to the active compound. Matrix devices may be biodegradable, i.e., they may slowly erode after administration (Kent et al. 1980; Wagner et al. 1984). Alternatively, matrix devices may be nonbiodegradable and rely on diffusion of the active compound through the walls or pores of the matrix (Hsieh et al. 1984). Although it is relatively easy to manufacture matrix type devices, it is extremely difficult to achieve constant release of the active compound from this type of device. The release rate is typically dependent on the dissolution, diffusion, and concentration of the active compound inside the matrices. Therefore, a nearly first-order release pattern is commonly observed from matrix type implants.

The other type of device is a reservoir type (Runkel and Lee 1991; Sommereville and Tarttelin 1983; Heitzman 1983). This type of device consists of a reservoir of active compound presented in a solution, suspension, gel, or solid and surrounded by a rate-controlling membrane. The rate-controlling membrane can be either porous or nonporous. It should be hydrated to some extent and yet able to maintain its integrity. Normally, the membrane is not biodegradable. The release of active compound from reservoir type devices depends on the dissolution of the active compound inside the reservoir, the diffusion of active compound through the membrane, the membrane thickness, and the total surface area of the rate-controlling membrane. The resulting release pattern is always zero-order kinetics, i.e., a constant release rate for a designated period of time. This type of device is generally easy to make. However, the sensitivity to rupture is sometimes a problem, since breaching the rate-controlling membrane can cause instantaneous dumping of the active compound.

Some sustained-release devices are hybrids, having a matrix core surrounded by a rate-controlling membrane (Hsieh et al. 1984; Robertson et al. 1983). By carefully designing the geometry, core matrix, and surrounding membrane, this type of device can give a true zero-order release pattern for a desired period of time. The rate-controlling membrane can smooth out the surface erosion release mechanism by limiting the release of active compound, and the center matrix core can be slowly depleted to form a reservoir of the active compound. The risk of dose dumping to a lethal level as a result of rupturing the controlling membrane of reservoir type devices is greatly reduced by this design.

Other sustained-release devices use a semipermeable membrane with mechanical action, such as electrical or osmotic pumps (Struyker-Bouldier and Smits 1978; Higuchi and Leeper 1973; Theeuwes and Higuchi 1974; Theeuwes 1975; Theeuwes and Yum 1976). This type of device normally requires the filling of solution, suspension, gel, semisolid, or solid into the reservoir, and the active compound is driven from the device by an osmotic engine. The osmotic engine is a compartment consisting of a center core tablet, which can generate high osmotic pressure upon dissolving, surrounded by a semipermeable membrane that can control the diffusion of water into the compartment. While this type of device may be capable of zero-order release, it is typically too expensive to compete economically with matrix and reservoir devices.

The most popular implant in the animal health field is the hormone pellet (Beck and Pope 1983). These pellets are used for the treatment of a disease in an animal, contraception, synchronization of estrus, or promotion or increase of weight gain and feed efficiency. The release of actives from the pellet implant needs to be sufficient to prevent disease from occurring, relieve the disease, or cause modification of the normal biological activities. The pellet implant is a matrix type solid formulation, and the in vivo dissolution rate can be described by the Noyes-Whitney dissolution rate law (Chien 1992):

$$R = kADS/d$$

where R is the release rate of actives from the pellet, k is a proportionality constant, A is the surface area of the pellet over the time of the dissolution, D is the diffusion coefficient of the actives, S is the solubility of actives, and d is the thickness of the aqueous diffusion layer around the pellet.

Since the active release mechanism from the disc or cylindrical pellet implants is surface erosion, the effective surface area directly in contact with tissue or body fluid is constantly changing with time. Therefore, the release rate of actives from the pellet implant is high initially and gradually decreases with time. This burst effect is normally observed as behavioral changes during the first few days after implantation when hormone implants are used for growth promotion.

Many implant depletion studies have suggested that the release rate of actives from the implant site can reach steady state 30 to 40 days after implantation. One major reason is the physiological changes at the implant site. Normally, a layer of tissue will grow around the pellet implants due to the body wound-repairing process protecting tissues from foreign substances. Depending on the physicochemical properties of the actives, the release mechanism of actives from the pellet implant can be switched from pellet dissolution control to the tissue membrane control. In other words, the tissue envelope surrounding the pellet implant can act as a rate-controlling membrane regulating the absorption of actives from the injection site if the dissolution rates of the actives are reasonably high.

Considerations in the design of an implant should include the total surface area of the implant, the hydrophilicity and lipophilicity of the actives, the particle size of the actives, the dissolution rate of the actives, the dissolution rate of the excipients, and the diffusivity of the actives through aqueous diffusion layers and tissue membranes.

From the above equation, the surface area of the implant directly in contact with the tissue and body fluid can greatly govern the release rate of actives from the pellet implant. Therefore, a defined surface area will be required to effect a flux of biologically active compound sufficient to achieve the desired physiological effect. The minimized surface area will be preferred due to the constraint in implantation techniques. For surface erosion pellet implants, the formulation can be modified by adding an adequate amount of water-soluble excipients that can create pores and channels inside the pellets during the residence period of the implant. The increased surface area produced by pores and water channels inside the pellet can compensate for the loss of outside surface area due to a reduction in the size of pellet. However, the balance between the type of excipients and amount of excipients used is critical for the pellet implant formulation design.

Modifying the diffusion layer thickness for the pellet implant can also increase the absorption of actives from the implant, since most long duration implants are composed of steroid hormones that are very water insoluble. Adding some surfactant to the formulation can help the dissolution of water-insoluble hormones as well as decrease the thickness of the aqueous diffusion layers surrounding the pellet at the implant site. All of these effects can benefit the dissolution of the pellet, thus increasing the absorption rate of the actives.

The particle size of the polymorphs of the actives can also greatly affect the release of actives from the implantation site. Some steroid hormones have several polymorphs. These polymorphs have different melting points and solubility in aqueous and nonaqueous solvents. By selecting the appropriate polymorph, the release rate of the actives from the pellet implant can be more controllable, and the physical stability of the implant can be greatly enhanced. The particle size of the crystals can affect the dissolution of actives from the pellet implant dramatically due to the effective surface area for active dissolution. Proper control of the crystal particle size used to manufacture the pellet implant can reduce the release rate variability to some extent.

In conclusion, most of the current veterinary implants sold on the market have prolonged release but not constant release of actives for a desired period of time. Efforts are needed to control the above discussed parameters to design sustained-release formulations not only with adequate duration but also a zero-order release pattern.

CURRENT MARKETED VETERINARY SUSTAINED–RELEASE PARENTERAL PRODUCTS

Most of the parenteral veterinary products on the market are solid implants, oil solutions, or suspensions. The product name, manufacturer, treatment classification, dosage, and sustained-release mechanism are listed in Table 12.1. Sustained-release preparations are primarily anabolic agents, antibacterials/antibiotics, and corticosteroids, since long-term treatment is a requirement in these areas.

Sustained-release solid implants for steroid hormones are popular because this type of product is used for increasing weight gain and feed efficiency for 70 to 180 days. Reimplantation is undesired due to the high cost and inconvenience for both feedlot and pasture animals. The release rate of steroid hormones from these implants depends on the dissolution rate of the steroids at the implantation site. Also, the release of the hormone into the systemic circulation is controlled by diffusion of the steroid molecules through the surrounding membrane. The cost of manufacturing implants is reasonably economical, the sustained-release effect is quite consistent, and clinical results are satisfactory.

Treatment with antibacterials and antibiotics normally requires a multiple regimen. By using the salt form of penicillin G, the oil solution of erythromycin, and the salt form of ceftiofur, the multiple day treatment can be dosed in one shot. It is very convenient and effective to treat infectious diseases. A challenge in manufacturing these types of preparations is possible instability, both physical and chemical.

Corticosteroids are another popular use of sustained-release technology to treat inflammation, since the use of standard formulations requires multiple and frequent dosing. Suspensions are most commonly formulated for this type of product to achieve prolonged pain relief.

Other areas such as arthritis treatment, allergenic injection, and vitamin supplementation also use sustained-release technology to provide the best clinical results.

CONCLUSION

The use of sustained-release technology in developing veterinary parenteral formulations is focused on simplicity, economy, and efficacy. Most of the currently available, sustained-release parenteral products in Table 12.1 show the benefit of extending the duration of the treatment by applying the basic principles of the sustained- or controlled-release technologies. Modifying traditional injectable preparations by using salt forms, stable aqueous suspensions, and oil solutions and suspensions results in new preparations that can offer prolonged action with reasonable cost and provides an advantage in reducing the frequency of administration, especially for pasture animals. Also, additional benefits such as achieving a more constant blood level, reduced side effects, reduced drug dose, and more efficient use of the drug generally result from the use of these sustained-release products.

ACKNOWLEDGMENTS

The authors would like to thank Dr. Richard A. Schiltz for his comments and corrections of this book chapter. Dr. Schiltz is currently a Consultant for Veterinary Product R&D; he was formerly the vice president of Animal Health R&D at Syntex Corporation.

REFERENCES

Barrios, S., J. H. Sorenson, and R. G. W. Spickett. 1975. Bioavailability of Cephalexin after intramuscular injection of its lysine salt. *J. Pharm. Pharmacol.* 27:711–712.

Bates, T. R., D. A. Lambert, and W. H. Johns. 1969. Correlation between the rate of dissolution and absorption of salicylamide from tablet and suspension dosage form. *J. Pharm. Sci.* 58:1468–1470.

Beck, L., and V. Z. Pope. 1983. Controlled-release delivery systems for hormones: A review of their properties and current therapeutic use. *Drugs* 27:528–547.

Cady, S. M., and R. Langer. 1992. Overview of protein formulations for animal health applications. *J. Ag. Food Chem.* 40:332–336.

Cady, S. M., W. D. Steber, and R. Fishbein. 1989. Development of a sustained release delivery system for bovine somatotropin. *Proceed. Int. Symp. Control. Rel. Bioact. Mater.* 16:22–23.

Carter, D. H., M. Luttinger, and D. L. Gardner. 1988. Controlled release parenteral systems for veterinary applications. *J. Control. Rel.* 8:15–22.

Chien, Y. W. 1992. Implantable controlled-release drug delivery systems. In *Novel drug delivery systems*, 2nd ed., edited by Y. W. Chien. New York: Marcel Dekker.

Chow, Y. P., and A. J. Repta. 1972. Complexation of acetaminophen with methyl xanthines. *J. Pharm. Sci.* 61:1454.

Dingle, J. T., J. L. Gordon, B. C. Hazelman, C. G. Knight, D. P. Page Thomas, N. C. Phillips, I. H. Shaw, F. J. Fildes, J. E. Oliver, G. Jones, E. H. Turner, and J. S. Lowe. 1978. Novel treatment of joint inflammation. *Nature* 271:372–373.

Dziuk, P. J., B. Cook, G. D. Niswender, C. C. Kaltenbach, B. B. Doane. 1968. Inhibition and control of estrus and ovulation in ewes with a subcutaneous implant of silicone rubber impregnated with a progestogen. *Am. J. Vet. Res.* 29:2415–2417.

Evans, E. F., J. D. Proctor, M. J. Frantkin, J. Valandia, and A. J. Wasserman. Blood flow in muscle groups and drug absorption. *Clin. Pharmacol. Ther.* 17:44–47.

Gower, P. E., C. H. Dash, and C. H. O'Callaghan. 1973. Serum and blood concentration of sodium Cephalexin in man given single intra muscular and intra venous injections. *J. Pharm. Pharmacol.* 25:376–381.

Hagedorn, H. C., B. N. Jensen, and N. B. Kramp. 1936. Protamine insulinate. *J. Amer. Med. Ass.* 106:177.

Heitzman, R. 1983. The absorption, distribution and excretion of anabolic agents. *J. Animal Sci.* 57:233–238.

Higuchi, T., and H. Leeper. 1973. Improved osmotic dispenser employing magnesium sulphate and magnesium chloride. U.S. Patent 3,760,804.

Hsieh, D., W. Rhine, and R. Langer. 1983. Zero-order controlled release polymer matrices for micromolecules and macromolecules. *J. Pharm. Sci.* 72:17–22.

Hsieh, D. S. T., N. Smith, and Y. W. Chien. 1987. Subcutaneous controlled delivery of estradiol by compudose implants in-vitro and in-vivo evaluations. *Drug Dev. Ind. Pharm.* 13:2651–2660.

Hussain, M. A., B. J. Aungst, M. B. Maurin, and L.-S. Wu. 1991. Injectable suspensions for prolonged release nalbuphine. *Drug Dev. Ind. Pharm.* 17:67.

Juliano, R. L., and D. Stamp. 1975. The effect of particle size and charge on the clearance rates on liposomes and liposome encapsulated drugs. *Biochem. Biophys. Res. Comm.* 63:651–658.

Juliano, R. L., and D. Stamp. 1978. Pharmacokinetics of liposome encapsulated anti-tumor drugs studies with vinblastine actinomycin D cytosine arabinoside and daunomycin. *Biochem. Pharm.* 27:21–28.

Kataoka, T., and T. Kobayashi. 1978. Enhancement of chemo therapeutic effect by entrapping 1-beta-D arabinofuranosyl cytosine in lipid vesicles and its mode of action. In *Liposomes and their uses in biology and medicine*, edited by D. Papahadjopoulos. *Ann. N.Y. Acad. Sci.* 308:387–393.

Kent, J. S., B. H. Vickery, and G. I. McRae. 1980. The use of a cholesterol matrix pellet implant for early studies on the prolonged release in animals of agonist analogues of LHRH. *Proceedings of the 7th International Symposium on Controlled Release of Bioactive Materials* 7:67.

Kimelberg, H. K., H. K. Tracy, T. F. Watson, R. E. Kung, F. L. Reiss, and R. S. Bourke. 1978. Distribution of free and liposome-entrapped [3H] methotrexate in the central nervous system after intracerebroventricular injection in a primate. *Cancer Res.* 38:706–712.

Marshall, A. B., and G. H. Palmer. 1980. Injection sites and drug bioavailability. In *Trends in veterinary pharmacology and toxicology*, edited by F. van Miert and F. W. van der Kreek. Amsterdam: Elsevier, pp. 54–60.

Mayhew, E., D. Papahadjopoulos, Y. M. Rustum, and C. Dave. 1978. Use of liposomes for the enhancement of the cytotoxic effects of cytosine arabinoside. *Ann. N.Y. Acad. Sci.* 308:371–386.

McCullough, H. N., and R. L. Juliano. 1979. Organ-selective action of an antitumor drug: Pharmacologic studies of liposome-encapsulated beta-cytosine arabinoside administered via the respiratory system of the rat. *J. Natl. Cancer Inst.* 63:727–731.

Nash, H., D. N. Robertson, Y. A. J. Moo, L. E. Atkinson. 1978. Steroid release from silastic capsules and rods. *Contraception* 18:367–394.

Nicholas, P., B. R. Meyers, and S. Z. Hirschman. 1973. Cephalexin pharmacologic evaluation following oral and parenteral administration. *J. Clin. Pharmacol.* 13:463–468.

Ober, S. S., H. C. Vincent, D. E. Simon, and K. J. Frederick. 1958. A rheological study of Procaine Penicillin G depot preparations. *J. Am. Pharm. Assoc. (Sci. Ed.)* 47:667–676.

Pope, D. G. 1983. Specialized dose dispensing equipment. In *Formulation of veterinary dosage forms*, edited by J. Blodinger. New York: Marcel Dekker, Inc., pp. 71–134.

Preston, R. L., and J. R. Rains. 1993. Response dynamics evaluated for estradiol/TBA implantation. *Feedstuffs* (January):18–20.

Rahman, Y. E., W. R. Hanson, J. Bharucha, E. J. Ainsworth, and B. N. Jaraslow. 1978. Mechanism of reduction of antitumor drug toxicity by liposome encapsulation. *Ann. N.Y. Acad. Sci.* 308:325–341.

Rasmussen, F. 1980. Tissue damage at the injection site after intramuscular injection of drugs in food-producing animals. In *Trends in veterinary pharmacology and toxicology*, edited by F. van Miert and F. W. van der Kreek. Amsterdam: Elsevier, pp. 27–33.

Rasmussen, F., and O. Svendsen. 1976. Tissue damage and concentration at the injection site after intra muscular injection of chemo therapeutics and vehicles in pigs. *Res. Vet. Sci.* 20:55–60.

Robertson, D. N., I. Sivin, H. Nash, J. Braun, and J. Dinh. 1983. Release rates of levonorgestrel from silastic capsules, homogeneous rods and covered rods in humans. *Contraception* 27:483–495.

Runkel, R. A., and J. C. Lee. 1991. Controlled release subcutaneous implant. U.S. Patent 5,035,891.

Schou, J. 1961. Absorption of drugs from subcutaneous connective tissue. *Pharmacol. Rev.* 13:441–464.

Sommreville, E., and M. F. Tarttelin. 1983. Plasma testosterone levels in adult and neonatal female rats bearing testosterone propionate-filled silicone elastomer capsules for varying periods of time. *J. Endocr.* 98:365–371.

Spiegel, A. J., and M. M. Noseworthy. 1963. Use of nonaqueous solvents in parenteral products. *J. Pharm. Sci.* 52:917–927.

Struyker-Bouldier, H., and J. Smits. 1978. The osmotic minipump: A new tool in the study of steady-state kinetics of the drug distribution and metabolism. *J. Pharm. Pharmacol.* 30:576–578.

Sund, R. B., and J. Schou. 1964. The determination of absorption rates from rate muscles: An experimental approach to kinetic descriptions. *Acta Pharmacol. Toxicol. (Copenh.)* 21:313–325.

Theeuwes, F. 1975. Elementary osmotic pump. *J. Pharm. Sci.* 64:1987–1991.

Theeuwes, F., and T. Higuchi. 1974. Osmotic dispensing device for releasing beneficial agent. U.S. Patent 3,845,770.

Theeuwes, F., and S. I. Yum. 1976. Principles of the design and operation of generic osmotic pumps for the delivery of semi-solid or liquid drug formulations. *Ann. Biomed. Eng.* 4:343–353.

Tice, T. R., and S. E. Tabibi. 1992. Parenteral drug delivery injectables. In *Treatise on controlled drug delivery: Fundamentals, optimization, applications,* edited by A. Kydonieus. New York: Marcel Dekker, pp. 315–339.

Visscher, G. E., J. E. Pearson, J. W. Fong, G. J. Argentieri, R. L. Robison. 1988. Effect of particle size on the in-vitro and in-vivo degradation rates of poly(dl-lactide-co-glycolide) microcapsules. *J. Biomed. Mater. Res.* 22:733–746.

Wagner, J. G. 1961. Biopharmaceutics: Absorption aspects. *J. Pharm. Sci.* 50:359–387.

Wagner, J., H. Brown, N. W. Bradley, W. Dinusson, and W. Dunn. 1984. Effect of monensin estradiol controlled release implants and supplement on performance in grazing steers. *J. Animal Sci.* 58:1062–1067.

13

Applications in Vaccine and Gene Delivery

Maninder Hora
Ramachandran Radhakrishnan
Chiron Corporation
Emeryville, California

Vaccine and gene delivery are two large areas of research within the umbrella of sustained-release injectable products. While vaccines are one of the oldest biologicals approved for use in humans, genes represent the newest cutting edge research in biologicals. The historical gap between the development of the two technologies has been bridged by the recent resurgence in research on vaccine delivery. The revival of efforts in vaccines is due to the emergence of recombinant vaccine antigens and a realization of the need for new adjuvants (agents used to enhance in vivo immunogenicity; explained later in chapter) for the development of safer, more effective vaccine dosage forms. Despite these major visible differences, certain aspects of these two biological agents, especially pertaining to their in vivo delivery, are quite similar. Both systems represent interesting classes of compounds for which sustained release may not be the only requirement for their meaningful in vivo delivery. This is partly due to a lack of sufficient understanding of how these agents work (for conventional vaccines) or how they are expected to work (for genes and subunit vaccines) in the body. For these and other reasons described more fully in the chapter, the authors prefer to think of sustained-action systems for vaccines and genes instead of sustained-release systems. We have chosen to emphasize this modified thinking to be consistent with the intended purpose of this book—to instruct the reader in practical and pertinent pharmaceutical issues relating to the subject matter. The first part of this chapter deals with vaccines; genes are discussed in the second part of this chapter.

VACCINES

Vaccination is a deliberate attempt to protect humans against disease. During the last 200 years, since the time of Edward Jenner, vaccination has controlled many major diseases such as smallpox, diphtheria, tetanus, polio, measles, and rubella (Plotkin and Mortimer 1988). Relatively inexpensive doses of vaccines administered at various stages of human growth can prevent devastating disease conditions while greatly containing the healthcare costs of society. The goal of a vaccine is to elicit a prolonged (preferably lifelong) immune response similar to that in a real infection while avoiding the risk of infection. Several strategies for vaccines have been employed over the years. Traditional forms of vaccines are prepared by inactivating the real bacteria or viruses. In this case, the vaccine *antigen* tries to simulate the actual virus and thus provides a strong immune reaction. These vaccines suffer from the fact that they can sometimes revert to the active viral or bacterial form and cause a real infection. Subunit vaccines derived from certain key parts of a virus or bacterium are generally safer. However, they may not produce a strong enough immune reaction to inactivate the whole invading organism. In recent years, greater understanding of these phenomena has lead to the development of subunit antigens produced by recombinant DNA (deoxyribonucleic acid) methodology. While the recombinant vaccines suffer from some of the same drawbacks as the traditional subunit vaccines, they are far safer due to their much greater purity and inability to revert to an infective form. Recombinant vaccines are more reproducible than their conventional counterparts, as the processes employed are better controlled and more consistent.

When an invading organism first enters the body, it is taken up by macrophages and is exposed to helper T cells. The immune system responds in two ways: a cellular response (associated with killer T cells) and an antibody response (associated with B cells and natural killer cells) are generated. The cellular response begins by activating the T cells. The T cells act as helper T cells and coordinate other immune cells, memory T cells that remember the specific invading pathogen, and killer T cells that are primed to attack and eliminate the invader. The helper T cells signal the B cells to make antibodies against parts of the invader. Both the cellular and humoral (i.e., antibody) responses are key to the design of effective vaccines. The reader is referred to the May 1998 supplement of *Nature Medicine* for an excellent set of reviews on vaccines.

Challenges in Vaccine Delivery

The immune response to subunit vaccines is limited by several factors. First, a response to the subunit in question may not be able to neutralize the entire organism. Second, the immune system may not recognize the subunits if they are too small in size. To overcome these problems, *adjuvants* are used. Adjuvants are immune stimulators and act by presenting the subunit vaccine antigen in a way that results in an enhanced immune response. Thus, adjuvants have become an integral part of many vaccines available today.

An important aspect of vaccination is the length of immunological memory created and its relationship to the duration of protection provided. For a number of vaccines, usually those derived from inactivated viruses, the memory is often good enough to last a lifetime. However, with many others, boosters are needed every few years. This is particularly true of subunit vaccines. For viruses that mutate frequently but in a predictable way, vaccines are needed regularly. For example, a new cocktail of influenza antigens is used for the flu vaccination every year. In addition to the purely scientific challenges, formidable sociologic and economic tasks also remain. To eradicate (or at least contain) the many deadly diseases that afflict masses of people around the world, global vaccination is orchestrated by the World Health Organization (WHO). Therefore, issues such as the ability to combine several vaccines in a single shot, reduction in the frequency of administration to enhance compliance, and cost become overriding factors.

Requirements for Sustained-Action Vaccine Dosage Forms

Vaccines are unique because they elicit prolonged immune response in vivo after a single or multiple (usually two or three) applications. For example, the administration of alum-adsorbed hepatitis B (HB) vaccine in humans results in the generation of anti-HB antibodies that persist for long periods of time (West and Callandra 1996). The level of anti-HBs wanes after vaccination rapidly within the first year and more slowly thereafter. Vaccinees respond much more rapidly to a booster vaccination due to immunologic memory residing in memory B lymphocytes sensitized through initial exposure to the antigen. For HB vaccine, this memory lasts for 5–12 years in healthy vaccinees. This time period is different for different vaccines. Therefore, each vaccine has unique requirements for delivery to afford full protection to the subject.

In this chapter, we have deliberately used the term *sustained action* in lieu of sustained release, as the key issue with vaccines is not sustained levels of antigen or adjuvant, not even persistency of antibody levels, but rather the sustained protection against infection. For a commercially viable, sustained-action vaccine dosage form, there are two issues of paramount importance.

Immunological Considerations

An ideal vaccine dosage form would deliver the antigen (and adjuvant if required for the desired immune response) in such a way that after a single injection, the subject is conferred lifelong immunity. While this qualitative statement describes the desired outcome from a clinical point of view, it is unclear what this means to a product development scientist. The ambiguity in this regard is due to the current lack of a complete understanding of how various vaccines work in vivo. For example, the following questions are important from a product development perspective.

- Does the antigen need to be delivered to the subject continuously over long periods of time? If so, for how long?

- Does the antigen and adjuvant need to be codelivered at all times or is it acceptable for the antigen to be delivered without the adjuvant after the primary vaccination?

- Are both cellular and humoral responses important for protection?

Answers to these and other questions specific to the vaccine under consideration are critical for determining the immunological properties of the proposed sustained-action dosage form.

Product Development Considerations

A successful sustained-action dosage form useful for global mass vaccination must meet a number of fairly demanding product criteria. The following is a partial list:

- *Easy and painless administration:* Since a majority of vaccines are intended for pediatric use, the vaccinee is unaware of the benefits. Even the parents or guardians of the child may not fully appreciate the advantages of vaccination at the time. Therefore, vaccine administration must be easy and painless. Some subjects may consider even mild side effects, e.g., headache and injection site pain and inflammation, as too severe.

- *Frequency:* A single injection is preferred over multiple administrations, as the subject may not come back for boosters for a variety of sociological or cultural reasons. A single injection combining several vaccines may be even more preferable.

- *Cost:* Since the risk of no vaccination is not clearly understood by the general population, the cost of a vaccine must be low. In Third World countries where mass vaccination is even more critical for disease prevention, this issue is of added significance.

- *Stability:* At the very least, the antigen should be delivered in a conformationally active form, and the entire system should be stable for at least 18–24 months under refrigeration conditions. In tropical climates, the system should be unaltered by short exposures (from days to weeks) to ambient temperatures (e.g., 30°C).

Sustained-Action Technologies for Vaccine Delivery

While a number of attributes of the vaccine (i.e., antigen + adjuvant) would be important for developing an ideal sustained-action dosage form, the following sections describe how various adjuvants have been used to fulfill this intended purpose. From a rather simplistic point of view, technologies that are suitable for the sustained delivery of proteins by subcutaneous (SC) or intramuscular (IM) routes are generally expected to be useful for the sustained release of vaccines because vaccine antigens are macromolecular in nature. As discussed in the following sections, the protein sustained-release technologies confer other attributes to the delivered vaccine and partially fulfill the sustained-action vaccine requirements to varying degrees.

Suspensions

Suspensions of alum are the only adjuvants that are currently present in licensed vaccine products. Of the 41 injectable vaccines that are marketed in the United States, 15 are adjuvanted with alum salts (PDR 1998). Of these, a majority (30) are administered intramuscularly and a few subcutaneously (10). One is administered orally. In these preparations, antigens are adsorbed to alum particles (~100 nm to 10 μm in size) and filled into vials as a suspension. The most commonly employed salts are aluminum hydroxide and aluminum phosphate, which have been extensively characterized for their crystallinity, morphology, and adsorption capacity. Aluminum adjuvants are believed to convert the soluble protein antigen into a particulate form by adsorption to the surface of the precipitated aluminum salt (Shirodkar et al., 1990). The adsorbed antigen depot allows the slow release of antigen, thus prolonging the time for interaction between antigen and antigen-presenting cells and lymphocytes. The aluminum-antigen beads become the focus of inflammation and attract immunocompetent cells, thereby forming granulomas that contain antibody-producing plasma cells. Aluminum suspensions do not normally generate cell-mediated immunity. Alum also plays a secondary but important role in the development of B cell memory.

Antigen release from alum-adsorbed vaccines has not been characterized in detail. However, the release duration is not believed to be greater than 1–2 weeks (Cox and Coulter 1997). For subunit glycoprotein antigens, in vitro desorption experiments revealed that the interaction between alum and the antigen was weak and that the protein was released from the alum in a matter of hours (Weissburg et al. 1995). With alum, the prolonged immunity is granted primarily by the vaccine, as it induces active synthesis of antibodies accompanied by immunological memory for the virus or bacterium that affords ongoing protection.

Recent studies on in vivo absorption of aluminum adjuvants were reported by Flarend et al. (1997). Using [26]Al-labeled adjuvant suspensions, they were able to show that Al levels in blood and urine were detected at 1 h and up to 28 days after IM injection in rabbits. The Al concentrations produced in the blood 1 h after injection were similar to those obtained 2–28 days after injection. A comparison of the crystalline aluminum oxyhydroxide (AH) with amorphous aluminum phosphate (AP) adjuvant indicated that AP was absorbed to a $3\times$ greater extent than AH. These studies provide a general idea of the rate and extent of degradation of aluminum adjuvants in vivo.

A suspension containing particles of γ-inulin embedded with alum (known as Algammulin) has also been evaluated for adjuvant activity and found to be more potent than alum in early studies (Cooper et al. 1991).

Emulsions

Extemporaneously prepared oil emulsions have been used as adjuvants for vaccines for many decades. Perhaps, the best known examples in this regard are Freund's complete and incomplete adjuvants (Gupta et al. 1993). Freund's complete adjuvant (FCA) consisted of a mineral oil–in-water emulsion containing killed myobacteria and was found to be too toxic for use in humans. FCA without the myobacteria is known as Freund's incomplete adjuvant (FIA). FIA is a less potent adjuvant compared to FCA but is much more tolerable. FIA was tested in human clinical studies but the reactogenicity profile was found to be not acceptable, and a recommendation was made to discontinue the use of mineral oil adjuvants in humans (Gupta et al. 1993).

Recent progress has been made toward the development of both water-in-oil (W/O) (Scalzo et al. 1995) and oil-in-water (O/W) (Ott et al. 1995) emulsions as adjuvants. In particular, more attention has been focused on manufacturing these products on an industrial scale using equipment and processes that yield consistent and reproducible adjuvant emulsions. Both the Chiron MF59 (Ott et al. 1995) and Syntex SAF (Lidgate et al. 1989) adjuvant emulsions are produced using high energy "microfluidization" processes and are physicochemically well defined and stable. The SAF emulsion has been shown to enhance the immunogenicity of many vaccines in animal studies (Hjorth et al. 1997; Byars et al. 1991). The MF59 emulsion has been administered in more than 6,000 healthy volunteers and was highly immunogenic, having no significant systemic or local side effects (Traquina et al. 1996). This emulsion adjuvant is currently in late clinical trials with many experimental vaccines. Finally, MF59 became the first emulsion adjuvant to be approved in a Western country: FluAd, an MF59-adjuvanted influenza vaccine, was approved in Italy in April 1997.

Several factors, such as antigen presentation and moderate immunomodulation through passive targeting of the oil particles to the lymph nodes, are believed to contribute to the adjuvant activity of emulsions. However, an important mechanism for the adjuvant action of emulsions is thought to be through slow release of antigen from the emulsion base. This is thought to be especially relevant for antigens that are integral membrane proteins, which readily associate with the oil phase of emulsions. The slow release properties of adjuvant emulsions have not been characterized in detail. Further understanding of this property may perhaps help in designing more powerful emulsion adjuvants.

Liposomes

Twenty years ago, liposomes were shown to function as immunological adjuvants (Allison and Gregoriadis 1974). From a presentation perspective, membrane-soluble viral subunits reconstituted into liposomes could elicit immune responses in a fashion similar to the native virus, provided that the orientation of the viral protein in liposomes was identical. Liposomes composed of natural phospholipid components are nonimmunogenic, biodegradable, and nontoxic. Even though the particulate structures must render a certain amount of adjuvancy to liposomes, these pseudomembranes or artificial bilayers are comparatively inferior to bacterial cell wall components. Aqueous encapsulation of soluble antigens (recombinant or otherwise), electrostatic adsorption of the polypeptide to the liposome surface, or incorporation of a hydrophobic viral protein into a membrane bilayer have been shown to contribute to the adjuvant property of liposomes (Shek 1984; Therien and Shahum 1989). The composition of the liposomes, as a general rule, has to be varied to accommodate the specific antigen(s) for encapsulation or membrane anchoring. It is possible that liposomes could act as a depot (Shek 1984), releasing the antigen to antigen-presenting cells (APCs) such that macrophages and/or liposome migration to lymph nodes results in elicitation of humoral immunity (Velinova et al. 1996). Liposomes have also been shown to induce cell-mediated immunity (CMI) by the activation of cytotoxic T lymphocytes (CTLs) (Lopes and Chain 1992).

Even though there has been a considerable controversy regarding CTL responses, vesicles with pH-sensitive lipids have been shown to fuse with endosomes and release the antigens into the cytosol. Processing of this protein and MHC Class I presentation then leads to the observed CTL responses (Reddy et al. 1992). Receptor-mediated targeting to macrophages using mannosylated ligands has been reported in further attempts to improve the adjuvancy of liposomes (Nair et al. 1992). Coentrapment of a cytokine such as interleukin-2 together with the antigen has been shown to enhance immune responses to the antigen (Ho et al. 1992).

In novel genetic vaccination approaches to improve immune responses, cationic liposomes with encapsulated antigen genome have been shown to elicit 100-fold higher antibody titers compared to surface complexes of DNA and naked DNA (Gregoriadis et al. 1997). Cationic liposomes complexed with human immunodeficiency virus (HIV-1) DNA and containing cholesterol-mannan as an integral component of the liposomes, when administered in mice, elicited significant HIV specific CTL activity (Toda et al. 1997). Mannan liposome-mediated DNA vaccination has thus been shown to augment Th-1 mediated immunity. It appears that this field will gain prominence and momentum with concomitant advances in gene therapy using similar systems.

Polymeric Systems

The principles of controlled drug and protein delivery are increasingly being applied to the delivery of vaccines. Nano- and microparticles of polymers have been employed as carriers for the preparation of prolonged action vaccines. These can be divided into two broad categories: nanoparticles (i.e., < 1 μm size) and microparticles. The distinction between these is necessary because their sustained-release attributes and adjuvant action are based on quite different mechanisms.

Whereas nanoparticles are more like other particulate systems (such as emulsions and liposomes) in their characteristics, microparticles are true sustained-delivery systems. In general, nanoparticles exhibit a modest depot effect, are transported from the injection site to other locations such as the lymph nodes, and exert their adjuvancy through strategic presentation of the antigen. Microparticles, on the other hand, display controlled release of the antigen over much greater time periods and remain in local areas around the injection site.

Nanoparticles. The mode of action of nanoparticles is very similar to those of classical vaccine adjuvants. Typically, antigens are adsorbed to polymeric particles < 1 μm in size and injected as a suspension. In studies employing influenza antigens adsorbed to polyacrylate nanoparticles, Kreuter and Haenzel (1978) showed that smaller-sized particles produced a much greater antibody response than particles larger than 0.5 μm. The use of nanoparticles resulted in a more reproducible response than with alum. This superiority is believed to be due to the ability to manufacture uniform nanoparticles in a physicochemically reproducible manner compared to alum. Similar results were obtained with a nanoparticulate-adsorbed HIV-2 whole-virus vaccine, in which greater antibody titers than with alum were maintained in mice for longer periods (Steineker et al. 1991). In both of these studies, nonbiodegradable polymers have been used to make vaccine nanoparticles, thereby limiting their future application to humans.

Microparticles. Significant advances in the controlled release of proteins from microspheres have been made in the past decade (Hora et al. 1990; Cohen et al. 1991; Johnson et al. 1996). These developments have spurred simultaneous interest in the use of microspheres for vaccines (O'Hagan et al. 1991; Eldridge et al. 1991; Esparza and Kissel 1992; Aguado and Lambert 1992). All of these studies have employed the biodegradable poly(lactide-co-glycolide) (PLGA) copolymer as the carrier for antigens. PLGA has been used as resorbable suture material in humans for approximately 30 years and possesses controllable in vivo degradation properties in addition to an excellent safety profile.

Eldridge and coworkers (1991) administered small (1–10 μm) PLGA microparticles loaded with Staphylococcal enterotoxin B (SEB) antigen by injection and oral routes in mice. A single injection produced a strong potentiation of the circulating antibody response, which appeared to be function of both a depot effect and the rapid phagocytosis of the microspheres by the APCs. Preliminary data also indicated that orally administered microspheres were adsorbed by the M cells overlaying the Peyer's patches and passed to the immune inductive environment of both the Peyer's patches and systemic lymphoid organs and produced concurrent low-level systemic and mucosal antibody responses. Eldridge et al. (1991) also postulated a vaccination strategy in which PLGA microspheres with various degradation times would be injected to produce distinct pulses of antigen release. By manipulation of a number of variables, including copolymer ratio and size, four discrete release pulses of antigen over a 120-day period were achieved.

More recently, Cleland et al. (1997) applied the same strategy and obtained two pulses of release (in vitro) of an experimental HIV-1 antigen MN gp120: the first pulse during the first day and the second after a few weeks or months. Cleland's group also encapsulated the adjuvant QS-21 in PLGA microspheres for coadministration with MN gp120 microspheres. Thomasin et al. (1996) showed that the second burst of tetanus toxoid (TT) from PLGA microspheres seen in vitro was much weaker in vivo. Finally, Singh et al. (1997) showed that a single shot of a hepatitis B surface antigen (HBsAg) was able to maintain an antibody response for 1 year, which was similar to that obtained with a conventional 3-injection schedule (at 0, 1, and 6 months) using HBsAg adsorbed to alum.

The studies with vaccine microparticles are extremely promising as indicated by the preclinical investigations cited above. However, the field is slightly behind the protein delivery field in terms of product development, where clinical testing of similar systems is already under way (Langer 1996).

Unmet Needs and Future Directions in Sustained Vaccine Delivery

A simplified comparison of various adjuvants for vaccine delivery is presented in Table 13.1.

By now, the reader should have gathered that the field of sustained-action vaccine delivery is deficient in many aspects. Since progress in vaccinology and vaccine delivery has been made in an empirical manner, many fundamental observations remain unexplained today. To a large degree, these "holes" are present due to our lack of a complete understanding of how vaccines work. Vaccinologists are still trying to comprehend the type, schedule, and duration of immune response desired for

Table 13.1. *Comparison of Various Vaccine Delivery Adjuvants*

Attribute	Suspensions	Emulsions	Liposomes	Nanoparticles	Microparticles
Particle size	1–100 μm	< 1 μm	< 50 μm	< 1 μm	1–200 μm
Duration of sustained release	days	days	days	days	months
Antigen presentation	no	yes	yes	no	no
Targeting	no	yes	yes	yes	no
Immunological action	Strong Th-2[1], IgE	Weak Th-1 and Th-2	none	none	none
CTL[2] induction	no	no	yes	no	no
Current clinical status	Only adjuvants approved currently in the U.S.	In advanced Phase III clinical trials in the U.S.	In early clinical studies	In preclinical evaluation	In preclinical evaluation

[1]Immunological activity as it relates to activation of the two CD4 T cell subpopulations Th-1 or Th-2. Th-1 responses usually induce complement fixing antibody and strong, delayed-type hypersensitivity reactions and are associated with the generation of interferon-γ, interleukin-2, and interleukin-12, while Th-2 responses generate high circulatory and secretory antibody levels, mainly IgE and interleukins -4, -5, -6, and -10.

[2]Cytotoxic T lymphocyte induction

various vaccines. A complicating factor is the diversity in antigen structure and their likely mode of action and limitations imposed by recombinant technology. A significant gap in knowledge is the memory aspect of vaccines. It is clear that fundamental advances in the vaccine field will be necessary before one can define precisely the requirements for a vaccine sustained-action system.

The proactive development of sustained-action dosage forms is considered highly desirable. So far, the sustained-action properties of vaccines have been obtained by default in many cases. A more detailed characterization of the currently applied sustained-action methodologies is required. An important aspect to sort out in systems such as suspensions and emulsions is the contribution of the depot effect versus other factors such as antigen presentation and passive targeting. Nanoparticles from biodegradable polymers need to be evaluated as vaccine adjuvants to expedite the evaluation of this technology in humans. In controlled delivery systems such as microspheres, a better control of the in vivo release profile is necessary to facilitate timely "boost" effects. Also, it needs to be determined if coadministration of antigen and adjuvant is required for immune protection. In this context, more sophisticated ways to provide antigens and adjuvants need to be devised. Novel ways to confer long-term stability of antigens and adjuvants when placed in the human body environment need to be investigated. New systems for vaccine delivery need to be continually developed to meet the ever-changing requirements for delivering vaccines to counter rapidly mutating viruses. Finally, all vaccine sustained-action systems need to be made more affordable, as cost remains a significant impediment to the application of these technologies to the nations of the Third World, where in many cases these systems are most needed.

GENE DELIVERY

A gene is a stretch of DNA and is the carrier of genetic information. Fidelity of genetic information is maintained by an accurate copying of the DNA sequence in a complementary ribonucleic acid (RNA) molecule and the decoding of the triplet sequence of the RNA (and therefore of the parent DNA) during the transcription and translation processes. Compartmentalization of cell types is achieved by the differential transcription of the genes during embryogenesis and organogenesis. Any impairment in this scheme of events leads to genetic defects and may be manifested in a number of pathological conditions. For example, cystic fibrosis, which afflicts 1 in 2,000 births primarily in the Caucasian population, has been shown to be due to a point mutation in the two defective copies of the cystic fibrosis transmembrane receptor (CFTR) gene that encodes a chloride channel in the airway epithelial cells. This transmembrane protein is required for electrolyte homeostasis (Collins 1992). It is conceivable that there may be partial or defective expressions in the levels of other proteins.

Historical Perspectives

Phenylketonuria (PKU) has long been recognized as a disease stemming from the loss of a gene that leads to an accumulation of the toxic metabolites of the amino acid phenylalanine. Since an understanding of the therapeutic approaches to the correction of genetic defects was not realized until the 1970s, the only available historic medical treatment for PKU is a dietary restriction. With advanced knowledge of the molecular genetics of bacteria and viruses and the biology of cellular processes, the concept of efficiently using viruses for gene delivery for therapeutic purposes was proposed (Friedmann and Roblin 1972; Roemer and Friedmann 1992). By 1988, more than 4,500 human diseases were classified as genetic in origin.

A detailed understanding of viruses provided clues for introducing pieces of useful DNA into the human genome. The existence of episomal plasmids such as the ones that confer drug resistance to bacteria and the knowledge that sufficient size plasmids with genetic information can synthesize proteins independent of the chromosomes gave impetus to the rational design of DNA transporters or vectors. These vectors or constructs include, in addition to the complementary DNA encoding of the protein of interest (cDNA), other segments to turn on and regulate the expression.

A milestone was achieved toward gene delivery with advances in recombinant DNA technology and genetic engineering, which permitted gene splicing, cloning of DNA segments into desired plasmids, and protocols for the introduction of such plasmids into organisms to alter their genetic basis. Ten years ago, an efficient procedure for the introduction of DNA using cationic liposomes as carrier systems was demonstrated in cell lines (Felgner et al. 1987). This process referred to as lipofection has become the method of choice for numerous experimental investigations for the delivery of useful therapeutic genes in vivo in animals and humans.

Concurrent with all of these advances has been the remarkable discovery that naked DNA or recombinant plasmid vectors can be introduced into cells in vivo (Wolff et al. 1990). IM injection of plasmid DNA is currently being explored for genetic vaccination (Ulmer et al. 1993; Hilleman 1995) and secretion of therapeutic proteins (Mumper and Rolland 1998) into systemic circulation, especially in cases where low circulating levels of these polypeptides are beneficial for the alleviation of diseases. Further efforts are directed toward sustained gene action, i.e., expression and development of sustained-release particulate carriers such as liposomes, PLGA microspheres, dendrimers, and so on, for DNA delivery.

Desired Attributes of a Sustained-Release Gene Delivery System

Unlike vaccines, the field of gene therapy is in its infancy. There have been several vaccines on the market for many decades; however, not a single gene product has reached the marketplace yet. As with any emerging technology, much needs to be understood about genes themselves before one could make a list of ideal attributes for their delivery. The characteristics of genes that are important for their inclusion in sustained-action dosage forms are their large size (tens of millions daltons) and strong negative charge. In general, the requirements for these systems can be divided into two categories: (1) those common to any drug delivery technology and (2) those specific to gene therapy only.

General Requirements for Gene Delivery Systems

To be successful, new gene therapy products will need to compete for clinical and patient acceptance against existing pharmaceutical and biological therapies. Like the existing pharmaceutical drugs, gene delivery products must meet the requirements of effectiveness, safety, quality, purity, stability, compliance, and cost (Ledley 1995). For example, the effective dose of gene products should conform to the cost-effectiveness of the therapy compared to existing recombinant protein pharmaceuticals. If liquid plasmids do not have adequate stability, freeze drying should render them sufficiently stable without loss of activity. In addition, gene products should be capable of being mixed and administered with carriers such as cationic liposomes (see below) from two vials or from a dual compartment syringe. Once these criteria are met, gene therapy products should easily fit into established clinical practice.

Requirements Specific to Gene Delivery

The efficiency and persistence of gene expression depends on the nature of the particular target cell type in vivo. For example, exogenously delivered genes have been reportedly expressed for several months resulting from the persistence of plasmid DNA in muscle that is associated with a slow turnover of myofiber nuclei (Wolff et al. 1990; Jiao et al. 1992). However, the clinical utility of direct gene transfer, even in a tissue such as muscle, is hampered by low and variable levels in expression, DNA degradation, or drainage into the local lymph nodes (Manthorpe et al. 1993). Sustained delivery or release in a simple case such as plasmid DNA delivery, therefore, will depend on the choice of the formulation that would enable the dispersal of DNA at the site of administration, protection of DNA from nucleases, site retention, and methods to facilitate the uptake of DNA by the regiospecific cells. The therapeutic application of genes in general requires noninvasive, cost-effective carriers that are safe for repetitive use, resulting in consistent levels of gene expression. Other desirable attributes for sustained gene delivery systems are the long in vivo residence times of the carriers and the tissue targetability.

An ideal delivery system containing all of the elements mentioned above is outlined in the cartoon representation presented in Figure 13.1. Assuming that the carrier/DNA complex is delivered intact into the cytoplasm or the complex exits the endosome intact, the nuclear localization sequence (NLS) built into the structure of the delivery system (see Figure 13.1) will guide it into the nuclear compartment.

Figure 13.1. *An ideal plasmid delivery system.*

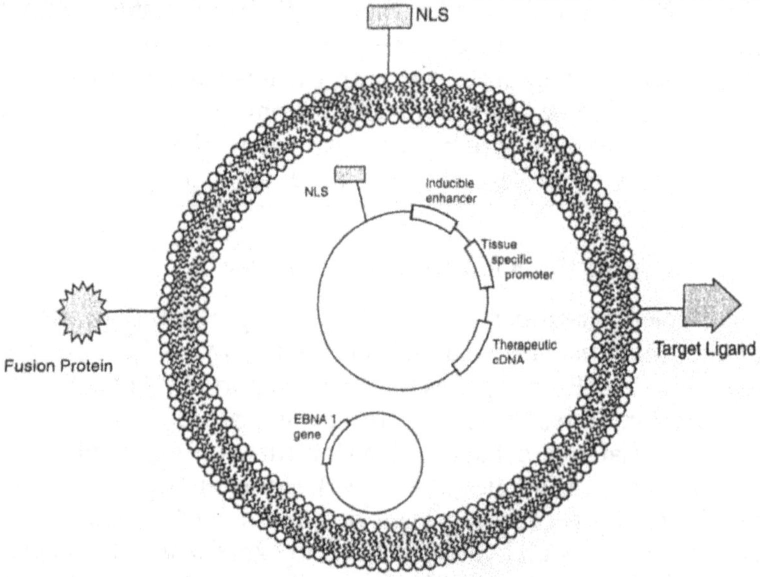

This cartoon depicts a liposome carrying the therapeutic cDNA with a tissue-specific promoter and an "on" and "off" enhancer. The nuclear localization sequence (NLS) on the plasmid would guide it to the nucleus. If the particle were to be delivered intact to the cellular interior, the NLS may be located on the particle exterior. A fusion peptide/protein (similar to a viral sequence) or a fusogenic lipid will allow the fusion of the particle with the cellular plasma membrane. Alternatively, a targeting ligand such as a monoclonal antibody may permit cell-specific interaction and endocytosis. Another plasmid intended to express Epstein-Barr nuclear antigen (EBNA-1), which would subsequently interact with the therapeutic plasmid, may allow longer expression of the functional plasmid episomally.

Sustained-Action Systems for Gene Delivery

Lipids and Liposomes

Felgner and coworkers (1987) published the first paper in the transfection of cells using cationic lipid complexes of DNA in culture. This has evolved into an area of intense investigation involving both synthesis of better cationic lipids and preparative methods for liposomes and their targeting. A comprehensive treatise on gene delivery applications using liposomes has appeared recently (Lasic 1997; Chonn and Cullis 1998).

Typically, the cationic lipid contains a positively charged head group(s), a spacer of varying length, a linker bond, and a hydrophobic anchor. A review of the literature shows that the first generation cationic lipids introduced for gene delivery included lipids with chemically stable ether linkages, nondegradable and resistant to cellular metabolism, inhibited protein kinase C activity, and exhibited direct cytotoxicity (Balasubramaniam et al. 1996). However, synthetic efforts from numerous laboratories have resulted in ester-linked lipids and other structures with an acceptable safety profile; the structural parameters have been reviewed by Gao and Huang (1995). A novel series of cationic lipids called lipitoids have recently been proposed for the cellular delivery of plasmid DNA (Huang et al. 1998).

The safety and short-term toxicity of a N-[1-(2,3-dimyristyloxy)propyl]-N,N-dimethyl-N-hydroxyethylammonium bromide (DMRIE) mixed with dioleoyl phosphatidylethandamine (DOPE) as a cationic liposome complex with HLA-B7 plasmid was studied in mice by tail vein injection (0.15 μmol lipid/dose × 3), and organ toxicities were monitored (San et al. 1993). Examination of serum biochemical enzymes and tissue pathologic studies revealed no abnormalities. In the same study, plasmid DNA–lipid complexes were administered directly into the arteries of pigs without any change in myocardial function, although vasculitis was evident. In a toxicological study with cationic liposome-plasmid complex by repeat intravenous (IV) injections in mice or cynomolgus monkeys, no significant toxicity was found based on clinical chemistry, hematology, or organ pathology (Parker et al. 1995). No abnormal immune response to the MHC class I HLA-B7 vector was observed. No clinical toxicity was recorded in a Phase I clinical trial in humans using the cationic 3β[N-N′,N′-dimethylaminoethane-carboxyl] cholesterol (DC-chol) DNA complex (Nabel et al. 1993).

The expression of genes delivered to the lungs in animals and humans has been reported (Caplen et al. 1995) as well as the tolerability of the DNA–cationic lipid complex delivered even in large doses. Lung expression can also be achieved by an IV administration of cationic liposome–DNA complex (Liu et al. 1997a, b) presumably due to the large particles being arrested in the pulmonary bed during circulation. Thus, it is clear that suitable nonviral delivery systems can be designed for repeat administrations in vivo, with a safety profile reasonable for clinical administration in humans. Sustaining the action of the DNA–lipid complex is feasible despite the moderate to poor expression levels realized with these systems.

Very recently, cationic liposomes complexed with DNA were targeted to specific cells in vitro by means of monoclonal antibodies (MAbs) or ligands associated with Stealth® lipid component (Kao et al. 1996). A significant increase in expression levels of the reporter genes was observed in appropriate cell lines.

In a line extension of the receptor-based targeting concept proposed as a "ternary-complex" approach, polylysine-antibody conjugates were bound to DNA, and the rest of the charges on the DNA were neutralized by cationic liposomes (Trubeskoy et al. 1992) added simultaneously.

Despite several issues that still exist, significant progress has been made with liposome carrier systems. Polysorbate 80 was included as a substitute for DOPE in cationic liposome formulations and emulsions that minimized the aggregation problem seen typically with DNA–liposome complexes and the inhibition of transfection activity owing to serum instability (Liu et al. 1997). Another cationic liposome formulation composed of a lipopolyamine and DOPE claimed DNA protection, slow plasma clearance, and effective cellular uptake following IV administration in mice (Thierry et al. 1997). Supercoiled DNA presumed to promote enhanced gene expression compared to relaxed or linear DNA was detected in blood up to 1 h after injection and resulted in significant transgene expression.

In a suicidal gene therapy approach, nude mice bearing pancreatic tumors were treated with a cationic liposome composed of dioctadecylamidoglycylspermine (DOGS) and DOPE, complexed with herpes simplex virus Thymidylate kinase (HSV TK) plasmid followed by ganciclovir administration (Aoki et al. 1997). Eight of the 14 mice treated were free of tumors, and the remaining 6 showed significant shrinkage. This is a powerful demonstration of the success achievable with cationic liposomal delivery coupled with a knowledge of virology. Finally, the safety record seen in the two early clinical trials (Nabel et al. 1993; Caplen et al. 1995) and the fact that several others are ongoing or planned attest to the potential of the liposome gene delivery approach.

Polymeric Systems

A classic example of a polymer used to facilitate and augment gene expression has been the use of polyethyene glycol (PEG) in transfections. In recent experiments to assess (β-galactosidase expression using the plasmid DNA in polyvinyl pyrrolidone (PVP) formulations, a linear dose response for the expression was demonstrated (Mumper et al. 1996). Lipopolyamines (Behr 1986; Behr et al. 1989) have been shown to condense DNA into small, discrete, multimolecular particles and deliver the gene in mammalian cells. Polycations such as poly-L-lysine are efficient DNA–condensing agents quite similar to natural histones. However, these complexes are only effective in the presence of a lytic peptide, inactivated adenovirus particles, or lysosomotropic agents such as chloroquine in gene delivery (Wagner et al. 1992). A number of targeted delivery systems based on polylysine amino conjugation chemistry and charge neutralization of plasmid DNA have been developed with varying degrees of success in vitro and in vivo. Receptor mediated in vitro gene transformation by a soluble carrier system for DNA was demonstrated by a chemical modification approach (Wu and Wu 1987), and this has been followed by numerous examples.

A new class of synthetic polymers, Starburst™ dendrimers (Baker et al. 1996), belonging to the group of polyamidoamines, have been developed recently. These are less toxic than linear poly-L-lysines (Haenzler and Szoka 1993) and appear to be non-immunogenic and noncytotoxic to cells. Numerous in vitro transfection studies have shown the utility of these carriers, and the potential for conjugation to antibodies and other ligands offers additional opportunities for the targeted and sustained delivery of DNA for many different types of therapies.

Utilizing the bioadhesive potential of polyanhydride copolymers of fumaric and sebacic acid, β-galactosidase gene was incorporated into microspheres prepared from these polymers and administered to rats by oral intubation (Mathiowitz et al. 1997). Preliminary results appear promising for sustaining gene delivery to elicit mucosal immune responses to specific antigens.

Viral Vectors

Retroviruses, single-stranded RNA viruses, have the capacity to randomly insert their genome into the host chromosome. From the perspective of a sustained action of the gene, longer expression can certainly be achieved with retroviruses. Murine Moloney Leukemia virus has been used in human clinical protocols (Mulligan 1993). The remote possibility of a recombination event leading to oncogenic potential in the host coupled with the limitation that gene expression only takes place in dividing cells render this approach restricted in scope. Recombinant retroviruses (rRVs) designed with appropariate deletions should make them safer, and further applications of such systems are awaited.

The adenovirus genome (a linear, double-stranded DNA) can be manipulated to include a foreign gene of 7–7.5 kb length and offers an attractive platform for gene delivery. Recombination is rare, and a number of replication defective vectors have been prepared and tested in animals and humans (Crystal et al. 1994). Unfortunately, the utility of this approach is limited by the acute inflammatory responses seen with high doses of the virus presented in earlier clinical trials as well as the antibody responses to viral proteins generated following the first dose (Crystal 1995; Knowles et al. 1995). A number of modifications, especially in combinations with nonviral systems such as liposomes, are currently being explored to target the vectors as well as reduce doses (Kichler et al. 1996).

Adeno-associated virus serotype 2 (AAV2) is a single stranded ~4.7 kb DNA human parvovirus and integrates on human chromosome 19. This virus has thus far been known to be safe. The obvious limitations with this virus are that the foreign DNA can only be ~4.3 kb in size, and the production of particles must ensure complete removal of the helper adenovirus. Electrotransfection of mouse bone marrow stem cells with an AAV plasmid-containing globulin gene has demonstrated (Ohi and Kim 1996) the production of human hemoglobin in these cells, offering hope for gene therapy of hemoglobinopathies. Several laboratories are also investigating the AAV approach for correcting other blood disorders such as hemophilia.

Plasmid Modifications

The idea of using "transcriptionally regulated" vectors provides a safety insurance policy in the applications of gene delivery (Miller and Whelan 1997). Even though carriers such as liposomes with attached antibodies or ligands may provide a reasonable targeting of the genes to one tissue rather than others, safety remains a serious concern in the development of gene therapy. Proteins that are therapeutic in one tissue may be harmful in another.

For example, restricting the expression of a gene in a specific tissue will reduce nonspecific toxicities and provide enhanced and presumably persistent expression in the target organ. Recombinant adenoviruses have been designed in which the human low-density lipoprotein (LDL) receptor gene was incorporated under the transcriptional control of the hepatotropic hepatitis B virus (HBV) core promoter that was also linked to HBV enhancer I (Sandig et al. 1996). These viruses permitted moderate to strong expression of the marker gene in vitro and in mice in vivo. Expression in lung and skeletal muscle was weak compared with the cytomegalovirus (CMV) promoter control.

In a recent report, infection of mini-pigs with recombinant adenovirus via different routes of administration (including IV) resulted in luciferase marker gene expression in a number of tissues with only intermediate liver expression (Torres et al. 1996). However, the moderate to strong expression of LDL receptor in liver was demonstrated using recombinant adenovirus, where the human LDL receptor gene was under the transcriptional control of the HBV core promoter, which bodes well for the plasmid construction or design to modulate tissue selective gene expression.

An approach for pulsed "expression" was outlined in the ecdysone-inducible expression cassette reported recently. This steroid insect hormone, which apparently has no effects on mammals, can be administered to initiate the expression of the steroid receptor, a member of nuclear receptor superfamily. The synthesis of the modified receptor regulates an optimized ecdysone responsive promoter. The "on" switch augments the synthesis of the gene product of interest. Inductions reaching four orders of magnitude have been achieved with ecdysone (No et al. 1996). It is likely that a vector will soon be made where tissue-specific expression of synthetic transcription factor will be combined with regulatable transgene expression, thus enabling precise control in the production of therapeutic proteins in desired tissues by the administration of an inducing or repressing agent. This inducible system is an excellent example for sustaining the "pulsed" delivery of the therapeutic protein to the tissue of interest and controlling the levels of a protein pharmaceutical with a therapeutic window by the use of the regulator hormones or drugs, e.g., doxycycline.

Knowledge gained recently regarding the EBNA 1 protein of Epstein-Barr virus (EBV) pertaining to its interaction with DNA segments to trigger replication (Laine and Frappier 1995) may help in designing vectors for sustaining the expression while staying episomally. Codelivery of the gene of interest and the EBNA 1 gene as part of the same plasmid or different plasmids would be an approach toward this goal. Based on the structure of the Env protein of the Maloney Leukemia Virus (Fass et al. 1997) or the NLS of growth factors (Lin et al. 1996), targeted vectors could be designed by a chemical modification of plasmids.

A comparative evaluation of both viral and nonviral delivery systems is presented in Table 13.2. The comparisons are made with respect to key criteria, such as the duration of action, toxicity of carriers, and manufacturability, to provide a flavor for the pharmaceutical utility of these systems. It is clear that no single system for gene delivery satisfies all of the criteria at the present time.

Table 13.2. *Comparative Evaluation of Different Gene Delivery Systems*

Systems	Applications/ Utility	Toxicity	Ease of Manufacture	Cost of Goods	Stability	Repeat Dosing	Duration of Action
Viral							
Retro	Gene therapy, ex vivo	Integration associated oncogenic potential	Customized therapy; may be cumbersome	Moderate to expensive	Good	One time ex vivo application	Months to year(s)
Adeno	In vivo gene therapy	Acute inflammation	Moderate to cumbersome	Moderate	Good	One-shot therapy	Months
AAV	In vivo gene therapy	Low potential toxicity	Moderate to difficult scale-up	Moderate	Good	Antibodies may preclude second application	Weeks
Nonviral							
Naked DNA	Genetic vaccination	Very low toxicity, excipient selection important	High feasibility	Comparatively inexpensive	Good	None	Several weeks to months
Polymer conjugate	Limited gene therapy	Toxicity of polymer, e.g., polylysine	Easy	Inexpensive to moderate	Fair-good	Less likely	

Table 13.2 continued on next page.

Table 13.2 continued from previous page.

Systems	Applications/ Utility	Toxicity	Ease of Manufacture	Cost of Goods	Stability	Repeat Dosing	Duration of Action
PLGA microspheres	Vaccination and limited gene therapy	Low carrier toxicity	Moderate difficulty in scale-up; targeted systems need a lot of work	Moderate to expensive	Fair to good	Expected to be compatible for repeats	Days to weeks
Liposomes							
DNA surface bound	Targeted gene therapy/limited vaccination	Minimal toxicity but selection of degradable lipids is key	Surface complexation easier	Moderate	Fair	No immunogenicity seen in clinic	Days to week(s)
DNA encapsulated	Targeted gene therapy/limited vaccination	Low toxicity with natural or biocompatible lipids	Encapsulated and targeted systems difficult to prepare and scale-up	Expensive	Fair-good	Expected to be good for repeat dosing	Days to week(s)

Future Directions in Gene Delivery

The explosive growth in the literature with publications ranging in scope from the ex vivo gene therapy for homozygous familial hypercholesteroalemia (Grossman et al. 1995) to retroviral gene transfer to correct ocular disorders (Dunaief et al. 1995) and discussions of strategies to achieve targeted gene delivery via receptor-mediated endocytosis pathways (Michael and Curiel 1994) all attest to the fact that we are moving away from the realm of mere speculation toward reality. Eleven clinical protocols were ongoing five years ago (Anderson 1992), and the human trials have increased exponentially since then. As would be evident to the reader from the foregoing presentation of the issues associated with the delivery vehicles, such as aggregation, storage stability, and tissue distribution, there are still numerous obstacles to tackle. Basic research should continue in the direction of optimizing retroviral vectors that can sustain the gene action without even a remote possibility of oncogenic potential. If adequate safety can be assured, lentiviruses, which have the ability to infect dividing and nondividing cells (Verma and Somia 1997), may be clinically useful. Work should continue in reducing inflammatory and immune responses to recombinant adenoviruses. The manufacture of sufficient quantities of non-pathogenic AAV should assist the entry of these systems to clinical trials and establish the true potential of this viral delivery system.

In view of the unique properties of some DNA carriers, such as cationic liposomes, effort should continue in the direction of developing biocompatible products, i.e., stable in the presence of biological fluids such as serum (Hashida et al. 1996). Even though liposome–DNA surface complexes can be prepared in water extemporaneously and tested in animals immediately, formulations thus generated will never be useful clinically. The numerous cellular and molecular barriers to gene transfer using cationic liposomes have been examined critically (Zabner et al. 1995).

The encapsulation of DNA in neutral or negatively charged liposomes using currently available techniques is abysmal, while cationic liposomes tend to bind DNA by electrostatic association on the surface, and true encapsulation in the aqueous interior in small liposomes is difficult to achieve. More work should go toward understanding the DNA condensation processes in the hope that such studies would provide clues to encapsulating native DNA in particulate carriers. The relevance of the various forms of DNA in expression and understanding ways to maintain the native DNA conformation even after encapsulation are key parameters that deserve careful study. It is worth mentioning that cationic liposome-encapsulated DNA has been shown to induce a greater than 100-fold increase in antibody titers compared to surface complexes (Gregoriadis et al. 1997).

Finally, tissue selective plasmid expression will remain a myth unless suitable guided missiles in the form of targeted carriers are developed for in vivo applications. Besides MAbs, knowledge gained regarding integrin-binding cyclic peptides (Hart et al. 1995) and the NLS of mitogenic growth factors (Lin et al. 1996) may be exploited in devising such systems. Robust and rapid manufacture of these new line pharmaceuticals based on cutting-edge technology is certainly awaited (Durland and Eastman 1998). Hopefully, the coming years will witness a rapid pace in technology and a revolution in terms of precise diagnosis of genetic disorders and the availability of an armamentarium of DNA–based injectable biopharmaceuticals to treat a myriad of diseases.

ACKNOWLEDGMENTS

The authors gratefully acknowledge Dr. Chen-Yi Huang for drawing the cartoon depiction of an ideal plasmid delivery system.

REFERENCES

Aguado, M. T., and P. H. Lambert. 1992. Controlled release vaccines–biodegradable polylactide/polyglycolide (PL/PG) microspheres as antigen vehicles. *Immunobiology* 184:113–125.

Allison, A. C., and G. Gregoriadis. 1974. Liposomes as immunological adjuvants. *Nature* 252: 252.

Anderson, W. F. 1992. Human gene therapy. *Science* 256:808–813.

Aoki, K., T. Yoshida, N. Matsumoto, H. Ide, K. Hosokawa, T. Sugimura, and M. Terada. 1997. Gene therapy for peritoneal dissemination of pancreatic cancer by liposome mediated transfer of Herpes Simplex Virus Thymidylate Kinase gene. *Hum. Gene Ther.* 8:1105–1113.

Baker Jr., J. R., A. Bielinska, J. Johnson, R. Yin, and J. F. Kukowska-Latallo. 1996. Efficient transfer of genetic material into mammalian cells using polyamidoamine dendrimers as synthetic vectors. In *Artificial self-assembling systems for gene delivery,* edited by P. L. Felgner, M. J. Heller, P. Lehn, J.P. Behr, and F. C. Szoka Jr. Washington, D.C.: American Chemical Society, pp. 129–145.

Balasubramaniam, R. P., M. J. Bennett, A. M. Aaberle, J. G. Malone, M. H. Nantz, and R. W. Malone. 1996. Structural and functional analysis of cationic transfection lipids: The hydrophobic domain. *Gene Ther.* 3:163–72.

Behr, J. P. 1986. DNA strongly binds to micelles and vesicles containing lipopolyamines or lipointercalants. *Tet. Lett.* 27:5861–5864.

Behr, J. P., B. Demeneix, J. P. Loeffler, and J. Perez-Mutul. 1989. Efficient gene transfer into mammalian primary endocrine cells with lipopolyamine-coated DNA. *Proc. Natl. Acad. Sci. USA* 86:6982–6986.

Byars, N. E., G. Nakano, M. Welch, D. Lehman, and A. C. Allison. 1991. Improvement of hepatitis B vaccine by the use of a new adjuvant. *Vaccine* 9:309–318.

Caplen, N. J., E. W. F. W. Alton, P. G. Middleton, J. R. Dorin, B. J. Stevenson, X. Gao, S. R. Durham, P. K. Jeffrey, M. E. Hodson, C. Coutelle, L. Huang, D. J. Porteous, R. Williamson, and D. M. Geddes. 1995. Liposome-mediated CFTR gene transfer to the nasal epithelium of patients with cystic fibrosis. *Nature Med.* 1:39–46.

Chonn, A., and P. R. Cullis. 1998. Recent advances in liposome technologies and their applications for systemic gene delivery. *Adv. Drug. Del. Rev.* 30:73–83.

Cleland, J. L., A. Lim, L. Barron, E. T. Duenas, and M. F. Powell. 1997. Development of a single-shot subunit vaccine for HIV-1: Part 4. Optimizing microencapsulation and pulsatile release of MN rgp120 from biodegradable microspheres. *J. Control. Rel.* 47:135–150.

Cohen, S., T. Yoshioka, M. Lucarelli, L. H. Hwang, and R. Langer. 1991. Controlled delivery systems for proteins based on poly(lactic-glycolic) microspheres. *Pharm. Res.* 8:713–720.

Collins, F. S. 1992. Cystic fibrosis: Molecular biology and therapeutic implications. *Science* 256:774–779.

Cooper, P. D., C. McComb, and E. J. Steele. 1991. The adjuvancity of Al-gammulin, a new vaccine adjuvant. *Vaccine* 9:408–415.

Cox, J. C., and A. R. Coulter. 1997. Adjuvants–a classification and review of their modes of action. *Vaccine* 15:248–256.

Crystal, R. G. 1995. Transfer of genes to humans: Early lessons and obstacles to success. *Science* 270:404–410.

Crystal, R. G., N. G. McElvany, M. A. Rosenfeld, C.-S. Chu, A. Mastrangeli, J. G. Hay, S. L. Brody, H. A. Jaffe, N. J. Eijja, and C. Danel. 1994. Administration of an adenovirus containing the human CFTR cDNA to the respiratory tract of individuals with cystic fibrosis. *Nat. Genet.* 8:42–51.

Dunaief, J. L., R. C. Kwun, N. Bhardwaj, R. Lopez, P. Gouras, and S. P. Goff. 1995. Retroviral gene transfer into retinal pigment epithelial cells followed by transplantation into rat retina. *Hum. Gene Ther.* 6:1225–1229.

Durland, R. H., and E. M. Eastman. 1998. Manufacturing and quality control of plasmid-based gene expression systems. *Adv. Drug. Del. Rev.* 30:33–48.

Eldridge, J. H., J. K. Staas, J. A. Muelbroek, J. R. McGhee, T. R. Tice, and R. M. Gilley. 1991. Biodegradable microspheres as a vaccine delivery system. *Molec. Immunol.* 28:287–294.

Esparza, I., and T. Kissel. 1992. Parameters affecting the immunogenicity of microencapsulated tetanus toxoid. *Vaccine* 10:714–720.

Fass, D., R. A. Davey, C. A. Hamson, P. S. Kim, J. M. Cunningham, and J. M. Berger. 1997. Structure of a murine leukemia virus receptor-binding at 2.0 Angstrom resolution. *Science* 277:1662–1666.

Felgner, P. L., M. J. Heller, P. Lehn, J. P. Behr, and F. C. Szoka Jr. 1996. *Artificial self-assembling systems for gene delivery.* Washington, D.C: American Chemical Society.

Felgner, P. L., T. R. Gadek, M. Holm, R. Roman, H. W. Chan, M. Wenz, J. P. Northrop, G. M. Ringold, and M. Danielson. 1987. Lipofection: A highly efficient, lipid-mediated DNA-transfection procedure. *Proc. Natl. Acad. Sci. USA* 84:7413–7417.

Flarend, R. E., S. L. Hem, J. L. White, D. Elmore, M. A. Suckow, A. C. Rudy, and E. A. Dandashli. 1997. In vivo absorption of aluminum-containing vaccine adjuvants using ^{26}Al. *Vaccine* 15:1314–1318.

Friedmann, T., and R. Roblin. 1972. Gene therapy for human genetic disease? *Science* 175:949–955.

Gao, X., and L. Huang. 1995. Cationic liposome-mediated gene transfer. *Gene Ther.* 2:710–722.

Gregoriadis, G., R. Saffie, and J. Brian de Souza. 1997. Liposome-mediated DNA vaccination. *FEBS Lett.* 402:107–110.

Grossman, M., D. J. Rader, D. W. M. Muller, D. M. Kolansky, K. Kozarsky, B. J. Clark III, E. A. Stein, P. J. Lupien, H. Bryan Brewer Jr., S. E. Raper, and J. M. Wilson. 1995. A pilot study of ex vivo gene therapy for homozygous familial hypercholesterolaemia. *Nature Med.* 1:1148–1154.

Gupta, R. K., E. H. Relyveld, E. B. Lindblad, B. Bizzini, S. Ben-Efraim, and C. K. Gupta. 1993. Adjuvants–a balance between toxicity and adjuvanticity. *Vaccine* 11:293–306.

Haenzler, J., and F. C. Szoka. 1993. Polyamidoamine cascade polymers mediate efficient transfection of cells in culture. *Bioconjugate Chem.* 4:372.

Hart, S. L., R. P. Harbottle, R. Cooper, A. Miller, R. Williamson, and C. Coutelle. 1995. Gene delivery and expression mediated by an integrin binding peptide. *Gene Ther.* 2:552–554.

Hashida, M., R. I. K. Kawabata, T. Miyao, M. Nishikawa, and Y. Takakura. 1996. Pharmacokinetics and targeted delivery of proteins and genes. *J. Control. Rel.* 41:91–97.

Hilleman, M. R. 1995. DNA vectors: Precedents and safety. In *DNA vaccines: A new era in vaccinology,* edited by M. A. Liu, M. R. Hilleman, and R. Kurth. New York: New York Academy of Sciences, pp. 1–14.

Hjorth, R. N., G. M. Bonde, E. D. Piner, K. M. Goldberg, and M. H. Levner. 1997. The effect of Syntex adjuvant formulation (SAF-m) on humoral immunity to the influenza virus in the mouse. *Vaccine* 15:541–546.

Ho, R. J. Y., R. L. Burke, and T. C. Merigan. 1992. Liposome-formulated interleukin-2 as an adjuvant of recombinant HSV glycoprotein gD for the treatment of recurrent genital HSV-2 in guinea pigs. *Vaccine* 10:209–213.

Hora, M. S., R. K. Rana, J. H. Nunberg, T. R. Tice, R. M. Gilley, and M. E. Hudson. 1990. Controlled release of interleukin-2 from biodegradable microspheres. *Bio/Technology* 8:755–758.

Huang, C., T. Uno, J. E. Murphy, S. Lee, J. D. Hamer, J. A. Escobedo, F. E. Cohen, R. Radhakrishnan, V. Dwarki, and R. Zuckermann. 1998. Lipitoids–novel cationic lipids for cellular delivery of plasmid DNA in vitro. *Chem & Biol.* 5:3456–3454.

Jiao, S., P. Williams, R. K. Berg, B. A. Hodgeman, L. Liu, G. Repetto, and J. A. Wolff. 1992. Direct gene transfer into non-human primate myofibers in vivo. *Hum. Gene Ther.* 3:21–33.

Johnson, O. L., J. L. Cleland, H. J. Lee, M. Charnis, E. Duenas, W. Jaworowicz, D. Shepard, A. Shahzamani, A. J. S. Jones, and S. D. Putney. 1996. A month-long effect from a single injection of microencapsulated human growth hormone. *Nature Med.* 2:795–799.

Kao, G. Y., L.-J. Chang, and T. M. Allen. 1996. Use of targeted cationic liposomes in enhanced DNA delivery to cancer cells. *Cancer Gene Ther.* 3:250–256.

Kichler, A., W. Zauner, C. Morrison, and E. Wagner. 1996. Ligand-polylysine mediated gene transfer. In *Artificial self-assembling systems for gene delivery,* edited by P. L. Felgner, M. J. Heller, P. Lehn, J. P. Behr, and F. C. Szoka Jr. Washington, D.C.: American Chemical Society, pp. 120–128.

Knowles, M. R., K. W. Hohneker, R. N., Z. Zhou, J. C. Olsen, T. L. Noah, P-C. Hu, M. W. Leigh, J. F. Engelhardt, L. T. Edwards, K. R. Jones, M. Grossman, J. M. Wilson, L. G. Johnson, and R. C. Boucher. 1995. A controlled study of adenoviral-vector-mediated gene transfer in the nasal epithelium of patients with cystic fibrosis. *New Engl. J. Med.* 333:823–831.

Kreuter, J., and I. Haenzel. 1978. Mode of action of immunological adjuvants: Some physicochemical factors influencing the effectivity of polyacrylic adjuvants. *Infect. Immun.* 15:667–675.

Laine, A., and Frappier, L. 1995. Identification of Epstein-Barr virus nuclear antigen 1 protein domains that direct interactions at a distance between DNA-bound proteins. *J. Biol. Chem.* 270:30914–30918.

Langer, R. 1996. Controlled release of a therapeutic protein. *Nature Med.* 2:742–743.

Lasic, D. D. 1997. Liposomes in the delivery of antisense oligonucleotides. In *Liposomes in gene delivery,* edited by D. D. Lasic. Boca Raton, Fla., USA: CRC Press, pp. 207–226.

Ledley, F. D. 1995. Nonviral gene therapy: The promise of genes as pharmaceutical products. *Hum. Gene Ther.* 6:1129–1144.

Lidgate, D. M., R. C. Fu, N. E. Byars, L. C. Foster, and J. S. Fleitman. 1989. Formulation of vaccine adjuvant muramyldipeptides. 3. Processing optimization, characterization, and bioactivity of emulsion vehicle. *Pharm. Res.* 6:748–752.

Lin, Y.-Z., Y. Y. Song, and J. Hawiger. 1996. Role of the nuclear localization sequence in fibroblast growth factor-1 stimulated mitogenic pathways. *J. Biol. Chem.* 271:5305–5308.

Liu, F., H. Qi, L. Huang, and D. Liu. 1997a. Factors controlling the efficiency of cationic lipid-mediated transfection in vivo via intravenous administration. *Gene Ther.* 4:517–552.

Liu, Y., L. C. Mounkes, H. D. Liggitt, C. S. S. Brown, I. Solodin, T. D. Heath, and R. J. Debs. 1997b. Factors influencing the efficiency of cationic liposome-mediated intravenous gene delivery. *Nature (Biotechnol.)* 15:167–173.

Lopes, L. M., and B. M. Chain. 1992. Liposome-mediated delivery stimulates a class I restricted cytotoxic T cell response to soluble antigen. *Eur. J. Immunol.* 22:287–290.

Manthorpe, M., F. Cornefert-Jensen, J. Hartikka, J. Felgner, A. Rundell, M. Margalith, and V. Dwarki. 1993. Gene therapy by intramuscular injection of plasmid DNA: Studies on firefly luciferase gene expression in mice. *Hum. Gene Ther.* 4:419–431.

Mathiowitz, E., J. S. Jacob, Y. S. Jong, G. P. Carlino, D. E. Chickering, P. Chaturvedi, C. A. Santos, K. Vijayaraghavan, S. Montgomery, M. Bassett, and C. Morrell. 1997. Biologically erodable microspheres as potential oral drug delivery systems. *Nature* 386:410–414.

Michael, S. I., and D. T. Curiel. 1994. Strategies to achieve targeted gene delivery via the receptor-mediated endocytosis pathway. *Gene Ther.* 1:223–232.

Miller, N., and J. Whelan. 1997. Progress in transcriptionally targeted and regulatable vectors for genetic therapy. *Hum. Gene Ther.* 8:803–815.

Mulligan, R. C. 1993. The basic science of gene therapy. *Science* 260:926–932.

Mumper, R. J., and A. P. Rolland. 1998. Plasmid delivery to muscle: Recent advances in polymer delivery systems. *Adv. Drug Delivery Rev.* 30:151–172.

Mumper, R. J., J. G. Duguid, K. Anwer, M. K. Barron, H. Nitta, and A. P. Rolland. 1996. Polyvinyl derivatives as novel interactive polymers for controlled gene delivery to muscle. *Pharm. Res.* 13:701–709.

Nabel, G. J., E. G. Nabel, Z-Y. Yang, B. A. Fox, G. E. Plautz, X. Gao, L. Huang, S. Shu, D. Gordon, and A. E. Chang. 1993. Direct gene transfer with DNA-liposome complexes in melanoma: Expression, biological activity, and lack of toxicity in humans. *Proc. Natl. Acad. Sci. USA* 90:11307–11311.

Nair, S., F. Zhou, R. Reddy, L. Huang, and B. T. Rouse. 1992. Soluble proteins delivered to dendritic cells via pH-sensitive liposomes induce primary cytotoxic T lymphocyte response in vitro. *J. Exp. Med.* 175:609–612.

No, D., T.-Y. Yao, and R. M. Evans. 1996. Ecdysone-inducible gene expression in mammalian cells and transgenic mice. *Proc. Natl. Acad. Sci. USA* 93:3346–3351.

O'Hagan, D. T., H. Jeffery, M. J. J. Roberts, J. P. McGee, and S. S. Davis. 1991. Controlled release microparticles for vaccine development. *Vaccine* 9:768–771.

Ohi, S., and B. C. Kim. 1996. Synthesis of human globin polypeptides mediated by recombinant adeno-associated virus vectors. *J. Pharm. Sci.* 85:274–281.

Ott, G., G. L. Barchfeld, D. Chernoff, R. Radhakrishnan, P. van Hoogevest, and G. Van Nest. 1995. MF59: Design and evaluation of a safe and potent adjuvant for human vaccines. In *Vaccine design: The subunit and adjuvant approach,* edited by M. F. Powell and M. J. Newman. New York: Plenum Press.

Parker, S. E., H. L. Vahlsing, L. M. Serfilippi, C. L. Franklin, S. G. Doh, S. H. Gromkowski, D. Lew, M. Manthorpe, and J. Norman. 1995. Cancer gene therapy using plasmid DNA: Safety evaluation in rodents and non-human primates. *Hum. Gene Ther.* 6:575–590.

PDR. 1998. *Physicians' desk reference.* Montvale, N.J., USA: Medical Economics Company, Inc.

Plotkin, S. A., and E. A. Mortimer. 1988. *Vaccines.* Philadelphia: W.B. Saunders Co.

Reddy, R., F. Zhou, S. Nair, L. Huang, and B. T. Rouse. 1992. In vivo cytotoxic T lymphocyte induction with soluble proteins administered in liposomes. *J. Immunol.* 148:1585–1589.

Roemer, K., and T. Friedmann. 1992. Concepts and strategies for human gene therapy. *Eur. J. Biochem.* 208:211–225.

San, H., Z-Y. Yang, V. J. Pompili, M. L. Jaffe, G. E. Plautz, L. Xu, J. H. Felgner, C. J. Wheeler, P. L. Felgner, X. Gao, L. Huang, D. Gordon, G. J. Nabel, and E. G. Nabel. 1993. Safety and short-term toxicity of a novel cationic lipid formulation for human gene therapy. *Hum. Gene Ther.* 4:781–788.

Sandig, V., P. Loser, A. Lieber, M. A. Kay, and M. Strauss. 1996. HBV-derived promoters direct liver-specific expression of an adenovirally transduced LDL receptor gene. *Gene Ther.* 3:1002–1009.

Scalzo, A. A., S. L. Elliott, J. Cox, J. Gardner, D. J. Moss, and A. Suhrbier. 1995. Induction of protective cytotoxic T cells to murine cytomegalovirus by using a nonapeptide and a human-compatible adjuvant (Montanide ISA 720). *J. Virol.* 69:1306–1309.

Shek, P. N. 1984. Applications of Liposomes in immumoprotection. In *Immunotoxicology: A current perspective of principles and practice,* edited by P. W. Mullen. Berlin: Springer-Verlag, pp. 103–125.

Shirodkar, S., R. L. Hutchinson, D. L. Perry, J. L. White, and S. L. Hem. 1990. Aluminum compounds used as adjuvants in vaccines. *Pharm. Res.* 7:1282–1288.

Singh, M., X-M. Li, J. P. McGee, T. Zamb, W. Koff, C. Y. Wang, and D. T. O'Hagan. 1997. Controlled release microparticles as a single dose hepatitis B vaccine: Evaluation of immunogenicity in mice. *Vaccine* 15:475–481.

Steineker, F., J. Kreuter, and J. Iower. 1991. High antibody titers in mice with polymethylmethacrylate nanoparticles as adjuvants for HIV vaccines. *AIDS* 5:431–435.

Therien, H. M., and E. Shahum. 1989. Importance of physical association between antigens and liposomes in liposomal adjuvanticity. *Immunol. Lett.* 22:253–258.

Thierry, A. R., P. Rabinovich, B. Peng, L. C. Mahan, J. L. Bryant, and R. C. Gallo. 1997. Characterization of liposome-mediated gene delivery: Expression, stability and pharmacokinetics of plasmid DNA. *Gene Ther.* 4:226–237.

Thomasin, C., G. Corradin, Y. Men, H. P. Merkle, and B. Gander. 1996. Tetanus toxoid and synthetic malaria antigen containing poly(lactide)/poly(lactide-co-glycolide) microspheres: Importance of polymer degradation and antigen release for immune response. *J. Control. Rel.* 41:131–145.

Toda, S., N. Ishi, E. Okada, K.-I. Kusakabe, H. Arai, K. Hamajima, I. Gorai, K. Nishioka, and K. Okuda. 1997. HIV-1 specific cell-mediated immune responses induced by DNA vaccination were enhanced by mannan-coated liposomes and inhibited by anti-interferon-g antibody. *J. Immunol.* 92:111–117.

Torres, J. M., C. Alonso, A. Ortega, S. Mittal, F. Graham, and L. Enjuanes. 1996. Tropism of human adenovirus type 5-based vectors in swine and their ability to protect against transmissible gastroenteritis corona virus. *J. Virol.* 70:3770–3780.

Traquina, P., M. Morandi, M. Contorni, and G. Van Nest. 1996. MF59 adjuvant enhances the antibody response to recombinant hepatitis B surface antigen vaccine in primates. *J. Infect. Dis.* 174:1168–1175.

Trubeskoy, V. S., V. P. Torchilin, S. Kennel, and L. Huang. 1992. Cationic liposome enhanced targeted delivery and expression of exogenous DNA mediated by N-terminal modified poly-L-lysine-antibody conjugate in mouse endothelial cells. *Biochim. Biophys. Acta.* 1131:311–313.

Ulmer, J. B., J. J. Donnelly, S. E. Parker, G. H. Rhodes, P. L. Felgner, V. J. Dwarki, S. H. Gromkowski, R. Randall Deck, C. M. DeWitt, A. Friedman, L. A. Hawe, K. R. Leander, D. Martinez, H. C. Perry, J. W. Shiver, D. L. Montgomery, and M. A. Liu. 1993. Heterologous protection against influenza by injection of DNA encoding a viral protein. *Science* 259:1745–1749.

Velinova, M., N. Read, C. Kirby, and G. Gregoriadis. 1996. Morphological observations on the fate of liposomes in the regional lymph nodes after footpad injection into rats. *Biochim. Biophys. Acta.* 1299:207–215.

Verma, I. M., and N. Somia. 1997. Gene therapy—promises, problems and prospects. *Nature* 389:239–242.

Wagner, E. M., K. Zatloukal, M. Cotten, H. Kirlappos, K. Mechtler, D. T. Curiel, and M. L. Bernstiel. 1992. Coupling of adenovirus to transfermin-polylysine/DNA complexes greatly enhances receptor-mediated gene delivery. *Proc. Natl. Acad. Sci. USA* 89:6099–6103.

Weissburg, R. P., P. W. Burman, J. L. Cleland, D. Eastman, F. Farina, S. Frie, A. Lim, J. Mordenti, M. R. Peterson, K. Yim, and M. F. Powell. 1995. Characterization of the MN gp120 HIV-1 vaccine: Antigen binding to alum. *Pharm. Res.* 12:1439–1446.

West, D. J., and G. B. Calandra. 1996. Vaccine induced immunologic memory for hepatitis B surface antigen: Implication for policy on booster vaccination. *Vaccine* 14:1019–1027.

Wolff, J. A., R. W. Malone, P. Williams, W. Chong, G. Acsadi, A. Jani, and P. L. Felgner. 1990. Direct gene transfer into mouse muscle in vivo. *Science* 247:1465–1468.

Wu, G. Y., and C. H. Wu. 1987. Receptor-mediated in vitro gene transformation by a soluble DNA carrier system. *J. Biol. Chem.* 262:4429–4432.

Zabner, J., A. J. Fasbender, T. Moninger, K. A. Poellinger, and M. J. Welsh. 1995. Cellular and molecular barriers to gene transfer by a cationic lipid. *J. Biol. Chem.* 270:18997–19007.

14

Microspheres for Protein and Peptide Delivery: Applications and Opportunities

Paul A. Burke

Amgen
Thousand Oaks, California

Scott D. Putney

Eli Lilly and Co.
Indianapolis, Indiana

In many ways, proteins are ideal drugs. They are specific and exert their effects at low concentrations, and human monoclonal antibodies can be developed to bind to virtually any desired target. Their virtually limitless number enables their use to influence a large variety of biological processes, giving a large number of therapeutic candidates. These positive attributes, however, may ultimately be overshadowed by the difficulties with protein delivery. These include negligible oral and transdermal bioavailabilities (Lee 1995), which necessitates their delivery by injections or infusions. In addition, protein in vivo half-lives generally range from less than one to several hours (Table 14.1), thus requiring frequent administration. Sustained-release formulations offer numerous advantages, including protecting the protein over an extended period from degradation or elimination; increased patient comfort, convenience, and compliance due to fewer and less frequent injections; and the ability to deliver the protein locally, at relatively constant concentrations, to a particular site or body compartment, thereby lowering overall systemic exposure.

There are several, well-established methods to accomplish continuous administration of low molecular weight drugs. These include transdermal patches; injectable depots such as polymeric matrices or vesicles (e.g., liposomes); and externally worn, implantable, or orally administered pressure-driven pumps (Langer 1990). Several of the many currently marketed examples include the orally administered pump that releases the antihypertensive nifedipine, the transdermal delivery of nitroglycerin for heart disease, scopolamine for motion sickness and for the treatment of nicotine addiction, and the ethylene–vinyl acetate or polyanhydride polymeric systems for the intrauterine release of levonorgestrel or for the delivery of 1,3-bis (2-chloroethyl)-1-nitrosourea (BCNU) to treat brain tumors (Arky 1997). In some

Table 14.1. *Half-Lives in Humans and Approved Dosing Frequencies of Several Recombinant Proteins*

Protein	Serum Half-Life (h)	Weekly Dosing Frequency (Arky 1997)
α-interferon (Zhi et al. 1995)	5.6	3
γ-interferon (Mordenti et al. 1993)	0.6	3
hGH (Harvey 1995)	0.3	7
EPO (Cohen 1993)	6	3
G-CSF (Cohen 1993)	5	7
GM-CSF (Stute et al. 1995)	2	7

hGH, human growth hormone

EPO, erythropoeitin

G-CSF, granulocyte colony stimulating factor

GM-CSF, granulocyte macrophage colony stimulating factor

cases, sustained-release systems have enabled drugs to be used in new ways that were not possible with conventional formulations (Langer 1990).

The difficulty in developing injectable, sustained-release protein systems has been due principally to the instability of the protein, both during incorporation of the protein into the delivery device and after administration. In general, processes that produce sustained-release forms of smaller, more stable compounds are generally too harsh for proteins. Moreover, after injection, the protein must remain stable in a hydrated form at physiological temperatures for up to several weeks. Unlike smaller compounds, proteins are large structures with complex architecture, possessing secondary, tertiary, and, in some cases, quaternary structure with labile bonds and chemically reactive groups on their side chains. Disruption of these structures (e.g., denaturation or aggregation) or the modification of side chains can occur readily with many proteins and lead to the loss of activity or immunogenicity. This may limit the efficacy of therapy or, worse, create safety concerns.

The early reports of injectable, sustained-delivery protein systems, formed by the microencapsulation of the protein, focused principally on demonstrating that the protein could be released in a sustained fashion; there was little emphasis on an analysis of protein integrity. More recently, as the development of human therapeutic proteins has matured and the need for better protein delivery systems has become more important, the emphasis has shifted to analyzing the effect of encapsulation methods on the protein. The goal is encapsulation and release of the protein in an unaltered form. Over the last several years, numerous therapeutic proteins have been microencapsulated (see Table 14.2); in some cases, proteins have been released with properties comparable to those of the unencapsulated protein. This work is paving the way for protein stabilization and microencapsulation technology to be applied to a wide variety of protein drugs, including those that are currently marketed, in development, or in preclinical research. These formulations will likely allow the development of proteins that, for reasons such as unacceptable pharmacokinetics (e.g., very short half-lives), excessive systemic toxicity, or the inability to access their site of action from systemic circulation, would never be developed in a traditional pharmaceutical dosage form.

Table 14.2. *Injectable, Biodegradable, Sustained-Release Formulations of Human Therapeutic Proteins*

Protein	Encapsulation Method (State of Protein During Encapsulation)	Duration of Release	Stabilization Strategy	Encapsulated Protein Integrity	Released Protein Integrity	Reference
erythropoietin	double emulsion (solution)	Protein released in vitro for 32 days	emulsion included hydroxypropyl-β-cyclodextrin and BSA	84%	> 40% covalent aggregates remaining at 32 days	Morlock et al. (1997)
erythropoietin	atomization (solid)	Protein released in vivo for 24 days (rats)	protein colyophilized with ammonium sulfate	100%	99%	Zale et al. (1996)
nerve growth factor	double emulsion (solution)	Protein released in vitro for 30 days	emulsion included carboxymethyl-dextran and BSA	< 60%	NR	Krewson et al. (1996)
nerve growth factor	double emulsion (solution)	Protein released in vitro for 35 days, and in vivo (rats) for ~ 2 months	emulsion included gelatin	NR	immunoreactive and bioactive protein released over 35 days	Mendez et al. (1997)

Table 14.2 continued on next page.

Table 14.2 continued from previous page.

Protein	Encapsulation Method (State of Protein During Encapsulation)	Duration of Release	Stabilization Strategy	Encapsulated Protein Integrity	Released Protein Integrity	Reference
γ-interferon	double emulsion (solution)	NR	maximized protein concentration in emulsion; included trehalose	> 90%	NR	Cleland and Jones (1996)
GM-CSF	double emulsion (solution)	Protein released in vitro for 1 week	none	NR	bioactive protein released	Pettit et al. (1996)
interleukin-1α	double emulsion (solution)	Protein released in vitro for up to 80 days, depending on formulation. Tumor-bearing mice survived longer than bolus-treated mice.	emulsion included BSA	NR	< 50% bioactivity released over 20 days	Chen et al. (1997)
insulin	melt press (solution)	NR	none	< 40%	NR	Tabata et al. (1993)

Table 14.2 continued on next page.

Table 14.2 continued from previous page.

Protein	Encapsulation Method (State of Protein During Encapsulation)	Duration of Release	Stabilization Strategy	Encapsulated Protein Integrity	Released Protein Integrity	Reference
insulin	melt press (solid)	NR	none	90%	NR	Tabata et al. (1993)
brain-derived neurotropic factor	double emulsion (solution)	Protein released in vitro for 30 days	emulsifier included in polymer phase	NR	bioactive protein released over 30 days	Mittal et al. (1994)
nerve growth factor	atomization (solid)	Protein released in vitro for 2 weeks	none	94%	NR	Cleland and Duenas (1997)
α-interferon	atomization (solid)	Protein released in vivo (rats) for 7 days	encapsulated protein complexed and coencapsulated with zinc	NR	immunoreactive protein released	Tracy et al. (1996)
growth hormone	double emulsion (solid)	NR	protein colyophilized with trehalose and mannitol	100%	NR	Cleland and Jones (1996)

Table 14.2 continued on next page.

Table 14.2 continued from previous page.

Protein	Encapsulation Method (State of Protein During Encapsulation)	Duration of Release	Stabilization Strategy	Encapsulated Protein Integrity	Released Protein Integrity	Reference
growth hormone	double emulsion (solution)	Protein released in vitro for 30 days	maximized protein concentration; trehalose in emulsion	92–98%	< 75% at 30 days	Cleland et al. (1997b)
growth hormone	atomization (solid)	Protein released in vivo (rats and monkeys) for 30 days	encapsulated protein complexed and coencapsulated with zinc	97%	98% at 28 days	Johnson et al. (1997)
β-interferon	spray drying (solid)	Protein released in vivo (mice) for 70 days	none	100%	50% bioactivity released at 5 weeks in vivo	Eppstein and Schryver (1990)
vascular endothelial growth factor	atomization (solid)	Protein released in vitro for up to 40 days	none	100%	93% at day 21	Cleland et al. (1997a)
transforming growth factor-b	solvent casting (solid)	Protein released in vitro for 25 days	protein colyophilized with demineralized bone and mannitol	NR	~85% bioactivity released	Chen et al (1994)

Table 14.2 continued on next page.

Table 14.2 continued from previous page.

Protein	Encapsulation Method (State of Protein During Encapsulation)	Duration of Release	Stabilization Strategy	Encapsulated Protein Integrity	Released Protein Integrity	Reference
interleukin-2	suspension emulsion (solid)	Protein released in vitro for 30 days	emulsion included HSA	NR	100% bioactivity recovered at 35 days	Hora et al. (1990)

All proteins were encapsulated with biodegradable block copolymers of lactide and glycolide. Percent integrity listed is percent monomer (or native dimer) (Cleland and Duenas 1997, Cleland et al. 1997a; Cleland and Jones 1996; Mendez et al. 1997) determined by size exclusion chromatography, except in Krewson et al. (1996). Morlock et al. (1997), and Tabata et al. (1993) where gel electrophoresis and ELISA were used. Integrity of released protein was determined following hydration of the microspheres. Integrity of encapsulated protein was determined by extracting the protein from the microspheres and assaying under nondenaturing conditions; in Cleland et al. (1997b) and Cleland and Jones (1996). encapsulated integrity was inferred from analysis of protein released immediately upon hydration. In addition to the examples listed, insulin (Creque et al. 1980), TGF-β (Dinbergs et al. 1996; Silberstein and Daniel 1987), bFGF (Chen et al. 1994), and EGF (Taniguchi et al. 1991) have been encapsulated as solids using nonbiodegradable ethylvinyl acetate polymers. NR: not reported.

BSA, bovine serum albumin

GM-CSF, granulocyte macrophage colony stimulating factor

HSA, human serum albumin

ELISA, emzyme-linked immunosorbent assay

TGF-β, transforming growth factor-β

bFGF, basic fibroblast growth factor

EGF, epidermal growth factor

OVERVIEW OF PROTEIN ENCAPSULATION

Efforts to develop sustained-release formulations of proteins have principally focused on incorporating the protein into microspheres made of biodegradable polymers, generally the homo- and copolymers of lactic acid and glycolic acid (PLGA) (for reviews, see Cleland [in press], Pitt [1990], and Schwendeman et al. [1996]). PLGA polymers degrade by hydrolysis to ultimately give the acid monomers, hence the microspheres do not have to be removed. They are chemically unreactive under the conditions used to prepare the microspheres and, therefore, do not modify the protein.

After subcutaneous (SC) or intramuscular (IM) injection, the protein is released by a combination of diffusion and polymer erosion (Figure 14.1). By varying the polymer composition and

Figure 14.1. *Microspheres composed of polymers of lactic acid and glycolic acid.*

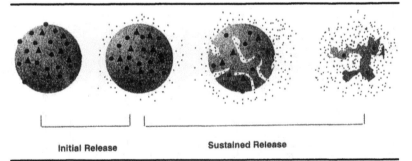

Initial Release   Sustained Release

The microspheres, about 50 to 100 μm in diameter, contain protein particles (denoted as large filled circles). After injection, protein release occurs in two ways. The initial release, which lasts about a day, results from dissolution of the protein at or near the surface of the microsphere (to give soluble protein, denoted as small circles). Sustained release occurs as the protein diffuses from the interior of the microsphere through pores created by polymer degradation or by dissolution of encapsulated solids. The duration of release can range from days to months. Degradation rates are determined by polymer molecular weight, the lactide:glycolide ratio, and the presence of hydrophobic end groups. Excipients can be included to stabilize the protein (triangles).

molecular weight, and by adding excipients that stabilize the protein or slow its release, the release rate can be adjusted to maintain the serum concentration within the therapeutic window (Figure 14.2). In contrast, following administration of solution formulations, protein levels quickly rise and decline, much of the time falling outside the therapeutic window.

Numerous methods have been used to encapsulate proteins into PLGA microspheres, because these polymers are not water soluble, essentially all encapsulation processes expose the protein to an organic solvent. These processes can be categorized by the physical state of the protein during encapsulation: either a solution or a solid. These two forms have significantly different stability profiles during encapsulation.

Figure 14.2. *Idealized serum profile (in arbitrary units) of a one-month microsphere formulation of a hypothetical protein.*

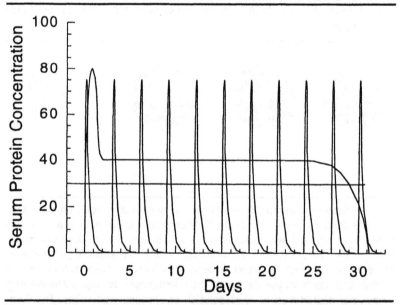

The solution formulation given every three days induces widely fluctuating serum levels that spike and then fall below the therapeutic concentration (horizontal line). The serum levels induced by the microsphere formulation peak (due to the initial release) and then remain constant above the therapeutic concentration.

Figure 14.3. *Processes for producing PLGA microspheres.*

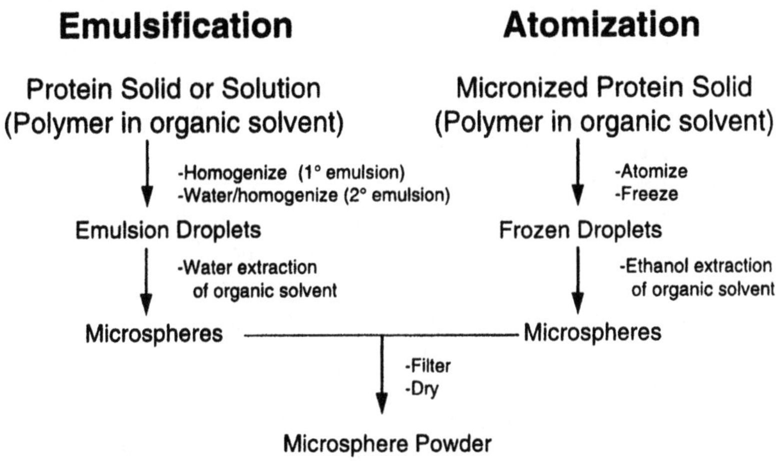

Emulsification

Protein Solid or Solution
(Polymer in organic solvent)

↓ -Homogenize (1° emulsion)
  -Water/homogenize (2° emulsion)

Emulsion Droplets

↓ -Water extraction
  of organic solvent

Microspheres ——————————

Atomization

Micronized Protein Solid
(Polymer in organic solvent)

↓ -Atomize
  -Freeze

Frozen Droplets

↓ -Ethanol extraction
  of organic solvent

—————— Microspheres

↓ -Filter
  -Dry

Microsphere Powder

Both processes begin with the polymer dissolved in an organic solvent (e.g., methylene chloride), in which the protein is insoluble. In the emulsification process (termed solvent evaporation or double emulsion), the protein, in an aqueous solution or as a solid, is emulsified (by sonication or homogenization) with the polymer solution. This gives small droplets or particles of drug contained in a larger droplet of polymer solution (the primary emulsion). The addition of water, which is followed by a second emulsification and extraction of the organic solvent, causes the PLGA to precipitate around the drug particles, resulting in hardened microspheres that are then filtered and dried. (Variations of the emulsification method shown are the single emulsion [Tabata et al. 1993] and phase separation [Lewis 1990] processes). In the atomization process, the micronized solid protein is suspended in the PLGA solution, which is then atomized using sonication or air atomization. This produces droplets that are then frozen in liquid nitrogen. The addition of ethanol (at –80°C), in which both the protein and the PLGA are insoluble, extracts the organic solvent from the microspheres.

With regard to encapsulation of protein solutions by emulsification, this method is used to produce PLGA microspheres of small molecular weight drugs and peptides, such as the 10 amino acid luteinizing hormone releasing hormone (LHRH) agonist (Ogawa et al. 1988; Sanders et al. 1986) (Figure 14.3). Although this type of process has been used to produce microspheres that release proteins in a sustained fashion, it is generally detrimental to the protein (Table 14.2). The main problem is the aqueous–organic solvent interface that is created during emulsification and present during the solvent extraction step. Proteins are generally amphipathic (i.e., contain both hydrophilic and hydrophobic surfaces) and often denature at such interfaces (Costantino et al. 1994; Hageman et al. 1992; Lu and Park 1995a, b; Tabata et al. 1993).

Despite these problems, approaches have been taken to stabilize proteins during emulsion processes. One is to reduce the fraction of drug in the surface-denatured layer by including surfactants in the primary emulsion (Krewson et al. 1996; Morlock et al. 1997; Mittal et al. 1994). In addition, carrier proteins such as gelatin or albumin have been included (Krewson et al. 1996; Chen et al. 1997; Mendez et al. 1997; Morlock et al. 1997). Similarly, the therapeutic protein concentration in the encapsulated solution has been maximized (> 100 mg/mL) (Cleland and Jones 1996). Overall, these approaches suggest that the protein in the interface represents a fixed loss and that exchange between surface-denatured and the remaining native protein is reversible or nonexistent. Another approach is to add small molecule osmolytes such as trehalose and mannitol to the protein phase (Cleland and Jones 1996). These act via preferential hydration of the native form (Arakawa and Timasheff 1985) and during the removal of water from the encapsulated protein solution (Pikal 1994).

To avoid the problems inherent in emulsion processes, a fundamentally different encapsulation method (atomization method, Figure 14.3) has been developed (Gombotz et al. 1991; Johnson et al. 1996, 1997). In this process, rather than protein in solution, lyophilized or precipitated dehydrated protein particles are exposed directly to the polymer solvent. Spectroscopic studies on the effect of organic solvents on protein powders have been done (e.g., using chymotrypsin, bovine pancreatic trypsin inhibitor, and lysozyme), and little additional denaturation results beyond that caused by lyophilization prior to solvent exposure (Griebenow and Klibanov 1996; Burke et al. 1992; Desai and Klibanov 1995). In addition, the effect of organic solvents on enzymes has been thoroughly characterized (Klibanov 1989). Additional stability can be conferred by lyoprotectants such as sucrose and ammonium sulfate that stabilize protein powders against organic solvent exposure (Burke et al. 1992; Cleland and Jones 1996).

Compared to emulsion processes, the atomization process maintains a higher degree of protein integrity for two reasons. First, other than that bound to the lyophilized protein, there is no water present and hence no aqueous–organic solvent interfaces. This reduces denaturation and aggregation, and the absence of water prevents reactions such as deamidation or hydrolysis. Second, during the atomization and solvent extraction steps, the protein remains at low temperatures. In addition to its benignity toward the protein, because there are no immiscible phases, the process requires no surfactants. Also, because the protein is not soluble in the extraction solvent, loss of the protein is minimal, thus increasing the encapsulation efficiency.

Regardless of the encapsulation method, microsphere properties and performance such as size, protein load, and duration of release can be adjusted to suit a particular application. The duration of release is determined principally by the polymer molecular weight and hydrophobicity. The longer the chain length of the polymer, the longer it takes to degrade to water-soluble lactide-glycolide oligomers. In addition, the more hydrophobic the polymer (e.g., a higher lactide content or the presence of a hydrophobic end group at the polyester terminus), the more slowly water is absorbed by the polymer and the longer the hydrolysis process. The protein load can vary from 0.1 percent to 30 percent, and lower or higher loading values are used for proteins requiring lower or higher doses, respectively. In addition, although the typical diameter of the microspheres is 50 to 100 μm, by modifying the emulsification (Jeffery et al. 1991) or atomization (Burke and Putney, unpublished) conditions, microspheres as small as a few micrometers can be produced for applications such as uptake by antigen-presenting cells or for injection through very thin needles.

With regard to the toxicity and safety profiles of PLGA microspheres, the polymer is nonimmunogenic and has a long safety record. In addition to its use for drug delivery, it is used to make biodegradable sutures, bandages, and bone plates (Austin et al. 1995; Pihlajamaki et al. 1992; Winde et al. 1993). Studies in rodents of protein- or peptide-containing PLGA microspheres have shown that the reaction to the depot is generally mild

(Anderson and Shive 1997). Studies of hGH–containing PLGA microspheres in rhesus monkeys showed that there is a mild foreign body reaction at the site of injection that resolves in about one month (Lee et al. 1997), and initial clinical studies of these microspheres showed no visible irritation at the site of injection (Vance et al. 1997). With regard to residual solvents, the level of methylene chloride in microspheres, made with either the emulsion or atomization process, is in the low ppm range, which is well below the limit for a pharmaceutical product (EMEA 1996).

USES OF MICROSPHERES TO DELIVER PROTEINS

The applications of sustained delivery of proteins can be grouped into four general types (Table 14.3):

1. *Systemic delivery*–examples of these range from proteins for replacement therapy to those given for autoimmune disorders, cancer, and antiviral therapy

Table 14.3. *Applications of Sustained-Release Protein Formulations*

Application	Examples
Systemic Delivery	Replacement (e.g., hGH, EPO, G-CSF)
	Cancer (cytokines)
	Antiviral (e.g., α-interferon)
	Autoimmune (e.g., β-interferon)
Local Delivery	Tumor immunotherapy (cytokines, tumor-associated antigens)
	Bones (growth factors)
	Joints (cytokine antagonists)
	Vasculature (growth factors, anti-inflammation proteins)
	Tumors (antiangiogenesis factors)
Sites Protected by Barriers	Brain (neurotrophic factors, cytokines)
	Eye (cytokines, antiangiogenesis proteins)
Vaccines	Viral
	Bacterial
	Tumor

2. *Local delivery* in which the site of action is a particular organ or tissue

3. *Delivery to sites that cannot be accessed by the bloodstream,* such as the brain or eye

4. *Vaccines.*

Systemic Delivery

The majority of the work with sustained protein delivery has been done with proteins that are administered systemically (Table 14.2). In addition to being administered less frequently, microencapsulated formulations provide relatively constant serum levels. The levels can be adjusted to fall within the therapeutic window by choosing the right dose and polymer degradation rate. Figures 14.4 and 14.5 show examples of serum levels

Figure 14.4. *Serum profiles of recombinant α-interferon in rats (Tracy et al. 1996).*

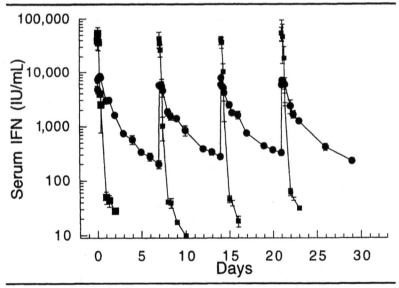

The serum concentration induced by the solution formulation (■) quickly falls to predose levels, whereas the serum levels induced by the microsphere formulation (●) remain elevated significantly longer.

Figure 14.5. *Serum profiles of hGH in juvenile rhesus monkeys (Johnson et al. 1996).*

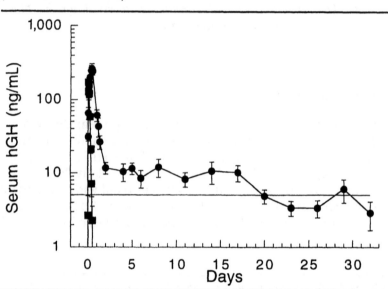

The serum concentration induced by the solution formulation (■) quickly falls to predose levels, whereas the serum levels induced by the microsphere formulation (●) remain elevated significantly longer. The horizontal line indicates the level of hGH (5 ng/mL) that induces elevated IGF-I levels.

induced by microencapsulated formulations of α-interferon and hGH relative to the same amount of protein given in solution.

One advantage of continuous delivery is the possibility of requiring a lower cumulative dose, and there are several examples of this phenomenon with peptide drugs. One is the injectable depot formulation of an LHRH agonist (Lupron Depot®, TAP Pharmaceuticals) in which fourfold less drug is required with the depot formulation than with conventional bolus therapy (0.25 mg/day versus 1.0 mg/day) (Arky 1997). This has also been seen with the continuous administration of analogues of somatostatin (Harris et al. 1995).

Likewise, the possibility of a cumulative dose reduction with continuous administration of protein drugs is now being observed. For example, with hGH, the biological effect of the sustained-release form (as measured by the increase in insulin-like growth factor-I, which mediates the growth effect) is significantly greater than when the same dose of protein solution is given as daily injections (Figure 14.6). Another example is bFGF, with which sustained release is threefold more effective in increasing the proliferation of vascular endothelial and smooth muscle cells than the same dose of the solution form (Dinbergs et al. 1996). It is interesting that in both cases, sustained delivery more closely mimics natural administration than bolus injections. In the case of hGH, it is released from the pituitary in a series of many frequent pulses (Daughaday 1995). Although this does not provide a continuous serum level, it does so more than large boluses given daily or three times a week. Moreover,

Figure 14.6. *The serum levels of IGF-I induced by the levels of hGH in Figure 14.5.*

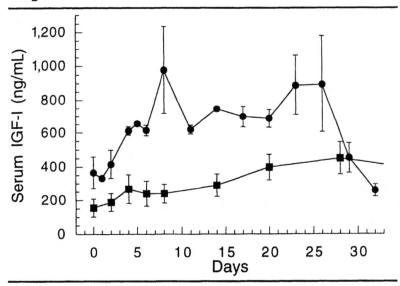

IGF-I mediates the effect of hGH, and the levels induced by the microspheres (●) are greater than those induced by daily injections of the same total amount of hGH as a solution (■). This suggests that a comparable therapeutic benefit would be elicited by an overall lower dose of hGH when delivered in a continuous fashion.

clinical studies have shown that continuous infusion of hGH via a pump results in growth velocity IGF-I levels comparable to that induced by daily injections (Laursen et al. 1995; Tauber et al. 1993; Jorgensen et al. 1990); one study (Tauber et al. 1993) suggested that the IGF-I levels induced were higher with continuous administration. In the case of bFGF, this protein remains bound to the extracellular matrix from which it is slowly released (Dinbergs et al. 1996). These results suggest that the natural mode of delivery and metabolism of the protein will determine whether treatment in a sustained fashion will require a lower cumulative dose.

Local Delivery

In many cases, proteins are administered systemically even though the desired site of action is a particular organ or tissue. It is generally not possible or practical to administer protein solutions directly to these sites because this would require invasive procedures and the effective concentration would quickly dissipate. Microsphere formulations would allow the effective concentration to be maintained locally for days to months, greatly reduce systemic exposure, and require significantly less protein.

Immunotherapy of Cancer

The goal of tumor immunotherapy is to induce or stimulate a CD8+ T cell (cytotoxic T lymphocyte, CTL) response against cells expressing tumor-associated antigens. Because cytokines (e.g., IL-2, the interferons, and GM-CSF) play a role in the activation of CTL precursors, the most straightforward approach to generating tumor-specific CTLs would be intralesional administration of the cytokine. This would provide local but transient levels to activate the developing immune response and avoid the significant dose-limiting toxicities resulting from systemic administration of these proteins. Unfortunately, clinical results with intralesional injections of cytokines have been limited to occasional responses (Cortesina et al. 1991; Ozzello et al. 1992; Vlock et al. 1994; Wussow et al. 1988), most likely due to the rapid clearance of the protein.

This has led to the more current approach in which autologous tumor cells (to provide the source of the tumor antigen) are cultured ex vivo, transfected with the gene for the cytokine using either recombinant viral or nonviral vectors, irradiated, and returned to the host as the vaccine. Studies in several animal models of cancer using this procedure have demonstrated induction of antitumor immunity able to prevent challenge with tumor cells or eradicate an existing tumor (Tepper and Mule 1994). Simply mixing the soluble cytokine with the irradiated cells provides little or no effect.

Cytokine-containing microspheres offer advantages over both local bolus administration and gene therapy approaches. First, microspheres would allow a precise dose of one or more cytokines to be mixed with irradiated tumor cells ex vivo. In ths case, the microspheres rather than the transfected gene would provide the source of the cytokine. The ability to control dose is currently difficult using gene transfer methods, both because the transfection efficiency is low and because the ability to precisely regulate the level of gene expression in different cell types is problematic (Mulligan 1993). With the finding that the efficacy of IL-2 as a vaccine against melanoma is dependent on the dose (Schmidt et al. 1995), the ability to deliver a precise amount of cytokine may be critical. Second, microspheres would allow the possibility of in situ delivery; this is difficult with gene therapy due to low transfection efficiencies. Third, because some of the cytokines currently under investigation for tumor vaccination are currently marketed for other indications (e.g., IL-2, γ-IFN, and GM-CSF), and thus their safety profiles are known, cytokine-containing microspheres would avoid the numerous safety and other regulatory concerns currently facing gene therapy.

Initial studies using microencapsulated cytokines rather than gene transfection have been promising. Tumor immunity can be induced using microspheres containing either IL-1 (Chen et al. 1997) or GM-CSF (Golumbek et al. 1993). The IL-1 microspheres were made using an emulsion method, and the GM-CSF microspheres were formed by cross-linking chondroitin sulfate with glutaraldehyde. Despite the use of these encapsulation methods, which caused loss of activity of the IL-1 and likely modified the GM-CSF by the reaction of glutaraldehyde with the free amino groups, in each case vaccination induced a protective immune response against subsequent tumor challenge. Mixing the irradiated cells with unencapsulated cytokine showed no effect.

Another approach to tumor immunotherapy is to deliver the tumor antigen directly to antigen-presenting cells (e.g., macrophages), which process the proteins to peptides and present them on the surface in the context of the MHC class I complex. Vaccination with recombinant vaccinia or fowlpox vectors using tumor antigens induces antigen-specific CTLs (Hodge et al. 1995; Wang et al. 1995). Although immunization with microspheres containing tumor-specific antigens has not yet been reported, antigens linked to beads about 10 microns in diameter are internalized by antigen-presenting cells and prime CD8+ CTLs several orders of magnitude more efficiently than soluble antigen (Kovacsovics-Bankowski et al. 1993).

Bones and Joints

Growth factors for engineering bone and cartilage is another class of proteins where sustained delivery has been investigated. Of specific interest are bone morphogenic (osteogenic) proteins (BMPs), of which two of the seven identified to date are in commercial development (Reddi 1994). Biodegradable polymers are attractive for the delivery of BMPs and other growth factors such as TGF-β (Gombotz et. al. 1993). Their action is primarily local, and the rate of matrix degradation can be adjusted to proceed at the same rate as new bone formation. Furthermore, microspheres can be combined with other biocompatible agents such as gels or other biopolymers to form pastes or composites, which are easily molded and can provide a scaffold for cell growth. For example, human BMP-2 administered as a paste of PLGA microspheres in a biopolymer solution is effective in repairing osseous defects in rats (Lee et al. 1994; Kenley et al. 1994). BMPs have also shown promise for periodontal regeneration (Ripamonti and Reddi 1994).

A related application is the treatment of joint disorders. In rheumatoid arthritis, the joints are sites of chronic inflammation, mediated by a variety of proinflammatory cytokines. Several cytokine antagonists, e.g., soluble p55 TNF receptor (Eliaz et al. 1996), an antitumor necrosis factor (TNF) monoclonal, an IL-1 receptor antagonist, and soluble IL-1 receptor (Feldmann et al. 1996), are being pursued in clinical trials as potential therapies. All of these proteins share the dual need for chronic delivery (to neutralize perpetual proinflammatory signals arising from the synovium) and for levels sufficient to inhibit the target cytokine (which can require a large molar excess). Rheumatoid arthritis may prove susceptible to antiangiogenic therapy as well. New vasculature is required to support the pathological growth of synovial cells in rheumatoid arthritis, and levels of the vascular endothelial growth factor (VEGF) are elevated in rheumatoid synovial tissue (Fava et al. 1994; Paleolog 1996).

Vasculature

In addition to being the most common avenue of drug delivery, the vasculature itself is a potential target of treatment. In an extrapolation of conventional intra-arterial drug administration, polymeric microspheres have been tested as chemoembolization anticancer agents. Targeted, endovascular delivery results in occlusion, followed by release of the chemotherapeutic from the emboli. First demonstrated in humans over 15 years ago when mitomycin was delivered from ethylcellulose microcapsules (Kato et al. 1981), this approach to cancer treatment is still an active area of research.

Small microspheres (nanospheres) are also being explored for the prevention of postangioplasty restenosis. PLGA nanoparticles containing the anti-inflammatory dexamethasone were administered intraluminally by a brief infusion in a rat carotid model of balloon injury. The nanospheres effectively penetrated the arterial wall, persisting for up to two weeks (Guzman et al. 1996). Several additional agents have subsequently been tested (Song et al. 1997), and a slow-release pluronic gel formulation of an antisense c-myb oligonucleotide inhibited smooth muscle cell accumulation in the carotid artery (Simons et al. 1992).

Microspheres containing angiogenic factors are also being considered for the treatment of coronary ischemia. bFGF was delivered from alginate microspheres when stabilized by the ligand heparin (Edelman et al. 1991). Heparin-alginate beads containing bFGF were administered extraluminally in a porcine model of chronic ischemia, resulting in improved coronary flow and reduction in infarct size (Harada et al. 1994). A related approach to revascularization of the heart is currently being tested in humans.

Cancer Therapy Using Angiogenesis Inhibitors

From the earliest days of the sustained release of proteins from polymeric matrices, this technology has been pursued for the delivery of anticancer proteins. An angiogenesis inhibitor, which deprives the tumor of growth-supporting vasculature, was found to inhibit tumor growth when delivered in a sustained, localized fashion (Langer et al. 1980; Langer and Murray 1983). To date, many such factors have been identified (Fidler and Ellis 1994; Folkman 1995), and several have been tested (with frequent administration of a protein solution) in clinical trials.

Antibodies to angiogenic factors have also been used (Kim et al. 1993). With the concept that angiogenic factors, such as VEGF, are not simply growth enhancers but are required for maintenance of existing vasculature, such antibodies may hold additional promise for cancer therapy. This is evidenced by the regression of preformed blood vessels following the administration of an antibody to VEGF (Yuan et al. 1996). This indicates a potential threat to normal vasculature by systemic delivery of these proteins and suggests that local delivery may not only be desirable but also necessary to ensure adequate safety.

Delivery to Protected Sites

Protein Delivery to the Brain

Proteins are unable to access some tissues from the bloodstream. An example is the brain, which is protected from the vasculature by the blood-brain barrier. In this case, the protein (or its gene) must be delivered using invasive techniques, and significant research has been done with gene delivery using recombinant viruses (Ridet and Privat 1995; Blomer et al. 1996) or transfected cells (Lindner et al. 1996; Kordower et al. 1994). An alternative would be to deliver a sustained-release microsphere formulation, and encapsulation and brain delivery of nerve growth factor (NGF) has been demonstrated (Camarata et al. 1992; Krewson and Saltzman 1996). One report showed prevention of cholinergic degeneration in the nucleus basalis after cortical devascularization (Maysinger et al. 1992). Recent studies of brain implants of PLGA have verified the biocompatibility of these polymers (Menei 1996; Kou et al. 1997).

The ability to control the duration of therapy by choosing different polymers is important. Short-term treatment would be required for acute neurodegeneration such as stroke or traumatic brain injury, where in vivo studies have shown that neurotrophic factors or fibroblast growth factor (FGF) are required for only several weeks (Dietrich et al. 1996; Mocchetti and Wrathall 1995). For example, treatment of adult paraplegic rats with acidic fibroblast growth factor (aFGF) mixed in a fibrin glue to give sustained release showed restoration of hind limb function after three weeks (Cheng et al. 1996). A longer duration of release would be needed to treat chronic neurodegeneration. Although permanent replacement of the protein cannot be achieved using microsphere formulations, it has been shown that short-term treatment with NGF results in sustained effects (Frick et al. 1997), in one case lasting up to six months (Liberini et al. 1993).

Intraocular Delivery

Another protected site is the retina, which is protected by the blood-retinal barrier. Consequently, numerous eye disorders, e.g., proliferative vitreoretinopathy, cytomegalovirus (CMV) retinitis, and endophthalmitis, must be treated by direct injection of drug into the posterior (vitreal) chamber (Peyman and Ganiban 1995). Due to risks of infection, retinal detachment, and the invasive nature of intravitreal injections, a strong incentive exists for sustained-release dosage forms. In 1996, the first such product, Vitrasert™, a nonerodible implant releasing ganciclovir over six months, was approved for the treatment of CMV retinitis (Smith et al. 1992). This device is placed out of the line of vision and requires eventual retrieval. Initial reports of intravitreal injection of PLGA microspheres have been made (Moritera et al. 1991; Giordano et al. 1995); although they would presumably be free floating within the vitreous, preliminary testing in rabbits suggests that microspheres are unlikely to interfere with vision (Giordano et al. 1995).

Whereas the majority of research into sustained intraocular drug delivery involves small drugs, numerous opportunities exist for proteins. Inhibitors of neovascularization have long been considered as potential therapies for diabetic and proliferative retinopathy (Patz et al. 1978), and cytokines are potential therapeutics for choroidal neovascularization (Chan et al. 1994) and herpes simplex virus infection (Geiger et al. 1994). A wide variety of ophthalmological pathologies result from uncontrolled vascularization (Aiello 1996), and retinal neovascularization has been inhibited in vivo with intravitreal injection of a chimeric receptor of VEGF (Aiello et al. 1995). A human anti-CMV antibody is being developed for CMV retinitis; although effective vitreal levels have been achieved in rabbits with systemic administration (Ostberg and Queen 1995), direct administration to the posterior vitreal chamber would both greatly reduce systemic exposure and reduce the risk of antibody formation.

Vaccines

It has long been recognized that depot formulations of antigens induce the most vigorous and long-lasting immunity (Janeway and Travers 1994), and immune responses to numerous PLGA–encapsulated viral and bacterial antigens have been demonstrated (see also Chapter 13). Humoral [immunoglobulin G and M (IgG and IgM)] responses have been elicited by parenteral administration of PLGA–encapsulated diphtheria toxoid, tetanus toxoid, and HIV gp120, and mucosal immunity [immunoglobulin (IgA)] was induced with *Bacillus pertussis* after intranasal dosing and with simian immunodeticiency virus (SIV) and Staphylococcal enterotoxin after oral administration (Cleland 1995). These microsphere formulations were made by encapsulation of a solution using an emulsion process, which often alters the integrity of the encapsulated protein (Table 14.2). However, because antibody formation is the goal, the need to maintain protein integrity is less stringent than with therapeutic proteins, where minimal or no immunogenicity of the released protein can be tolerated. Unless the principal antigenic determinants are discontinuous, and hence depend on a particular tertiary structure, a considerable degree of denaturation can be tolerated.

The most vigorous and long lasting immunity is believed to be induced if antigen exposure is not only continuous but if it occurs in two or more pulses (boosters) separated by several weeks or months (Janeway and Travers 1994). Achieving such pulsatile delivery is considerably more challenging because this requires release at a predetermined time as well as maintaining the unreleased protein in an antigenic form in vivo for weeks or months. This has recently been achieved with tetanus toxoid, in which pulses of release at three and seven weeks was achieved by using different microsphere formulations consisting of PLGA that degrade at different rates (Sanchez et al. 1996).

It is increasingly being recognized that to vaccinate against many viruses e.g., HIV (Hoth et al. 1994; Bolognesi 1996), it will be necessary to elicit cellular immunity (CD8+ CTL). This is difficult if not impossible to do with unencapsulated antigen (Kovacsovics-Bankowski et al. 1993) because the protein must be engulfed by antigen-presenting cells, protolysed, and presented on the surface with the MHC-I complex (Kovacsovics-Bankowsli and Rock 1995). As mentioned above in the context of tumor vaccines, this can be done using antigens encapsulated into small (< 10 μm) particles. Both emulsion and atomization processes can make microspheres of this size, either could be used for this application because, after internalization, the protein is processed and, therefore, the requirement for unaltered integrity is less stringent.

CONCLUSIONS

Protein therapeutics are a long-standing focus of the biotechnology industry. Hundreds of protein therapeutics are in advanced stages of clinical testing or are marketed (Struck 1994; Talmadge 1993), and it is likely that they will continue to be developed at an increasing rate. Although work is ongoing to develop alternative protein delivery systems, such as oral (Lee et al. 1992; Wang 1996), pulmonary (Patton and Platz 1992; Adjei and Gupta 1994), rectal (deBore et al. 1992), vaginal (Richardson and Illum 1992), buccal (Ho et al. 1992), and dermal (Cullander and Guy 1992), delivery of these relatively complex and unstable molecules remains problematic. Although the first report of sustained release of a microencapsulated protein was published over 20 years ago (Langer and Folkman 1976), it was over 20 years later that the first sustained-release formulation of a protein (recombinant human growth hormone, rHGH) (Vance et al. 1997) entered clinical testing. This work, plus recent advances in protein stabilization necessary to produce these dosage forms, is paving the way for the use of sustained-release systems to be applied to a wide variety of protein drugs for an even larger number of different indications. Such advances may eventually result in proteins comprising a significant part of the pharmacopeia.

ACKNOWLEDGMENTS

The authors thank Steve Zale, Jim Wright, Mark Tracy, Funmi Johnson, Dwaine Emerich, Robert Langer, Robert Breyer, Richard Pops, Alex Klibanov, and Jeff Cleland for helpful discussions.

REFERENCES

Adjei, A., and P. Gupta. 1994. Pulmonary delivery of therapeutic proteins. *J. Control. Rel.* 29:361–373.

Aiello, L. P. 1996. Vascular endothelial growth factor and the eye: Past, present, and future. *Arch Ophthalmol.* 114:1252–1254.

Aiello, L. P., E. A. Pierce, E. D. Foley, H. Takagi, H. Chen, L. Riddle, N. Ferrara, G. L. King, and L. E. H. Smith. 1995. Suppression of retinal neovascularization in vivo by inhibition of vascular endothelial growth factor (VEGF) using soluble VEGF-receptor chimeric proteins. *Proc. Natl. Acad. Sci. USA* 92:10457–10461.

Anderson, J. M., and M. S. Shive. 1997. Biodegradation and biocompatibility of PLA and PLGA microspheres. *Adv. Drug. Del. Rev.* 28:5–24.

Arakawa, T., and S. Timasheff. 1985. The stabilization of proteins by osmolytes. *Biophys. J.* 47:411–414.

Arky, R. 1997. *Physician's desk reference.* Montvale, N.J., USA: Medical Economics Company.

Austin, P. E., K. A. Dunn, K. Elly-Cofield, C. K. Brown, W. A. Wooden, and J. F. Bradfield. 1995. Subcuticular sutures and the rate of inflammation in noncontaminated wounds. *Ann. Emerg. Med.* 25:328–330.

Blomer, U., L. Naldini, I. M. Verma, D. Trono, and F. Gage. 1996. Applications of gene therapy to the CNS. *Human Molec. Gen.* 5:1397–1404.

Bolognesi, D. P. 1996. Overview of HIV vaccine development. *Antib. & Chemother.* 48:63–67.

Burke, P. A., R. G. Griffin, and A. M. Klibanov. 1992. Solid-state NMR assessment of enzyme active center structure under nonaqueous conditions. *J. Biol. Chem.* 267:20057–20064.

Camarata, P. J., R. Suryanarayanen, D. A. Turner, R. G. Parker, and T. J. Ebner. 1992. Sustained release of nerve growth factor from biodegradable polymer microspheres. *Neurosurgery* 30:313–319.

Chan, C. K., S. J. Kempin, S. K. Noble, and G. A. Palmer. 1994. The treatment of choroidal neovascular membranes by alpha interferon: An efficacy and toxicity study. *Ophthalmology* 101:289–300.

Chen, B., T. Arakawa, C. F. Morris, W. C. Kenney, C. M. Wells, and C. G. Pitt. 1994. Aggregation pathway of recombinant human keratinocyte growth factor and its stabilization. *Pharm. Res.* 11:1581–1587.

Chen, L., R. N. Apte, and S. Cohen. 1997. Characterization of PLGA microspheres for the controlled delivery of IL-1 for tumor immunotherapy. *J. Control. Rel.* 43:261–272.

Cheng, H., Y. Cao, and L. Olson. 1996. Spinal cord repair in adult paraplegic rats: Partial restoration of hind limb function. *Science* 273:510–513.

Cleland, J. L. 1995. Design and production of single-immunization vaccines using polylactide polyglycolide microsphere systems. In *Vaccine design: The subunit and adjuvant approach,* edited by M. F. Powell and M. J. Newman. New York: Plenum Press, pp. 439–462.

Cleland, J. L. Protein delivery from biodegradable microspheres In *Protein delivery: Physical systems,* edited by L. Sanders and W. Hendren. New York: Plenum Publishing. In press.

Cleland, J. L. and E. T. Duenas. 1997. Controlled delivery of nerve growth factor for local treatment of neuronal disease. *Proc. Intl. Symp. Control. Del. Bioact. Mater.* 24:823–824.

Cleland, J. L., and A. J. S. Jones. 1996. Stable formulations of recombinant human growth hormone and interferon gamma for microencapsulation in biodegradable polymers. *Pharm. Res.* 13:1462–1473.

Cleland, J. L., E. T. Duenas, J. Kahn, A. Park, A. Daugherty, and A. Cuthbertson. 1997a. Local controlled delivery of vascular endothelial growth factor provides neovascularization. *Proc. Intl. Symp. Control. Del. Bioact. Mater.* 24:85–86.

Cleland, J. L., A. Mac, J. Boyd, J. Yang, E. T. Duenas, D. Yeung, D. Brooks, C. Hsu, H. Chu, V. Mukku, and A. J. S. Jones. 1997b. The stability of recombinant human growth hormone in poly(lactic-co-glycolic acid) PLGA microspheres. *Pharm. Res.* 14:420–425.

Cohen, A. M. 1993. Erythropoietin and G-CSF In *Therapeutic proteins: Pharmacokinetics and pharmacodynamics,* edited by A. H. C. Kung, R. A. Baughman, and J. W. Larrick. New York: W. H. Freeman, pp. 165–186.

Cortesina, G., A. DeStefani, E. Galeazzi, G. Cavallo, C. Jemma, M. Giovarelli, S. Vai, and G. Forni. 1991. Interleukin 2 injected around tumour draining lymph nodes in head and neck cancer. *Head and Neck* 13:125–131.

Costantino, H. R., R. Langer, and A. Klibanov. 1994. Moisture-induced aggregation of lyophilized insulin. *Pharm. Res.* 11:21–29.

Cullander, C., and R. H. Guy. 1992. Transdermal delivery of peptides and proteins. *Adv. Drug Del. Rev.* 8:291–329.

Daughaday, W. F. 1995. Growth hormone, insulin-like growth factors and acromegaly. In *Endocrinology,* vol. 2, edited by L. J. DeGroot. Philadelphia: W. B Saunders and Co., pp. 303–329.

deBore, A. G., E. J. van Hoogdalem, and D. D. Breimer. 1992. Rate of controlled rectal peptide drug adsorption. *Adv. Drug Del. Rev.* 8:237–251.

Desai, U. R., and A. M. Klibanov. 1995. Assesssing the structural integrity of a lyophilized protein in organic solvents. *J. Am. Chem. Soc.* 117:940–3945.

Dietrich, W. D., O. Alonso, R. Busto, and S. P. Finklestein. 1996. Posttreatment with intravenous basic fibroblast growth factor reduces histopathologic damage following fluid-percussion brain injury in rats. *J. Neurotrauma* 13:309–316.

Dinbergs, I. D., L. Brown, and E. R. Edelman. 1996. Cellular response to transforming growth factor-beta 1 and basic fibroblast growth factor depends on release kinetics and extracellular matrix interactions. *J. Biol. Chem.* 271:29822–29829.

Edelman, E. R., E. Mathiowitz, R. Langer, and M. Klagsbrun. 1991. Controlled and modulated release of basic fibroblast growth-factor. *Biomaterials* 12:619–626.

Eliaz, R., D. Wallach, and J. Kost. 1996. Long-term protection against the effects of tumor necrosis factor by controlled delivery of the soluble p55 TNF receptor. *Cytokine* 8:482–487.

Eppstein, D. A., and B. B. Schryver. 1990. Controlled release of macromolecular polypeptides. U.S. Patent 4,962,091.

EMEA. 1996. *Note for guidance on impurities: Residual solvents.* London: European Agency for the Evaluation of Medicinal Products.

Fava, R. A., H. M. Olsen, G. Spencer-Green, K.-T. Yeo, T.-K. Yeo, B. Berse, R. W. Jackman, D. R. Senger, H. F. Dvorak, and L. F. Brown. 1994. Vascular permeability factor/endothelial growth factor (VPF/VEGF): Accumulation and expression in human synovial fluids and rheumatoid synovial tissue. *J. Exp. Med.* 180:341–346.

Feldmann, M., F. M. Brennan, and R. N. Maini. 1996. Role of cytokines in rheumatoid-arthritis. *Ann. Rev. Immunol.* 14:397–440.

Fidler, I. J., and L. M. Ellis. 1994. The implications of angiogenesis for the biology and therapy of cancer metastasis. *Cell* 79:185–188.

Folkman, J. 1995. Angiogenesis in cancer, vascular, rheumatoid and other disease. *Nature Med.* 1:27–31.

Frick, K. M., D. L. Price, V. E. Koliatsos, and A. L. Markowska. 1997. The effects of nerve growth factor on spatial recent memory in aged rats persist after discontinuation of treatment. *J. Neuroscience* 17:2543–2550.

Geiger, K., E. L. Howes, and N. Sarvetnick. 1994. Ectopic expression of gamma interferon in the eye protects transgenic mice from intraocular herpes simplex virus type 1 infections. *J. Virology.* 68:5556–5567.

Giordano, G. G., P. Chevez-Barrios, M. F. Refojo, and C. A. Garcia. 1995. Biodegradation and tissue reaction to intravitreous biodegradable poly(D,L-lactic-co-glycolic)acid microspheres. *Cur. Eye Res.* 14:761–768.

Golumbek, P. T., R. Azhari, E. M. Jaffee, H. I. Levitsky, A. Lazenby, K. Leong, and D. M. Pardoll. 1993. Controlled release, biodegradable cytokine depots: A new approach in cancer vaccine design. *Cancer Res.* 53:5841–5844.

Gombotz, W., M. Healy, and L. Brown. 1991. Very low temperature casting of controlled release microspheres. U.S. Patent 5,019,400.

Gombotz, W. R., S. C. Pankey, L. S. Bouchard, J. Ranchalis, and P. Puolakkainen. 1993. Controlled release of TGF-beta 1 from a biodegradable matrix for bone regeneration. *J. Biomater. Sci. Ed.* 5:49–63.

Griebenow, K., and A. M. Klibanov. 1996. On protein denaturation in aqueous-organic mixtures but not in pure organic solvents. *J. Am. Chem. Soc.* 118:11695–11700.

Guzman, L. A., V. Labhasetwar, C. X. Song, Y. S. Jang, A. M. Lincoff, R. Levy, and E. J. Topol. 1996. Local intraluminal infusion of biodegradable polymeric nanoparticles. A novel approach for prolonged drug delivery after balloon angioplasty. *Circulation* 94:1441–1448.

Hageman, M. J., J. M. Bauer, P. L. Possert, and R. T. Darrington. 1992. Preformulation studies oriented toward sustained delivery of recombinant somatropins. *J. Agric. Food Chem.* 40:348–355.

Harada, K., W. Grossman, and M. Friedman. 1994. Basic fibroblast growth factor improves myocardial function in chronically ischemic porcine hearts. *J. Clin. Invest.* 94:623–630.

Harris, A. G., S. P. Kokoris, and S. Ezzat. 1995. Continuous vs. intermittent subcutaneous infusion of octreotide in the treatment of acromegaly. *J. Clin. Pharmacol.* 35:59–71.

Harvey, S. 1995. Growth hormone metabolism. In *Growth hormone,* edited by S. Harvey, C. G. Scanes, and W. H. Daughaday. Boca Raton, Fla., USA: CRC Press, pp. 285–301.

Ho, N. F. H., C. L. Barsuhn, P. S. Burton, and H. P. Merkle. 1992. Mechanistic insites to buccal delivery of proteinaceous substances. *Adv. Drug Del. Rev.* 8:197–235.

Hodge, J. W., J. Schlom, S. J. Donohue, J. E. Tomaszewski, C. W. Wheeler, B. S. Levine, L. Gritz, D. Panicali, and J. A. Kantor. 1995. A recombinant vaccinia virus expressing human prostate-specific antigen (PSA): Safety and immunogenicity in a non-human primate. *Int. J. Cancer.* 63:231–237.

Hora, M. S., R. K. Rana, J. H. Nunberg, T. R. Tice, R. M. Gilley, and M. E. Hudson. 1990. Controlled release of interleukin-2 from biodegradable microspheres. *Bio/Technology* 8:755–758.

Hoth, D. F., D. P. Bolognesi, L. Corey, and S. H. Vermund. 1994. HIV vaccine development: A progress report *Annals Int. Med.* 121:603–611.

Janeway, C. A., and P. Travers. 1994. *Immunobiology: The immune system in health and disease.* London Current Biology, Ltd.

Jeffery, H., S. S. Davis, and D. T. O'Hagan. 1991. The preparation and characterization of poly(lactide-co-glycolide) microparticles. I: Oil-in-water emulsion solvent extraction. *Int. J. Pharm.* 77:169–175.

Johnson, O. L., J. L. Cleland, H. J. Lee, M. Charnis, E. Duenas, W. Jaworowicz, D. Shepard, A. Shahzamani, A. J. S. Jones, and S. D. Putney. 1996. A month-long effect from a single injection of microencapsulated human growth hormone. *Nature Med.* 2:795–799.

Johnson, O. L., W. Jaworowicz, J. L. Cleland, L. Bailey, M. Charnis, E. Duenas, C. Wu, D. Shepard, S. Magil, T. Last, A. J. S. Jones, and S. D. Putney. 1997. The stabilization and encapsulation of human growth hormone into biodegradable microspheres. *Pharm. Res.* 14:734–739.

Jorgensen, J. O., N. Moller, T. Lauritzen, and J. S. Christiansen. 1990. Pulsatile vs. continuous intravenous administration of growth hormone (GH) in GH-deficient patients: Effects on circulating insulin-like growth factor-I and metabolic indicies. *J. Clin. Endocrinol. Metab.* 70:1616–1623.

Kato, T., R. Nemoto, H. Mori, M. Takahshi, Y. Tanakawa, and M. Horada. 1981. Arterial chemoembolization with microencapsulated anticancer drug. *J. Am. Med. Assoc.* 245:1123–1127.

Kenley, R., L. J. Marden, T. Turek, L. Jin, E. Ron, and J. O. Hollinger. 1994. Osseous regeneration in the rat calvarium using novel delivery systems for recombinant human bone morphogenetic protein-2 (rhBMP-2). *J. Biomed. Mater. Res.* 28:1139–1147.

Kim, K. J., B. Li, J. Winer, M. Armanini, N. Gillett, G. S. Phillips, and N. Ferrara. 1993. Inhibition of vascular endothelial growth factor-induced angiogenesis suppresses tumour growth in vivo. *Nature* 362:841–844.

Klibanov, A. M. 1989. Enzymatic catalysis in anhydrous organic solvents. *Trends in Biochem. Sci.* 14:141–144.

Kordower, J. H., S. R. Winn, Y. Liu, E. J. Mufson, J. R. Sladek, J. P. Hammang, E. E. Baetge, and D. F. Emerich. 1994. The aged monkey basal forebrain: Rescue and sprouting of axotomized basal forebrain neurons after grafts of encapsulated cells secreting human nerve growth factor. *Proc. Natl. Acad. Sci. USA* 91:10898–10902.

Kou, J. H., C. Emmett, P. Shen, S. Azwani, T. Iwamoto, R. Vaghefi, and L. Sanders. 1997. Bioerosion and biocomaptibility of poly(d,l-lactic-co-glycolic acid) implants in brain. *J. Control. Rel.* 43:123–130.

Kovacsovics-Bankowsli, M., and K. L. Rock. 1995. A phagosome-to-cytosol pathway for exogenous antigens presented on MHC class I molecules. *Science* 267:243–246.

Kovacsovics-Bankowski, M., K. Clark, B. Benacerraf, and K. L. Rock. 1993. Efficient major histocomplatibility complex class I presentation of exogenous antigen upon phagocytosis by macrophages. *Proc. Natl. Acad. Sci.* 90:4942–4946.

Krewson, C. E., and W. M. Saltzman. 1996. Transport and elimination of recombinant human NGF during long-term delivery to the brain. *Brain Res.* 727:169–181.

Krewson, C. E., R. Dause, M. Mak, and W. M. Saltzman. 1996. Stabilization of nerve growth factor in controlled release polymers and in tissue. *J. Biomater. Sci. Polymer Ed.* 8:103–117.

Langer, R. 1990. New methods of drug delivery. *Science* 249:1527–1533.

Langer, R., and J. Folkman. 1976. Polymers for the sustained release of proteins and other macromolecules. *Nature* 263:797–800.

Langer, R., and J. Murray. 1983. Angiogenesis inhibitors and their delivery systems. *Appl. Biochem. Biotech.* 8:9–24.

Langer, R., H. Conn, J. Vacanti, C. Haudenschild, and J. Folkman. 1980. Control of tumor-growth in animals by infusion of an angiogenesis inhibitor. *Proc. Natl. Acad. Sci. USA* 77:4331–4335.

Laursen, T., O. L. Jorgensen, G. Jakobsen, B. L. Hansen, and J. S. Christiansen. 1995. Continuous infusion versus daily injections of growth hormone (GH) for 4 weeks in GH-deficient patients. *J. Clin. Endocrinol. and Metab.* 80:2410–2418.

Lee, H. J. 1995. Biopharmaceutical properties and pharmacokinetics of peptide and protein drugs. In *Peptide-based drug design,* edited by M. Taylor and G. Amidon. Washington, D.C.: American Chemical Society, pp. 69–97.

Lee, H. J., G. Riley, O. Johnson, J. L. Cleland, N. Kim, M. Charnis, L. Bailey, E. Duenas, A. Shahzamani, A. J. S. Jones, and S. D. Putney. 1997. In vivo characterization of sustained-release formulations of human growth hormone. *J. Pharm. Expt. Ther.* 281:1431–1439.

Lee, S. C., M. Shea, M. A. Battle, K. Kozitza, E. Ron, T. Turek, R. G. Schaub, and W. C. Hayes. 1994. Healing of large segmental defects in rat femurs is aided by rhBMP-2 in PLGA matrix. *J. Biomed. Mater. Res.* 28:1149–1156.

Lee, V. H. L., S. Dodda-Kashi, G. M. Grass, and W. Rubas. 1992. Oral route of peptide and protein drug delivery. In *Protein and peptide drug delivery* edited by V. H. L. Lee. New York: Marcel Dekker, pp. 691–737.

Lewis, D. H. 1990. Controlled release of bioactive agents from lactide/glycolide polymers. In *Biodegradable polymers as drug delivery systems,* edited by M. Chasin and R. Langer. New York: Marcel Dekker, pp. 1–41.

Liberini, P., E. P. Pioro, D. Maysinger, F. R. Ervin, and A. C. Cuello. 1993. Long-term protective effects of human recombinant nerve growth factor and monosialoganglioside GM1 treatment on primate nucleus basalis cholinergic neurons after neocortical infarction. *Neuroscience* 53:625–637.

Lindner, M. D., C. E. Kearns, S. R. Winn, B. Frydel, and D. F. Emerich. 1996. Effects of intraventricular encapsulated hNGF-secreting fibroblasts in aged rats. *Cell Transpl.* 5:205–223.

Lu, W., and T. G. Park. 1995a. In vitro release profiles of eristostatin from biodegradable polymeric microspheres: Protein aggregation problems. *Biotech. Prog.* 11:224–227.

Lu, W., and T. G. Park. 1995b. Protein release from poly(lactic-co-glycolic acid) microspheres: Protein stability problems. *Pharma. Sci. Technol.* 49:13–19.

Maysinger, D., I. Jalsenjak, and A. C. Cuello. 1992. Microencapsulated nerve growth factor: Effects on the forebrain neurons following devascularizing cortical lesions. *Neuroscience Letters* 140:71–74.

Mendez, A., P. J. Camarata, R. Suryanarayanan, and T. J. Ebner. 1997. Sustained intracerebral delivery of nerve growth factor with biodegradable polymer microspheres. *Methods Neurosci.* 21:150–167.

Menei, P. M. 1996. Effect of stereotactic implantation of biodegradable 5-flurouracil-loaded microspheres in healthy and C6 glioma-bearing rats. *Neurosurgery* 39:117–123.

Mittal, S. A., A. Cohen, and D. Maysinger. 1994. In vitro effects of brain derived neurotropic factor released from microspheres. *Neuro Report* 5:2577–2582.

Mocchetti, I., and J. R. Wrathall. 1995. Neurotrophic factors in central nervous system trauma. *J. Neurotrauma* 12:853–869.

Mordenti, J., S. A. Chen, and B. L. Ferraiolo. 1993. Pharmacokinetics of interferon-gamma In *Therapeutic proteins: Pharmacokinetics and pharmacodynamics,* edited by A. H. C. Kung, R. A. Baughman, and J. W. Larrick. New York: W. H. Freeman, pp. 187–199.

Moritera, T., Y. Ogura, Y. Honda, R. Wada, S.-H. Hyon, and Y. Ikada. 1991. Microspheres of biodegradable polymers as drug delivery system in the vitreous. *Invest. Ophthal. Vis. Sci.* 32:1785–1790.

Morlock, M., H. Kroll, G. Winter, and T. Kissel. 1997. Microencapsulation of rh-erythropoietin using biodegradable poly(d,l-lactide-co-glycolide): Protein stability and the effects of stabilizing exipients. *Eur. J. Pharm. Biopharm.* 43:29–36.

Mulligan, R. C. 1993. The basic science of gene therapy. *Science* 260:926–932.

Ogawa, Y., H. Okada, M. Yamamota, and T. Shimamoto. 1988. In vivo release profiles of leuprolide acetate from microcapsules prepared with polylactic acids or copoly(lactic/glycolic) acids and in vivo degradation of these polymers. *Chem. Pharm. Bull.* 36:2576–2581.

Ostberg, L., and C. Queen. 1995. Human and humanized monoclonal antibodies: Preclinical studies and clinical experience. *Biochem. Soc. Trans.* 23:1038–1043.

Ozzello, L., D. Habif, C. De Rosa, and K. Cantell. 1992. Cellular events accompanying regression of skin recurrences of breast carcinomas treated with intralesional injections of natural interferons alpha and gamma. *Cancer Research.* 52:4571–4581.

Paleolog, E. M. 1996. Angiogenesis: A critical process in the pathogenesis of RA–A role for VEGF? *Br. J. Rheumatology* 35:917–919.

Patton, J. S., and R. M. Platz. 1992. Pulmonary delivery of proteins and peptides for systemic action. *Adv. Drug. Del. Rev.* 8:179–196.

Patz, A., G. Lutty, A. Bennett, and W. R. Coughlin. 1978. Inhibitors of neovascularization in relation to diabetic and other proliferative retinopathies. *Trans. Am. Ophthalmol. Soc.* 76:102–107.

Pettit, D. S., S. Pankey, N. Nightlinger, M. Disis, and W. Gombotz. 1996. GM-CSF encapsulated in PLGA microspheres for vaccine adjuvancy. *Proc. Intl. Symp. Control. Del. Bioact. Mater.* 23:857–858.

Peyman, H. A., and G. J. Ganiban. 1995. Delivery systems for intraocular routes. *Adv. Drug Deliv. Rev.* 16:107–123.

Pihlajamaki, H., O. Bostman, E. Hirvensalo, P. Tormala, and P. Rokkanen. 1992. Absorbable pins of self-reinforced poly-L-lactic acid for fixation of fractures and osteotomies. *J. Bone Joint Surg.* 74:853–857.

Pikal, M. J. 1994. Freeze-drying of proteins: Process, formulation and stability. In *Formulation and delivery of proteins and peptides,* edited by J. Cleland and R. Langer. Washington, D. C.: American Chemical Society, pp. 120–133.

Pitt, C. G. 1990. The controlled parenteral delivery of polypeptides and proteins. *Int. J. Pharm.* 59:173–196.

Reddi, A. H. 1994. Symbiosis of biotechnology and biomaterials: Applications in tissue engineering of bone and cartilage. *J. Cellular Biochem.* 56:192–195.

Richardson, J. L., and L. Illum. 1992. The vaginal route of peptide and protein drug delivery. *Adv. Drug Del. Rev.* 8:341–366.

Ridet, J. L., and A. Privat. 1995. Gene therapy in the central nervous system: Direct versus indirect gene delivery. *J. Neurosci. Res.* 42:287–293.

Ripamonti, U., and A. H. Reddi. 1994. Periodontal regeneration: Potential role of bone morphogenetic proteins. *J. Periodont. Res.* 29:225–235.

Sanchez, A., R. K. Gupta, M. J. Alonso, G. R. Siber, and R. Langer. 1996. Pulsed controlled-release system for potential use in vaccine delivery. *J. Pharm. Sci.* 85:547–552.

Sanders, L. M., B. A. Kell, G. I. McRae, and G. W. Whitehead. 1986. Prolonged controlled release of nafarelin, a luteinizing hormone-releasing hormone analogue, from biodegradable polymeric implants: Influence of composition and molecular weight of polymer. *J. Pharm. Sci.* 75:356–360.

Schmidt, W., T. Scheeighoffer, E. Herbst, G. Maass, M. Berger, F. Schilcher, G. Schaffner, and M. L. Bernstiel. 1995. Cancer vaccines: The interleukin 2 dosage effect. *Proc. Natl. Acad. Sci. USA* 92:4711–4714.

Schwendeman, S. P., M. Cardamone, M. R. Brandon, A. Klibanov, and R. Langer. 1996. Stability of proteins and their delivery from biodegradable polymer microspheres. In *Microspheres/microparticles–characterization and pharmaceutical application,* edited by S. Cohen and H. Bernstein. New York: Marcel Dekker, Inc., pp. 1–49.

Simons, M., E. R. Edelman, J.-L. DeKeyser, R. Langer, and R. D. Rosenberg. 1992. Antisense c-myb oligonucleotides inhibit intimal arterial smooth muscle cell accumulation in vivo. *Nature* 359:67–70.

Smith, T. J., P. A. Pearson, D. L. Blandford, J. D. Brown, K. A. Goins, J. L. Hollins, E. T. Schmeisser, P. Glavinos, L. B. Baldwin, and P. Ashton. 1992. Intravitreal sustained-release ganciclovir. *Arch. Ophthalmol.* 110:255–258.

Song, C. X., V. Labhasetwar, H. Murphy, X. Qu, W. R. Humphrey, R. J. Shebuski, and R. J. Levy. 1997. Formulation and characterization of biodegradable nanoparticles for intravascular local drug delivery. *J. Control. Rel.* 43:197–212.

Struck, M. M. 1994. Biopharmaceutical R&D success rates and development times: A new analysis provides benchmarks for the future. *Bio/Technol.* 12:674–677.

Stute, N., W. L. Furman, M. Schell, and W. E. Evans. 1995. Pharmacokinetics of recombinant human granulocyte-macrophage colony-stimulating factor in children after intravenous and subcutaneous administration. *J. Pharm. Sci.* 84:824–828.

Tabata, Y., Y. Takebayashi, T. Ueda, and Y. Ikada. 1993. A formulation method using D,L-lactic acid oligomer for protein release with reduced initial burst. *J. Control. Rel.* 23:55–64.

Talmadge, J. E. 1993. The pharmaceutics and delivery of therapeutic polypeptides and proteins. *Adv. Drug Del. Rev.* 10:247–299.

Tauber, H., H. De Bouet du Portal, B. Sallerin-Caute, P. Rochiccioli, and R. Bastide. 1993. Differential regulation of serum growth hormone (GH)-binding protein during continuous infusion vs. daily injection of recombinant human GH in GH-deficient children. *J. Clin. Endocrinol. and Metab.* 76:1135–1139.

Tepper, R. I., and J. J. Mule. 1994. Experimental and clinical studies of cytokine gene-modified tumor cells. *Human Gene Therapy* 5:153–164.

Tracy, M. A., H. Bernstein, and M. A. Kahn. 1996. Controlled release of metal-stabilized interferon. U.S. Patent 5711968.

Vance, M. L., C. J. Woodburn, S. Putney, J. Grous, H. J. Lee, and O. L. Johnson. 1997. Effects of sustained release growth hormone on serum GH, IGF-I and IGFBP-3 concentration in GH deficient adults. *Endocrin. Metab.* 4 (Suppl. A):75.

Vlock, D., C. Snyderman, J. Johnson, E. Myers, R. J. Eibling, J. Kirkwood, J. Dutcher, and G. Adams. 1994. Phase 1b trial of the effect of peritumoural and intranodal injections of IL-2 in patients with advanced squamous cell carcinoma of the head and neck; An Eastern Co-operative Oncology Group Trial. *J. Immunol.* 15:134–139.

Wang, M., V. Bronte, P. W. Chen, L. Gritz, D. Panicali, S. A. Rosenberg, and N. P. Restifo. 1995. Active immunotherapy of cancer with a non-replicating recombinant fowlpox virus encoding a model tumor-associated antigen. *J. Immunol.* 154:4685–4692.

Wang, W. 1996. Oral protein drug delivery. *J. Drug Target.* 4:195–232.

Winde, G., B. Reers, H. Nottberg, T. Berns, J. Meyer, and H. Bunte. 1993. Clinical and functional results of abdominal rectopexy with absorbable mesh-graft for treatment of complete rectal prolapse. *Eur. J. Surg.* 159:301–305.

Wussow, P., B. Block, F. Hartmann, and J. Deicher. 1988. Intralesional interferon alpha in advanced malignant melanoma. *Cancer* 61:1071–1074.

Yuan, F., Y. Chen, M. Dellian, N. Safabakhsh, N. Ferrara, and R. K. Jain. 1996. Time-dependent vascular regression and permeability changes in established human tumor xenografts induced by an anti-vascular endothelial growth factor/vascular permeability factor antibody. *Proc. Natl. Acad. Sci. USA* 93:14765–14770.

Zale, S. E., P. A. Burke, H. Bernstein, and A. Brickner. 1996. Composition for sustained release of non-aggregated erythropoietin. U.S. Patent 5674534.

Zhi, J., S. B. Teller, H. Satoh, S. G. Koss-Twardy, and D. R. Luke. 1995. Influence of human serum albumin content in formulations on the bioequivalency of interferon alpha-2a given by subcutaneous injection in healthy male volunteers. *J. Clin. Pharmacol.* 35:281–284.

15

Toxicology and Biocompatibility Evaluation of Microsphere and Liposome Sustained-Release Drug Delivery Systems

M. Gary I. Riley
Alkermes, Inc.
Cambridge, Massachusetts

Scott D. Putney
Eli Lilly and Co.
Indianapolis, Indiana

The principal goal of the preclinical toxicity evaluation of any drug formulation, whether immediate or sustained release, is to demonstrate safety within a therapeutically useful dose range. In addition, sustained-release formulations must demonstrate that controlled-release claims are justified and that neither the components nor the operation of the controlled-release system impart any additional toxic liability. Toxicity and biocompatibility profiles of immediate and sustained-release formulations of the same agent may differ due to pharmacokinetic and drug distribution effects, and different toxicities may occur due to interactions between the host and the delivery system. This means that toxicity, bioavailability, and drug release studies used to support the registration of conventional formulations of a drug may be insufficient to characterize a sustained-release formulation.

In these studies, biocompatibility and systemic and local toxicity are each evaluated against a background of appropriate drug exposure. The comprehensive toxicological assessment of a sustained-release formulation must include data on the drug release system, the drug of interest, and all excipients. Frequently, toxicity data are already available for some components, especially excipients or polymers, and thus may be addressed by reference. Similarly, toxicity data for immediate-release formulations of the drug itself may be available before a sustained-release formulation is developed. Such data can be of significant value—they can be useful in planning studies to define the toxicity of the sustained-release formulation and to characterize long-term drug exposure.

Biocompatibility testing provides a measure of the extent to which sustained-release formulations produce local effects at the site of administration. These effects are the result of the interplay of drug, the sustained-release system, and contiguous tissue. As with toxicity testing, biocompatibility testing of sustained-release formulations involves more experimental variables than conventional formulations. Biocompatibility studies of sustained-release systems are also usually of longer duration than the local irritation studies performed for immediate-release formulations. Usually designed to characterize the tissue changes of at least one treatment cycle, biocompatibility studies should also include an evaluation of the injection site after drug release has ceased. It has been suggested that the scope of biocompatibility be enlarged to include the effect of biodegradable delivery systems on delivered agents (Gombotz and Pettit 1995). Although important, this invites confusion and does not contribute to an understanding of biocompatibility as conventionally defined.

The drug delivery systems to be discussed in this chapter are those intended for local drug release and include subcutaneous (SC), intramuscular (IM), and intralesional formulations (intravenous [IV] administration is not considered). Structurally, these systems are based on encapsulation within a matrix or a permeation barrier. Most are thus included in the categories of biodegradable and nonbiodegradable polymer-based formulations and liposomes. Encapsulation systems based on biodegradable polymers account for most of the injectable, sustained-release products currently marketed and in advanced stages of clinical development, and these promise to be significant areas of sustained-release drug delivery research in the foreseeable future. The other types of parenteral sustained-release systems based on dissolution, adsorption, or esterification mechanisms (Chien 1992) are not widely used and not addressed here.

BIOCOMPATIBILITY OF INJECTABLE, SUSTAINED–RELEASE PRODUCTS

The biocompatibility of a sustained-release device is determined over an appropriate time course by evaluating the sequence of local tissue changes that follow its administration. The sequence consisting of injection/implantation trauma, inflammation, tissue repair, and foreign body reaction is consistently observed with all implanted polymeric devices and is not restricted to drug delivery systems (Anderson and Shive 1997). Biocompatible substances react minimally with tissue at the implantation site, ideally causing no significant pain or discomfort and only low levels of inflammation, local fixed tissue destruction, and scar tissue formation. After the device degrades or is surgically removed, the optimal outcome is an absence of residual effects at the site of administration. Evaluation of biocompatibility is performed by macroscopic and microscopic examination of the local tissue response.

A distinction is made between inflammation resulting from trauma or cytotoxic effects of the implanted material and inflammation from the host response. The former reactions involve the destruction of local tissue elements and, depending on the site of administration, may potentially result in clinically significant complications or loss of function. The latter response, referred to as a foreign body reaction, is characterized by the actual or attempted engulfment of the foreign material by macrophages or their syncytical counterparts, foreign body giant cells. Foreign body reactions occur in response to the presence of even well-tolerated materials and may persist after active inflammation and wound healing is complete. These reactions reverse when the foreign material biodegrades or is removed surgically.

Biocompatibility of Biodegradable Polymeric Delivery Devices

Biodegradable polymers used for drug delivery include bulk eroding polymers such as polymers of lactide/glycolide; poly-cyanoacrylates; surface eroding polymers such as polyanhy-drides and polyorthoesters; and hydrogels, which include natural polymers such as collagen and synthetic polymers (e.g., pluronic polyols, polyvinyl alcohol, and polyvinyl pyrrolidone). Of these, polylactide-polyglycolide homo- and copolymers (PLGA) are most widely used in parenteral drug delivery systems. Other polymers may be suitable for this purpose, but it has been noted that there is a general reluctance to introduce new polymers, given the economics and delays inherent in establishing safety and efficacy. There are four sustained-release, biodegradable, polymer-based injectable or implantable systems currently approved for marketing in the United States: three PLGA–based luteinizing hormone releasing hormone (LHRH) peptide analogs (Lupron®, Takeda Abbott; Zoladex®, ICI; and Decapeptyl®, Ipsen Biotech) and Gliadel, a 1,3-bis (2-chloroethyl)-1-nitrosourea (BCNU) wafer for surgical implantation at sites of brain tumor resection (Guilford). Lupron® and Decapeptyl are microspheres, and Zoladex® is an implantable rod. In addition, a PLGA microsphere formulation of human growth hormone (hGH) is currently in clinical testing (Vance et al. 1997; Moshang et al. 1998), and the toxicology and biocompatibility of this formulation has been investigated (Lee et al. 1997).

Biodegradable delivery devices are generally injected subcutaneously or intramuscularly and, depending on the properties of the polymer, release encapsulated drug for a period of weeks or months. Biodegradation occurs in parallel with drug release and is necessary for release. Biodegradable polymers offer the significant advantage of not requiring removal from the patient when the drug load has been discharged. In addition, they are suitable for the delivery of drugs having a broader range of molecular size than are nonbiodegradable polymers (Langer 1990).

Numerous studies on biocompatibility and/or in vivo biodegradation of various polymer types have shown that the life span of the polymer in tissue, particle size, and surface characteristics affect the intensity of tissue reaction. This is exemplified by studies showing that persistence of the PLGA polymer increases with lactide content. In microsphere implantation studies in rats (Visscher et al. 1986, 1988), microspheres formulated from a polymer composed of 50:50 lactide:glycolide attracted foreign body giant cells in 4 days after implantation in skeletal muscle. Both the microspheres and the local tissue reaction had almost completely disappeared by day 56, with decreased intensity of local inflammation and fibrosis starting about day 30. This corresponded with morphological evidence of polymer degradation.

In contrast, microspheres fabricated from 100 percent lactide (PLA) evoked a foreign body giant cell response by day 11. Here the local reaction persisted for 480 days, consistent with the slower hydrolysis of this polymer. Similar patterns of early local myositis followed by local fibrosis were observed, and fibrosis was more obvious with the 50:50 formulation. Microspheres comprised of polymer with 65:35 lactide:glycolide degraded over a 150-day period (Yamaguchi and Anderson 1993). Persistence of polymer in tissue may also be affected by drug loading and excipients (Maulding et al. 1986; Zhang et al. 1995).

Biocompatibility of Nondegradable Polymeric Delivery Devices

Nonbiodegradable polymers are used for a wide range of structural devices, e.g., in orthopedic and cardiovascular medicine, and are used extensively in transdermal delivery systems. Their use in injectable, sustained-release systems is limited, however, and they are now largely superseded by biodegradable formulations. The only injectable, nonbiodegradable, sustained-release formulation currently marketed in the United States is Norplant®, a silicone rubber reservoir that releases the contraceptive agent levonergestrol. Medical nonbiodegradable polymers include silicones, polyesters, polyethylene, polyacrylamide, polystyrene, polyamides, polyurethanes, and acrylic acid poly (ε-caprolactone) (Kopecek and Ulbrich 1983). In practice, nonbiodegradable polymers exhibit low levels of biodegradation, but this degradation is minimal and usually does not contribute to the drug release properties of the product. Release is via diffusion rather than a combination of diffusion and degradation, which places limits on the size of the drug that can be delivered (e.g., molecules of greater than 600 Da are unable to diffuse through silicone) (Schwendeman et al. 1997).

Differences between the biocompatibility of biodegradable and nonbiodegradable polymer systems are chiefly limited to the persistence of the latter at the implantation site and, more importantly, to surgical trauma attending their explantation. In other respects, the biocompatibility issues—implantation trauma, inflammation, and tissue repair—are similar. Nonbiodegradable polymer implants evoke a foreign body reaction and varying degrees of fibrosis as long as they are present. The foreign body reaction abates promptly when the polymer is removed; local fibrosis generally diminishes but does not usually disappear completely (Yamaguchi and Anderson 1993).

Effect of Polymer Surface Features on Biocompatibility

With delivery devices made of both biodegradable and nonbiodegradable polymers, biocompatibility can be influenced by polymer surface features (Matlaga et al. 1976; Gombotz and Pettit 1995; Lam et al. 1995). Surfaces that minimize local tissue pressure and the shedding of polymer particles are associated with favorable biocompatibility. Conversely, sharply angular conformations and the tendency to form free fragments of polymer promote tissue irritation and inflammation. Fortunately, polymer drug delivery devices such as implantable rods, microspheres, or wafers generally lack the mass to exert excessive physical forces in tissue. Irregular surfaces with high porosity or pitting facilitate the attachment of monocytic phagocytes, which can enhance foreign-body reaction by exposing additional surface area rather than significantly impacting overall biocompatibility (Zhao et al. 1991).

Other properties of the implanted material, such as polymer purity and degradation products, also affect biocompatibility. Reactive substances that promote inflammation may be found on the surface of undegraded polymer or may not be brought into contact with tissue until degradation and solubilization of the polymer occurs. The presence of polymer impurities, such as unreacted monomeric polymer or unremoved solvents, may also affect biocompatibility (Kobayashi et al. 1992; Nakamura 1993).

Biocompatibility of Liposomes

Liposomes are biodegradable and are typically phagocytized by scavenger cells and degraded intracellularly, with little or no detectable residue (Kadir et al. 1993). Liposomes function by isolating the drug in membrane bound vesicles and subsequently releasing it in a delayed fashion. In general, liposomes provide favorable biocompatibility, low toxicity, low immunogenicity, and biodegradability. Liposomal formulations may be used for deposition in fixed tissues such as the subcutis, muscles, and tumors (Kadir et al. 1992).

The biocompatibility of liposomes is evaluated after administration of the formulation as a depot injection. Local degradation of void liposomes is associated with accumulation of macrophages at the margin of the depot, and minimal scarring (but no evidence of cytotoxicity) is occasionally noted (Kadir et al. 1992). The inclusion of active agents, and their subsequent local release by diffusion, may be associated with dose-limiting local irritation, although there have been few reports of tolerability of drug-loaded liposomes in normal tissue (Kadir et al. 1993) (see also Chapter 7).

The Design of Studies to Evaluate Biocompatibility

Establishing a biocompatibility profile requires evaluation in at least one animal species in a study that closely mimics the intended conditions of use. Study designs for biocompatibility will normally evaluate clinically relevant dose parameters, including intended route, volume, and concentration. By comparison, large dose multiples, such as those used in toxicity studies, may require drug concentrations, dose volumes, and osmolarity that are inconsistent with biocompatibility. In vivo biocompatibility studies of sustained-release polymer systems should span the period from dosing until the local tissue response abates. In certain circumstances, e.g., where the same site is subject to reuse, biocompatibility studies may span more than one cycle of treatment.

In vitro studies of biocompatibility using blood or tissue culture systems are useful screening approaches and can provide information on the passive tissue response to direct cytotoxicity of drug delivery systems. However, they do not measure the active responses of metabolic adaptation, inflammation, and healing; hence, definitive investigation of biocompatibility requires in vivo evaluation. Furthermore, because they are usually static, in vitro systems do not reflect diffusion and clearance mechanisms.

TOXICOLOGY OF INJECTABLE, SUSTAINED–RELEASE PRODUCTS

The toxicology assessment of an injectable, sustained-release formulation includes evaluations of local and systemic effects in at least two animal species at several doses (Majors and Friedman 1990). These doses should range from the no-effect level to a dose inducing obvious toxicity. If toxicity is not achievable, the maximum dose is limited by the dose volume. Toxicity studies provide a dose response profile that is used as a guide to establish the clinical dose and to predict clinical adverse effects.

Drug Exposure

The measurement of drug exposure, including characterization of the time course of release, is an important element of toxicity evaluation for sustained-release systems (Majors and Friedman 1990). This may be accomplished by direct pharmacokinetic (PK) methods, including the measurement of parent and metabolites in blood, urine, and feces and, where technically feasible, the measurement of unreleased drug. Alternatively, if techniques are available, measurement of pharmacodynamic (PD) response may be used (Ritschel 1988). PK or PD testing is necessary to confirm not only that drug release occurred but that it did so at the intended rate. The frequency of PK/PD monitoring is guided by the characteristics of the delivery system. Systems having more than one phase of release (e.g., an early burst phase followed by a sustained-release phase) require more frequent sampling than those with zero-order release.

Drug release studies of parenteral systems in animals are more likely to have a satisfactory clinical correlation when conducted with the clinical dose and via the intended route of administration. Animal models in which typical human doses can be administered at appropriate tissue sites are preferred for these studies. Generally, dogs, rabbits, and monkeys, in which sufficient SC or IM tissue mass usually exists, are suitable models. Changes in drug release profiles may occur with devices such as microspheres, where the dimensions of the deposited material vary according to administered dose volume or to disposition within tissues (Riley and Zale, unpublished observations). Discrete devices such as implants have fixed dimensions and are not subject to this effect.

Toxicity of Polymeric Delivery Devices

For practical purposes, the toxicological effects of the polymers commonly used for sustained-release delivery applications are entirely local. Systemic toxicity is not an issue due to low exposure and the low inherent toxicity of polymers and their degradation products. Nonbiodegradable polymers are inert or only slowly release trace amounts of degradation products; the degradants of most biodegradable polymers, while more abundant, pose little or no significant safety risk. An exception among biodegradable polymers is polycyanoacrylates, which release formaldehyde as a degradation product (Leonard et al. 1966; Pitt and Schindler 1983; Gombotz and Pettit 1995; Leong et al. 1995). Systemic toxic effects of drug-loaded polymer devices are due to the active agent; frequently the drug-based toxicity of the sustained-release formulation may be predicted from previous toxicology studies of an immediate-release formulation. Sustained-release formulations avoid the potential toxic consequences of frequent drug concentration spikes, which are associated with immediate-release formulations; however, the pharmacological response to the active agent may be enhanced. This is most notable with agents having short elimination half-lives and disproportionately long dosing intervals (a relatively common situation with many therapeutic proteins). Sustained release extends the duration of organ exposure and enhanced PD effects, including toxicity, may ensue. Onset of adverse effects may be immediate or delayed (Ritschel 1988).

Although sustained-release formulations are generally investigated thoroughly in vitro and in other preclinical studies, several differences between the performance of polymeric systems in vitro and in vivo underscore the importance of a thorough toxicological investigation. The first is differences in in vitro and in vivo release profiles from biodegradable polymer systems due to the influence of drug formulation and host variables. Eleven separate categories of formulation variables have been described that affect the in vivo polymer degradation rate of PLGA and PLA microspheres (Anderson and Shive 1997). In addition, injection site effects that result in differences in the polymer degradation rate of up to twofold have also been described (Kamei et al. 1992). Secondly, there is evidence that local inflammation can affect polymer degradation. The low pH (~ 3) generated by activated macrophages has been proposed to accelerate polymer degradation (Anderson and Shive 1997). Finally, microsphere interactions with the delivery vehicle may influence drug exposure, and, by inference, toxicity. Vehicles for use with microspheres generally contain viscosity-enhancing agents and surfactants (e.g., carboxymethyl cellulose and polysorbates). Whereas these excipients are usually safe, they do exhibit a low incidence of adverse effects, such as local irritation and allergic or pseudoallergic reactions (Weiner and Bernstein 1989). In addition to being clinically important, these events may also affect drug release.

Toxicity of Liposomal Delivery Devices

The toxicity of liposomes is determined by two principal factors: the components of the vesicles and the pharmacology of the encapsulated drug. Locally administered liposomes are, depending on their size, either retained at the injection site or removed. Following local (IM or SC) administration, lymphatic uptake of particles < 100 nm occurs (Osborne et al. 1979). Larger particles, remaining locally, are distributed extracellularly or are taken up by phagocytes (Kadir et al. 1993). Ultimately, these larger liposomes are locally degraded and ingested by phagocytic cells. The systemic effects of liposomes cleared via the lymphatics are similar to those seen following IV administration, although there has been some success in formulating small liposomes for retention in draining lymph nodes by manipulating the surface charge (Hirano et al. 1985).

The toxic effects of vesicles are chiefly limited to local irritation; significant systemic toxicity from liposomal lipids, at the doses usually employed for local delivery, is unlikely. Toxicity depends primarily on the lipid component, and a number of lipids have been identified for use in fabricating liposomes. For example, Phillips (1993) reported on the use of dipalmitoyl-phosphatidylglycerol and dimyristoyl-phosphatidylglycerol, which showed a low level of toxicity for cells of the reticuloendothelial system (RES). In vitro studies have linked cytotoxicity to charged lipids, whereas neutral lipids are less liable to exhibit toxicity. Stearylamine and phosphatidylserine, cationic and anionic, respectively, are both cytotoxic (Kadir et al. 1993).

Although the major route of liposome clearance following IV administration is phagocytic uptake by the RES, there is less evidence that direct RES uptake is a major clearance mechanism following local liposomal delivery (Arrowsmith et al. 1984). Certainly, there are anatomical reasons to believe that micrometer-sized liposomes will be denied access to the circulation and thus to the extensive RES lining the blood vessels in many organs. Studies of liposome lipid retention have shown significant persistence at the injection site and some local lymph node accumulation (Arrowsmith et al. 1984).

Four types of biological interactions leading to drug release are proposed for liposomes: cellular adsorption, lipid and protein exchange, membrane fusion, and endocytosis (Kadir et al. 1993). Within this framework, it seems reasonable to assume situational differences in the relative contribution of each mechanism to liposomal degradation and clearance. The cellular adsorption of liposomes may involve electrostatic or hydrophobic interactions or specific receptor binding, and adsorbed liposomes can deliver their contents directly into the cytoplasm. Alternatively, an exchange of lipids and proteins may occur between the lipid bilayer and the surrounding biological fluid environment. In addition to exchange mechanisms capable of altering the physical characteristics of liposomes, the vesicles may also be subject to degradative forces such as lipase action and phagocytosis by macrophages and neutrophils.

Liposome formulations are frequently used with drugs with significant dose-limiting toxicity (e.g., bleomycin, amphotericin, and platinum oncolytic agents) (Petrak 1993), and thus a safe and predictable time course of drug release is a primary safety concern. In a stable liposome formulation, net drug release is rate limiting and determines tissue and/or blood levels of the active agent. The rate of drug release may be influenced by the fluidity of the lipid bilayer, and interactions between liposomes and the environment may alter the fluidity, size, and permeability of the liposomes from their original state. For example, gentamicin release may be increased sevenfold when the phosphatidylcholine component is changed from egg to the less saturated soybean form (Kadir et al. 1993). Moreover, insulin formulated in liposomes produced a stronger pharmacodynamic response when the vesicles carried a negative charge than when neutral liposomes were used (Stevenson et al. 1982). The drug release rate also depends on the route of administration; in studies with liposomal chloroquine in mice, faster drug release was obtained with the SC route than by IM delivery (Titulaer et al. 1990) (see also Chapter 7).

Design of Studies to Evaluate Toxicity

Toxicity studies of sustained-release systems have some features not shared by conventional formulations. These include the use of additional control groups and, in certain cases, an extended study duration. The use of at least three doses of the active formulation is a customary practice in toxicology studies of both conventional and sustained-release formulations. A single vehicle control group is usual in studies of conventional formulations, whereas additional controls in studies of sustained-release formulations generally include the nondrug-containing device and another formulation of the same drug. This formulation can be a marketed immediate-release formulation or another sustained-release formulation, if available. Since one of the uses of the comparator formulation is to provide bridging data to previous toxicity studies, it is often advantageous to repeat a dose from a study with that formulation.

In toxicology evaluations of conventional formulations, multiple dose studies are generally performed to investigate steady state drug levels for at least the intended duration of treatment. Similarly, the approach to toxicologic dosing of sustained-release formulations is to use repeated doses at an interval that is governed by the duration of release of the formulation. The frequency of dosing is generally similar but not necessarily identical to the duration of drug release from the formulation. An alternative approach is to administer the formulation more frequently than the release interval. This leads to an escalation in drug concentrations, thereby potentially accentuating toxicity as well as masking the toxicological response to the actual drug delivery profile of the test formulation. With this regimen, the chance of detecting differences in toxicodynamics between successive iterations of a formulation may be diminished.

CONCLUSIONS

The safety evaluation of parenteral, sustained-release formulations requires characterization of systemic toxicity and biocompatibility in animals. Biocompatibility of depot formulations depends on the properties of both the delivery system and the drug, and the local tissue response generally represents a summation of drug and delivery system effects. To interpret any unexpected effects produced by the formulation, it is important to also study, preferably in parallel, the toxicity of the components of the sustained-release system as well as the immediate-release forms of the drug of interest. Studies should include parallel PK measurement and, if possible, PD measurement.

The biocompatibility of polymers used in sustained-release delivery systems is well established; however, it is influenced by polymer composition, the presence of process contaminants, altered degradation rate, and the size and shape of the delivery device. The usual response to degradable polymers is a benign local foreign body reaction that is reversible upon removal or dissolution of the polymer. Systemic toxicity of sustained-release polymeric systems is influenced chiefly by the effects of the drug and its release profile. Based on extensive prior use, the polymer components are not expected to contribute significantly to the systemic effects of the formulation.

Biocompatibility studies indicate that locally deposited liposomes are also well tolerated. Liposome vesicles are ingested, and ultimately degraded, by local accumulations of RES cells and produce little inflammatory reaction. In contrast to polymeric formulations, most liposomes can be administered intravenously as well as by SC and IM injection. When dosed intravenously, they may produce RES overload, potentially leading to blockade. The inclusion of nonnatural phospholipids, especially positively charged lipids, may be associated with adverse effects, most notably involving cellular components of blood and the RES and coagulation systems. With locally administered liposomes, the potential for these changes is minimized. The safety evaluation of drug-loaded liposome formulations should include PK and toxicologic studies of target organ toxicity and drug biodistribution.

REFERENCES

Anderson, J. M., and M. S. Shive. 1997. Biodegradation and biocompatibility of PLA and PLGA microspheres. *Adv. Drug Del. Rev.* 28:5–24.

Arrowsmith, M., J. Hadgraft, and I. W. Kellaway. 1984. The in vivo release of cortisone esters from liposomes and the intramuscular clearance of liposomes. *Int. J. Pharmaceutics* 20:347–362.

Chien, Y. W. 1992. *Novel drug delivery systems.* New York: Marcel Dekker.

Gombotz, W. R., and D. K. Pettit. 1995. Biodegradable polymers for protein and peptide drug delivery. *Bioconj. Chem.* 6:332–351.

Hirano, K., C. A. Hunt, A. Strubbe, and R. D. MacGregor. 1985. Lymphatic transport of liposome-encapsulated drugs following intraperitoneal administration: Effect of lipid composition. *Pharm. Res.* 2:271–278.

Kadir, F., W. M. C. Eling, D. Abrahams, J. Zuidema, and D. J. A. Crommelin. 1992. Tissue reaction after intramuscular injection of liposomes in mice. *Int. J. Clin. Pharm. Ther. Toxicol.* 30:374–382.

Kadir, F., J. Zuidema, and D. J. A. Crommelin. 1993. Liposomes as intramuscular and subcutaneous injection drug delivery devices. In *Pharmaceutical particulate carriers: Therapeutic applications,* edited by A. Rolland. New York: Marcel Dekker, pp. 165–198.

Kamei, S., Y. Inoue, H. Okada, Y. Yamada, H. Ogawa, and H. Toguchi. 1992. New method for analysis of biodegradable polyesters by high-performance liquid chromatography after alkali hydrolysis. *Biomaterials* 32:953–958.

Kobayashi, H., K. Shiraki, and Y. Ikada. 1992. Toxicity test of biodegradable polymers by implantation in rabbit cornea. *J. Biomed. Mat. Res.* 26:1463–1476.

Kopecek, J., and K. Ulbrich. 1983. Biodegradation of biomedical polymers. *Prog. Polym. Sci.* 9:1–58.

Lam, K. H., J. M. Schakenraad, H. Groen, H. Esselbrugge, P. J. Dijkstra, J. Feijen, and P. Nieuwenhuis. 1995. The influence of surface morphology and wettability on the inflammatory response against poly(L-lactic acid): A semi-quantitative study with monoclonal antibodies. *J. Biomed. Mat. Res.* 29:929–942.

Langer, R. 1990. New methods of drug delivery. *Science* 249:1527–1533.

Lee, H. J., G. Riley, O. Johnson, J. L. Cleland, N. Kim, M. Charnis, L. Bailey, E. Duenas, A. Shahzamani, A. J. S. Jones, and S. D. Putney. 1997. In vivo characterization of sustained-release formulations of human growth hormone. *J. Pharm. Expt. Ther.* 281:1431–1439.

Leonard, F., R. K. Kulkarni, G. Brandes, J. Nelson, and J. J. Cameron. 1966. Synthesis and degradation of poly(alkyl)α-cyanoacrylates. *J. Appl. Polym. Sci.* 10:259–265.

Leong, K. W., B. C. Brott, and R. Langer. 1995. Bioerodible polyanhydrides as drug-carrier matricies. I: Characterization, degradation, and release characteristics. *J. Biomed. Mat. Res.* 19:941–955.

Majors, K. R., and M. B. Friedman. 1990. Animal testing of polymer based systems. In *Polymers for controlled drug delivery,* edited by P. J. Tarcha. Boca Raton, Fla., USA: CRC Press, pp. 231–239.

Matlaga, B. F., L. P. Yasenchak, and T. N. Salthouse. 1976. Tissue response to implanted polymers: The significance of sample shape. *J. Biomed. Mater. Res.* 10:391–397.

Maulding, H. V., T. R. Tice, D. R. Cowsar, J. W. Fong, J. E. Pearson, and J. P. Nazareno. 1986. Biodegradable microcapsules: Acceleration of polymeric excipient hydrolytic rate by incorporation of a basic medicament. *J. Control. Rel.* 3:103–117.

Moshang, T., E. O. Reiter, B. L. Silverman, K. Ford, G. Downing, J. W. Frane, J. L. Cleland, S. D. Putney, and K. M. Attie. 1998. Treatment of growth hormone deficiency with a sustained-release growth hormone formulation. Abstract from the Meeting of The American Pediatric Society, 12–16 June in New Orleans, Louisiana.

Nakamura, A. 1993. Biological safety of biomaterials and devices. In *Biomedical applications of polymeric materials,* edited by T. Tsuruta, T. Hayashi, K. Kataoka, K. Ishihara, and Y. Kimura. Boca Raton, Fla., USA: CRC Press, pp. 429–468.

Osborne, M. P., J. J. Richardson, K. Jeyasingh, and B. E. Ryman. 1979. Radionuclide-labeled liposomes: A new lymph node imaging agent. *Int. J. Nucl. Med. Biol.* 6:75–83.

Petrak, K. 1993. Design and properties of particulate carriers for intravascular administration. In *Pharmaceutical particulate carriers: Therapeutic applications,* edited by A. Rolland. New York: Marcel Dekker, pp. 275–298.

Phillips, N. C. 1993. Subcutaneous liposomal tumor antigen vaccines: Preclinical and clinical studies. In *Pharmaceutical particulate carriers: Therapeutic applications,* edited by A. Rolland. New York: Marcel Dekker, pp. 199–226.

Pitt, C. G., and A. Schindler. 1983. Biodegradation of polymers. In *Controlled drug delivery,* edited by S. D. Bruck. Boca Raton, Fla., USA: CRC Press, pp. 54–80.

Ritschel, W. A. 1988. Pharmacokinetic and biopharmaceutical aspects in drug delivery. In *Drug delivery devices,* edited by P. Tyle. New York: Marcel Dekker, pp. 17–80.

Schwendeman, S. P., H. R. Costantino, R. K. Gupta, and R. Langer. 1997. Peptide, protein, and vaccine delivery from implantable polymeric systems. In *Controlled drug delivery: Challenges and strategies,* edited by K. Park. Washington D.C.: American Chemical Society, pp. 229–249.

Stevenson, T. W., H. M. Patel, J. A. Parsons, and B. E. Ryman. 1982. Prolonged hypoglycemic effect in diabetic dogs due to subcutaneous administration of insulin in liposomes. *Diabetes* 31:506–511.

Titulaer, H. A., W. M. Eling, D. H. Crommelin, P. A. Peeters, and J. Zuidema. 1990. The parenteral controlled release of liposome encapsulated chloroquine in mice. *J. Pharm Pharmacol.* 42:529–532.

Vance, M. L., C. J. Woodburn, S. Putney, J. Grous, H. J. Lee, and O. L. Johnson. 1997. Effects of sustained release growth hormone on serum GH, IGF-I and IGFBP-3 concentration in GH deficient adults. *Endocrin. Metab.* 4 (Suppl. A):75.

Visscher, G. E., J. E. Pearson, J. W. Fong, G. J. Argentieri, R. L. Robison, and H. V. Maulding. 1988. Effect of particle size on the in vitro and in vivo degradation rates of poly(DL-lactide-co-glycolide) microcapsules. *J. Biomed. Mat. Res.* 22:733–746.

Visscher, G. E., R. L. Robison, H. V. Maulding, J. W. Fong, J. E. Pearson, and G. J. Argentieri. 1986. Note: Biodegradation of and tissue reaction to poly (DL-lactide) microcapsules. *J. Biomed. Mat. Res.* 20:667–676.

Weiner, M., and I. L. Bernstein. 1989. *Adverse reactions to drug formulation agents: A handbook of excipients.* New York: Marcel Dekker.

Yamaguchi, K., and J. M. Anderson. 1993. In vivo biocompatibility studies of medisorb 65/35 D,L-lactide/glycolide copolymer microspheres. *J. Control. Rel.* 24:81–93.

Zhang, Y., S. Zale, L. Alukonis, and H. Bernstein. 1995. Effects of metal salts on PLGA hydrolysis. *Proceed. Intern. Symp. Control. Rel. Bioact. Mater.* 22:83–84.

Zhao, Q., N. Topham, J. M. Anderson, A. Hiltner, G. Lodoen, and C. R. Payet. 1991. Foreign-body giant cells and polyurethane biostability: In vivo correlation of cell adhesion and surface cracking. *J. Biomed. Mat. Res.* 25:177–183.

16

Scale-Up, Validation, and Manufacturing of Microspheres

Kunio Kawamura
Kinkii University
Higashi-Osaka, Japan

Recent advances in pharmaceutical manufacturing technology have enabled the development of significantly improved drug delivery systems for injectable products. In the early 1980s, controlled-release injectable microspheres containing potent LHRH (luteinizing hormone release hormone) agonists were prepared using a biodegradable polymer carrier, poly(lactic acid-co-glycolic acid) (PLGA), by two methods, phase separation and a water/oil/water (w/o/w) emulsion-solvent evaporation technique. These methods were investigated simultaneously and produced commercially successful one- and three-month release injectable products. These products have decreased the frequency of administration while minimizing patient discomfort. A brief summary of these products is shown in Table 16.1.

There are many published reports on sustained-release injectable products, most of which are manufactured at laboratory

Table 16.1. *Representative Characteristics of Biodegradable, Controlled-Release Injectable and Implantable Products*

Drug (Number of Amino Acids)	Polymer (Molecular Wt.)	Preparation Method	Size (Drug Content)	Length of Release In Vivo
Goserelin acetate (10)	PLGA (50/50) (20,000)	Melt extrusion	Cylindrical rod (21%)	28 days
Nafarelin acetate (10)	PLGA (50/50) (0.38 dL/g)	—	20–50 μm (1%)	24 days
Tryptorelin (10)	PLGA (53/47) (0.7 dL/g)	—	50 μm (2%)	30 days
Leuprorelin acetate (9)	PLGA (75/25) (14,000)	w/o/w	20–30 μm (8%)	4 weeks
	PLA (15,500)	w/o/w	20–30 μm (8%)	28 days
TRH (3)	PLGA (75/25) (14,000)	w/o/w	30 μm (7%)	4 weeks
Buserelin (9)	PLGA (50/50)	—	—	28 days

TRH, thyroid releasing hormone

PLA, poly(lactic acid)

scale. After obtaining successful results in animal studies in small-scale production, many technical challenges and objectives must be accomplished during scale-up, validation, manufacturing, and technology transfer to reach the commercial market. During the process from laboratory to commercial scale, there are various challenges that must be overcome. However, the difficulties faced in the scale-up of pharmaceuticals has unique differences from other industries. This scale-up and technology transfer must be performed without changes in the quality, safety, and efficacy of the products, since these are scrutinized and fixed at the time of a clinical trial. The following items must be addressed when developing a manufacturing process for sustained-release products for industrial production:

- *Scale-up:* Large-scale production capabilities need to be established.

- *Drug release:* The rate of drug release should be reproducible and meet the product requirements following large-scale production.

- *Quality control:* Reliable quality control test methods must be established to ensure product quality, especially the reproducibility of drug release.

- *Chemical stability:* Stability of the drug and the release pattern should be assured.

- *Biological activity:* The bioactivity of labile compounds, such as proteins, following incorporation into a sustained-release injectable system must be assured.

- *Physical properties:* The stability of the physical properties of the microspheres should be maintained (i.e., aggregation [stickiness] of the microspheres over time should be prevented).

- *Sterility:* Sterility assurance is one of the most important and challenging issues in the manufacture of sustained-release injectable products, since these products typically cannot be terminally sterilized and are usually manufactured via aseptic processing.

- *Impurities:* Impurities associated with the manufacturing process must be controlled within specific limits.

Residual solvent and/or degradation products must be below acceptable levels.

- *GMPs:* Good Manufacturing Practices must be followed to obtain approval for commercialization.

SCALE–UP AND TECHNOLOGY TRANSFER

In general, when undertaking the scale-up and technology transfer of a pharmaceutical product, the greatest challenge is to maintain the consistency and quality of the product, including bioequivalency and bioavailability. Although consistency and quality should always be considered, there will be differences in the extent of evaluation. If clinical trials are in late Phase II or later, careful attention must be applied to maintain the bioavailability and bioequivalency of the product prepared at large scale equivalent to that prepared at smaller scales. The same attention must be devoted to the scale-up of the manufacturing method after production is already at commercial scale. However, in the early stage of development in the laboratory, precise confirmation of bioequivalency and bioavailability in human subjects may not be necessary, if the consistency of quality is ensured by all other parameters. In performing the bioequivalency study, it is essential that a relationship between bioavailability and bioequivalency is established. It is preferable that the evaluation method of bioequivalency by in vitro methods is correlated to in vivo bioequivalency and bioavailability. Scale-up and technology transfer are completed by ensuring the consistency of the product by two methods before and after scale-up. All parameters relating to quality and manufacturing conditions, in addition to the bioavailability and bioequivalency study, must be well characterized at bench scale before proceeding to scale-up. Other important factors that must be considered are the assurance of sterility and nonpyrogenicity. Sustained-release injectable products are typically manufactured by incorporating an active pharmaceutical substance into a polymer matrix; the usual sterilization methods cannot always be applied.

In the early stages of the development of sustained-release injectable products, the following items must be considered when developing and establishing microsphere technology for commercial production:

- *Drug release:* The rate of drug release from a large-scale production lot should be reproducible and meet product requirements. This includes drug release characteristics. An active ingredient must be released in a zero-order fashion for over one month (one-month depot) or three months (three-month depot) with only a small initial burst.

- *Encapsulation efficiency:* During the manufacturing process, the active ingredient must be entrapped into the microspheres effectively, and loss of the active ingredient into the solvent must be minimized.

- *Residual solvent:* Undesirable residual solvent must be minimized.

- *Sterilization:* The method of sterilization must be determined.

MANUFACTURING METHODS

Microsphere Fabrication Techniques

Several methods of fabrication and polymers have been utilized in the preparation of sustained-release microspheres. The most widely used polymers used for sustained-release biodegradable microspheres have been poly(lactic acid) (PLA), poly(glycolic acid) (PGA), and/or the copolymer of lactic acid and glycolic acid (PLGA) (Kent et al. 1981, 1987, Ogawa et. al. 1988a, b, c; Okada et al. 1988; Okada 1989; Redding et al. 1984; Sanders et al. 1984; Schwope et al. 1975; Toguchi 1996; Wise et al. 1980). Injectable sustained-release microspheres of water-soluble drugs have usually been prepared by a phase separation method (Sanders 1984; Redding 1984; Ruiz and Benoit 1991) or an in-water drying method (Ogawa 1988a, b, c; Heya 1991a, b).

Phase Separation

The phase separation method requires large amounts of organic solvent, which presents challenges in large-scale production. In addition, the prepared microspheres have a tendency to agglomerate, and the residual organic solvent can be difficult and time-consuming to remove.

In-Water Drying

The in-water drying method is a w/o/w emulsification process and is not usually suitable for the production of microspheres containing a high concentration of a water-soluble drug. This method has many steps, and the temperature and viscosity of the w/o/w emulsion must be carefully controlled (Kent et al. 1981; Ogawa et al. 1988a, b, c; Yamamoto et al. 1985).

Spray Drying

The spray drying method may also be applicable, since the process is fast and is performed under much milder process conditions (Pavanetto et al. 1993). However, considerable loss of an active ingredient can occur during spray drying, depending on the degree of adhesion to the inside wall of the spray drying apparatus. The spray drying method also has some difficulties in assuring the sterility of the product (Nonomura et al. 1994; Takada et al. 1994).

Scale-Up

Regardless of the chosen manufacturing method and the conditions of laboratory preparation, the rationale of this selection must be fully reflected in scale-up and technology transfer. In the scale-up of the manufacturing method, the specifics of the equipment and manufacturing conditions should be designed by considering the critical points during development of the laboratory scale process. Successful scale-up and technology transfer can be achieved by comprehensive study in the laboratory. During development in the laboratory or small-scale manufacturing, critical parameters in the manufacturing method must be clarified. Among these critical parameters, some can be easily controlled throughout the scale-up process, while others are not easily controlled. For example, the temperature of a liquid or emulsion in a vessel can be easily controlled, while mixing or agitation of the liquid and/or emulsion will depend on the size of vessel. It is important to understand what parameters will be fixed from laboratory scale to commercial scale and what parameters are variable, so that they can be optimized in the scale-up of the manufacturing method. Examples of the relationship between laboratory-scale preparation and the scaled-up product are discussed here by reviewing a reported manufacturing and preparation method.

Case Study: Lupron Depot®

The manufacturing method of Lupron Depot®, a very successful sustained-release injectable product is discussed below. As discussed in other chapters of this book, Lupron Depot®, is a once-a-month or once-every-3-months injectable PLGA microsphere formulation containing leuprolide acetate. PLGA is a biodegradable polymer that has been used for many years in biodegradable surgical sutures and other medical applications. A preparation method using in-water drying through a w/o/w emulsion (Ogawa et al. 1988a, b, c; Okada 1987) is described in the following sections.

Laboratory Preparation

The emulsion method was selected based on a novel w/o/w emulsion technology for encapsulating water-soluble substances originally reported by Okada et al. in 1988. This method was selected because the preparation procedure is simple, medical grade materials are readily available, and organic solvents with a low boiling point are used to overcome the problems associated with residual solvent. The peptide and gelatin were dissolved in a small amount of distilled water. PLGA dissolved in dichloromethane (DCM) was mixed with the above solution and agitated vigorously with a homogenizer. This w/o emulsion was cooled to increase the viscosity of the inner water phase and the emulsion, and then poured into a chilled PVA (polyvinyl alcohol) solution through a long, narrow nozzle with stirring using a mixer. The resulting w/o/w emulsion was stirred gently to remove the organic solvent to form the microspheres. The semidry microspheres were sieved to remove larger particles, collected by centrifugation after washing with water, and lyophilized to complete the removal of organic solvent and water. In this procedure, only one volatile organic solvent (DCM) was used, and the residual concentration was adequately reduced by well-controlled heating under reduced pressure during the lyophilization process. The residual solvent level was far below the specification limit (500 ppm in USP 23 <467>).

Formation of a stable w/o/w emulsion is required in the solvent evaporation process. A stable w/o emulsion was obtained by controlling the temperature, concentration, and volume of each solution. The preparation procedure for one-month depot microspheres of leuprolide acetate using a w/o/w emulsion-solvent evaporation method is shown in Figure 16.1.

In general, water-soluble substances, especially those with a low molecular weight, are easily released to the outer water phase, even after the formation of a stable w/o/w emulsion during solvent evaporation over a long period. A high encapsulation efficiency was obtained by increasing the viscosity of the inner water phase (by adding gelatin to the water phase and lowering

Figure 16.1. *Preparation procedure of leuprolide acetate microspheres by an in-water drying method.*

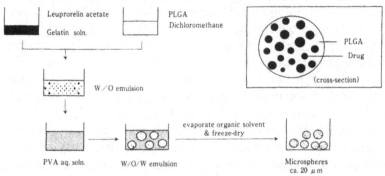

the temperature) and the w/o emulsion (by increasing the concentration of drug and polymer). Adding gelatin and lowering the temperature of the w/o emulsion have increased the encapsulation efficiency of a water-soluble peptide by using the in-water drying method.

The microspheres prepared by the emulsion method are polycore microspheres in which the drug is primarily dispersed throughout the interior of the polymer matrix. Mannitol is added to distilled water to redisperse the semidry microspheres before the lyophilization process.

It is reported that the same kind of technology was successfully applied to the other active ingredients, such as TRH (thyrotropin releasing hormone), TNP-470 [6-O-(N-chloroacetylcarbamoyl)-fumagillol], and vaccines in laboratory-

---

Figure 16.2. *Chemical structures of LHRH, leuprolide acetate, and TRH.*

5-Oxo-Pro-His-Trp-Ser-Tyr-Gly-Leu-Arg-Pro-Gly-NH$_2$

1 2 3   4 5 6 7 8 9 10

LHRH

5-Oxo-Pro-His-Trp-Ser-Tyr-Gly-Leu-Arg-Pro-NH-C$_2$H$_5$

1 2 3   4 5 6 7 8 9

Leuprolide
(Leuprorelin acetate: M.W. = 1,269.5)

5-Oxo-Pro-His-Pro-Gly-NH$_2$

1 2 3

TRH
(M.W. = 364.2)

scale preparations. TRH enhances the physiological activities of the brain and stimulates the release of TSH (thyroid stimulating hormone). TNP-470, a fumagillin derivative, has strong antiangiogenic activity (Ingber et al. 1990; Kamei et al. 1993; Okada et al. 1992). Although these products are not commercially available, the same technology transfer procedures would be utilized. The chemical structures of LHRH, leuprolide acetate, and TRH are shown in Figure 16.2. The description from the 1997 *Physician's Desk Reference* for Lupron Depot® 3.75 mg and 7.5 mg is shown in Figure 16.3.

Scale-Up Considerations

The ratio of the amount of each ingredient, the solvent, and the temperatures of the solutions must be fixed in the scale-up trial production, since these parameters have already been thoroughly studied in laboratory experiments and small-scale production.

Figure 16.3. *Description of Lupron Depot® 3.75 mg and 7.5 mg in 1997* Physician's Desk Reference.

Lupron Depot® 3.75 mg <PDR>
Lupron Depot® 7.5 mg <1997>
(leuprolide acetate for depot suspension)

DESCRIPTION
Leuprolide acetate is a synthetic nonapeptide analog naturally occurring gonadotropin-releasing hormone (GnRH or LH-RH). The analog possesses greater potency than the natural hormone. The chemical name is 5-oxo-L-prolyl-L-histidyl-L-tryptophyl-L-seryl-L-Tyrosyl-D-leucyl-L-argibyl-N-ethyl-prolinamide acetate (salt) with the following structural formula:

LUPRON DEPOT is available in a vial containing sterile lyophilized microspheres, which when mixed with diluent, become a suspension that is intended as a monthly intramuscular injection.

Variables in the manufacturing of the Lupron Depot® dosage form at scale-up production include the size of vessels and equipment. At this stage, the agitation or stirring of solution and emulsion, the conditions to remove the organic solvent, and the conditions of lyophilization for the complete removal of organic solvent and water should be carefully studied based on knowledge obtained in small-scale production. In order to obtain the same agitation or stirring effect with the small-scale vessel of solutions and/or emulsion, the shape of the vessel and the stirring apparatus must be considered.

RAW MATERIAL PROCUREMENT

In addition to the elements described in the previous section, there are several other essential factors that must be considered. The procurement of raw materials of consistent high quality with the correct degree of polymerization and distribution of molecular weight is important for the polymer used in the matrix. Even at bench scale or small-scale manufacturing, a procurement procedure for raw materials should be established. It is important to have a stable and continuous supply of quality raw materials. The most important raw material is the polymer used in the microencapsulation matrix. Therefore, the polymer should be manufactured similar to a finished product. It is important to procure endotoxin-free and nonpyrogenic raw materials, since endotoxin cannot be removed in a typical manufacturing process.

IN VITRO DRUG RELEASE STUDIES

For the evaluation of sustained-release injectable products, in vitro drug release studies and their relationship with in vivo drug release are the most important parameters to ensure a quality product.

In Vitro Drug Release

For the routine evaluation of the quality of sustained-release injectable products, in vitro drug release test methods have been developed. This test method is performed by suspending microspheres in the test solution and shaking the solution at $37 \pm 1°C$. The microspheres (50 mg) are suspended in 10 mL of release medium (33 mM phosphate buffer, pH 7.0, with Tween 80®). The residual active ingredient in the microspheres is periodically determined by high performance liquid chromatography (HPLC) after filtering the microspheres (Heya et al. 1991, 1994).

In Vivo Drug Release

In order to compare in vitro dissolution with in vivo dissolution, in vivo drug release was determined in rats by determining the amount of leuprolide acetate remaining at the subcutaneous injection site. The excised microspheres, surrounded by a thin layer of connective tissue, were homogenized and extracted with the same solution used in the in vitro method. After the homogenate was centrifuged at 3,000 rpm for 15 min, the drug content in the aqueous phase was analyzed by HPLC.

Correlation Between the In Vitro and In Vivo Drug Release Profile

The relationship between the in vitro and in vivo drug release profiles showed good agreement and indicated that the drug release rate and the release profile can be determined with an in vitro method for routine quality control. Figure 16.4 shows the in vitro and in vivo relationship of the TRH depot. For routine daily testing, accelerated dissolution testing with the same solution has also been developed.

Figure 16.4. *Comparison of in vitro and in vivo dissolution of TRH depot in PLGA matrix.*

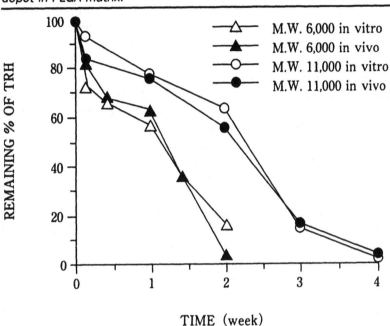

STERILITY ASSURANCE

Sterilization Methods for Sustained-Release Injectable Products

Any injectable medical products, including sustained-release injectable products, must be sterile. In order to obtain sterile sustained-release injectable products, two methods can be employed: aseptic processing or terminal sterilization. Aseptic processing of the entire microsphere manufacturing process has several challenges that can be met by careful process validation. Lupron Depot® is manufactured by aseptic processing. It may appear simple to manufacture nonaseptic microspheres and sterilize the nonaseptic microspheres by radiation sterilization at the end of the process.

When gamma irradiation, which is known to reduce the molecular weight of PLGA, is used as the sterilization process, the molecular weight of the L/G ratio in PLGA in the microspheres must be redetermined. The effect of gamma irradiation on the release rate must be demonstrated by in vivo and in vitro tests. The process also must be validated for its capability in producing uniform results on all samples in a single run, as well as consistent results from run to run. In addition to the above, radiation sterilization produced a variety of degradation products of both the polymer and the peptide drug (Chiu et al. 1989). Determining the toxicity and bioactivity of all the degradation products is practically impossible, especially for peptides and proteins. The molecular weight of the polymer is dependent on the dose of irradiation, and the polymer may contain oligomers that might promote polymer degradation by self-catalysis. An increase in initial peptide loss and shortening of the release period due to the presence of oligomers created by irradiation were reported for tryptorelin microspheres using PLGA (50/50) (Ruiz and Benoit 1991). The presence of oligomers also affects the phase separation of PLGA during the microencapsulation process. Because of the insurmountable challenges of terminal sterilization, aseptic processing is successfully applied in the production of sustained-release injectable products.

To ensure a sterile product following aseptic processing, sterile components and equipment must be utilized. Traditional steam and dry heat sterilization of the equipment are used on heat resistant equipment, containers, and closures. Filtration sterilization is used for all organic and aqueous solutions. The aseptic process must be validated to ensure a sterile product.

Stages of Sterility Assurance

Sterility assurance of product manufactured by aseptic processing consists of three major components: process validation, environmental and process monitoring, and intermediate and finished product testing.

Process Validation

Validation of the entire aseptic process in the manufacture of microspheres can be divided into four operations:

1. Simulation of the sterile intermediate materials manufacturing process

2. Simulation of the wet microsphere manufacturing process by aseptic filtration (Figure 16.1)

3. Simulation of the lyophilization process

4. Simulation of the filling process (media fill run)

These four processes are validated separately by simulating each process using suitable media to confirm that each process is completed aseptically. Completion of all four stages is necessary to achieve complete process validation (Tognchi et al. 1996).

Preparation of Sterile Raw Materials. The preparation process of sterile starting raw materials consists of filtration sterilization followed by lyophilization. All of the materials used, leuprolide acetate, mannitol, PLGA, distilled Water for Injection (WFI), and organic solvents, are sterilized by filtration just before manufacturing the microspheres. Filtration sterilization is validated by a biological indicator (BI) challenge in the same solvent system and simulation test.

The first validation in this process is performed by confirming that each filtration system has enough capacity for the filtration sterilization and is validated by using *Brevundimonas diminuta* or an equivalent-sized microorganism suspended in the same solvent. The second validation is performed by simulating each process using appropriate media at production scale. As in the simulation of the sterile mannitol manufacturing process, aseptically filtered mannitol solution, at a concentration that does not inhibit microbial growth, is placed in the lyophilizer and stored under vacuum without freezing. After collecting the solution, the number of microbes in the entire solution is counted after membrane filtration and incubation.

Other processes for the preparation of sterile starting raw materials have been validated by simulating each stage of the process with the mannitol process and the same amount of raw materials and solvent. The sterility of each component is confirmed by incubation of the filter following filtration of each solvent/solution containing the individual components.

Wet Microsphere Manufacturing Process. In order to assure the sterility of the processes, the emulsification and wet microsphere manufacturing processes are simulated by using the mannitol solution, which does not inhibit microbial growth with the same equipment and scale as an actual production run.

Lyophilization. The lyophilization and microsphere drying processes are simulated using the same amount of mannitol solution as used in actual production. The number of microbes that might potentially exist in the process is counted after membrane filtration and incubation.

Filling. The filling process is validated by three consecutive media fill runs, which consist of filling sterile mannitol followed by adding SCD (soybean casein digest) liquid medium and incubating. The relationship between the process flow of microsphere manufacturing and the four kinds of process simulation is shown in Figure 16.5.

In order to assure the integrity of aseptic processing, the operation should be automated or performed by robots as much as possible and should minimize the number of manipulations performed by manufacturing personnel.

Environmental Monitoring and Control of Facilities

In addition to the prospective validation of the process, facility, and equipment after installation, the critical processes and the environment are monitored continuously during production. Airborne microbes, airborne particulates, and pressure differentials are the most critical items to be intensively monitored in the manufacturing environment. A continuous monitoring system should be utilized.

Figure 16.5. *Media simulation test for Lupron Depot® production.*

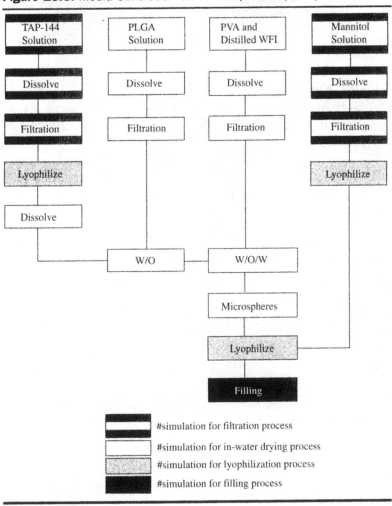

Validation (i.e., prospective validation) is important in establishing the manufacturing procedure and confirming that the procedure, facility, and equipment are adequate for their intended use. During production, critical processes and the environment should be monitored continuously or periodically. The following are examples of the intensive monitoring of the manufacturing process and the environment: computerized continuous and automatic pressure differential monitoring and control, continuous and automatic airborne particulate monitoring, periodical microbial monitoring, review and assessment.

Pressure Differential Monitoring. Among the various environmental monitoring methods and controls, monitoring of the air pressure differentials in the aseptic area is the most critical. Figure 16.6 shows a schematic drawing of the pressure differential cascade in the manufacturing of a sterile product, such as penicillin or hormonal pharmaceuticals. In this pressure differential cascade, aseptic processing should be performed in the area of the highest pressure against the surroundings. There must be a negative air pressure area in the facility in order to avoid the spread of drug powder, which can be controlled by continuous use of a computer monitoring and control system. The computer controller display indicates the air pressure differentials in each room and the directions of airflow. The air pressure differentials and the airflow are automatically controlled. An alarm system is also installed to alert manufacturing personnel of any changes in the pressure differential cascade.

Figure 16.6. *A schematic drawing of pressure differential cascade.*

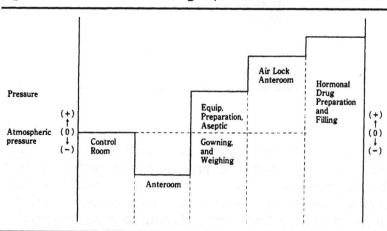

Airborne Particulate Monitoring. In order to maintain the aseptic condition of the working area, the level of airborne particulate matter must be below the specified limit. For this purpose, airborne particulate matter in each controlled area is monitored continuously by using probes installed at all critical locations. The monitoring display of the computer controller panel indicates the number of airborne particulates in every critical location and triggers an alarm if out of specification.

Review and Assessment. Data obtained in environmental monitoring and control should be statistically reviewed with daily, monthly, semiannual, and annual computer controls to maintain a high degree of quality assurance.

Sterility Testing of Intermediate and Final Product

Sterility testing, microbial monitoring, and pyrogen testing are performed on raw materials, intermediate product, and finished product. These tests are classified as follows:

- Bioburden and pyrogen testing for raw materials

- Sterility tests for intermediate raw materials

- Microbial testing of supernatant/waste solutions by filtration and cultivation of the filter

- Sterility testing of intermediate and finished product

Raw materials, intermediate product, and finished product are also tested to confirm sterility by taking samples from each step in the production process. Sterility assurance by microbial testing of the final product is often less reliable and is usually only for confirmation. However, in the testing of Lupron Depot® microsphere manufacturing, supernatant/waste solution obtained in the wet microsphere manufacturing process is subjected to the test of microbes by membrane filtration of the entire solution. The test of microbial counts in the supernatant and waste solution, which are collected from all processes, can verify the sterility of the product with more precision than testing of samples taken from the finished product alone. This is not a sampling inspection; it is a 100 percent inspection.

For the sterility test of the finished product of Lupron Depot®, the outer surface of the microsphere and the interior of the microsphere are tested by dissolving in organic solvent. The sterility assurance of the overall process is shown in the Table 16.2.

SPECIFICATIONS

Matrix Polymer (PLGA) Specifications

One of the most important specifications in the production of microspheres is the physicochemical characteristics of the matrix polymer, PLGA. PLGA is not a novel chemical compound; for use in sustained-release injectable products, however, it is considered as a new molecular entity because of the extent of polymerization and the ratio of lactic acid to glycolic acid

Table 16.2. *Major Factors in Sterility Assurance of the Microsphere Manufacturing Process*

Items	Specifications
Raw materials	Bioburden and pyrogen check
Filtration sterilization	Integrity test, *Brevandimonas diminuta* challenge
Lyophilization	Simulation test
Sterile raw materials	Sterility test
Production of microspheres	Microbial check of supernatant/waste solution by filtration (100% for all volume of filtrate)
Microspheres	Sterility test of outer surface of microsphere interior
Aseptic filling	Sterility test, media fills
	Usually 3 media fills each consisting of 5,000 units each or more

(L/G). Complete information on the chemistry, manufacturing, and controls is necessary. Particularly, the L/G ratio and the distribution of molecular weight are critical. Molecular weight distribution and monomer and oligomer impurities usually affect entrapment of the water-soluble active ingredient. The establishment of specifications for the molecular weight distribution of the matrix polymer and consistent procurement of high quality matrix polymer are critical in developing sustained-release products. In addition to standard tests for raw materials, every lot of PLGA must be tested for the molecular weight and the L/G ratio, which affect the release rate of the peptide or an active ingredient from the microspheres. Other tests, such as intrinsic viscosity, free acid, and residual solvents, should be considered as standard tests for polymers.

Product Specifications

The specifications of injectable microsphere dosage forms must be established to include test items such as the size of the microspheres, peptide loading, peptide release pattern, content uniformity, residual solvent, and sterility.

Sterility

As already described, sterility assurance is one of the most important and difficult issues in the manufacturing of sustained-release injectable products, since the products cannot be terminally sterilized—sterility is assured by aseptic processing. The method to ensure sterility has already been described in the manufacturing process. In addition to the assurance of sterility in the process, sterility tests should be established as one quality confirmation method. Because the microspheres are insoluble in culture media, a sterility test specified by a general method can only demonstrate sterility of the exterior of the microspheres. The microspheres should be dissolved in a suitable solvent to demonstrate sterility of the interior of the microspheres. Therefore, sterility testing of sustained-release injectable products should consist of the exterior and interior sterility tests.

Dissolution

Dissolution is based on the relationship between in vitro dissolution and the in vivo release test. In vivo release testing is the direct evaluation of sustained-release injectable products, but it is difficult to perform the in vivo test to evaluate routine manufactured products. In vitro dissolution tests have been established for the purpose of routine quality evaluation. However, for some types of sustained-release injectable products that release a drug for one or more months, conventional, in direct relation accelerated tests are also desired. For such purposes, a conventional dissolution test method may be applied for expedited feedback of the quality information for routine monitoring purposes. A short-term, conventional, accelerated dissolution test has been developed based on the study of in vivo and in vitro (full long term; one month) tests, and the in vitro (full long term) study and the short-term dissolution test. A short-term drug release test capable of producing a release rate profile that correlates with a long-term drug-release test is also accepted. Of course, lots rejected by the long-term test are also rejected by the short-term test.

Content Uniformity of the Active Ingredient

Lot-to-lot content uniformity of microspheres depends on the degree and consistency of entrapment during the microsphere manufacturing process. This is affected by the physicochemical properties of the matrix polymer and also by the microsphere manufacturing process. In order to monitor the entrapment ratio of the active ingredients, the amount of active ingredient in the microspheres is measured and must fall within the specified limit.

Appearance and Physical Properties

Inspection with a microscope is also important in evaluating the quality of the microspheres. The shape and surface of microspheres should be one of the most important indications observed. Particle distribution should also be determined and evaluated.

Residual Solvent

The removal of organic solvent used in the manufacturing process and reduction of residual solvent to below the specified limit are critical and sometimes difficult operations. The process should be carefully studied and established; thereafter, residual solvent should be monitored and carefully controlled. DCM, a known potential carcinogen, is often used during the manufacturing of microspheres. The residual DCM present in the microspheres must be within the permitted levels.

Entrapment of the Active Ingredient

There is a close relationship between entrapment of the active ingredient and the content or yield of the final product. Entrapment and content are good indicators for evaluating the process.

Stability

Stability studies should be performed at both accelerated conditions and long-term storage conditions.

In Vivo Bioequivalency

It is desirable and necessary to perform an in vivo bioequivalency study between the laboratory-scale product and the commercial-scale product in humans.

Microsphere Size

An acceptable range of particle sizes should be established based on the release rate profiles. These data are determined from various sizes of microspheres obtained by sieving.

Summary

The specifications developed for confirming scale-up effects have been applied to the manufacture of products at each stage of production. Sample specifications of sustained-release injectable products are shown in Table 16.3.

Table 16.3. *Specifications for a Sustained-Release Injectable Product*

Test Items	Specifications and Comment
Appearance	Visual appearance
	Microscopic appearance
Identification	Positive
Aggregation	Passes through a needle easily (the absence of aggregation or the sticking of microspheres should be carefully checked.)
Endotoxin	Within the FDA's guidelines
Sterility	Interior: sterile
	Exterior: sterile
Content uniformity	Within the specification (meets the USP's requirements)
Water content	Within the specification (Excess water causes aggregation or sticking of microspheres.)
Residual solvent	Within the specification (The removal of organic solvent used in the manufacturing process should be carefully checked.)
Dissolution pattern	Meets the specification (This is one of the most important quality items. Routine testing is performed by the short-term conventional dissolution test method.)
Content of active ingredient	Within limits
Content of additives	Within limits (Additive content is significant because it affects the ratio of active ingredient to additives.)

CONSTRUCTION OF AN
ASEPTIC PROCESSING FACILITY

A facility for the production of microspheres by an aseptic process must meet stringent specifications. The major construction features of an aseptic processing room (area) for manufacturing sustained-release injectable products are summarized below.

- Air supplied to the aseptic processing room should be of Class 100 (air cleanliness, in which the number of particles, 0.5 μm and larger, per cubic feet is less than 100. Class 100 corresponds to M3.5 (SI units) or higher quality.

- An area where products are exposed to the open air should be covered by Class 100 unidirectional flow or higher quality air and separated from personnel by barriers.

- The air pressure of the aseptic processing room should be kept positive against the surrounding area.

- Outlet of exhaust air is through a HEPA (high efficiency particulate air) filter to prevent backflow of contaminated air. The amount of exhausted air is adjusted to keep the inside air at a positive pressure.

- All of the supplied water should be distilled WFI with a hot water circulation system. The water system is periodically steam sterilized.

- All materials and equipment are sterilized through a double-ended sterilizer.

- Personnel must be well trained for aseptic processing. They should not manipulate the product, equipment, and materials directly.

- There should not be any space between the equipment installation and the floor and/or wall.

- The walls, ceiling, and floors should be smooth, without any space or cracks.

- There should be no floor drain.

These major construction features are shown in Figure 16.7. After completing the construction of the aseptic processing room, all parts of the processes should be carefully validated. Installation qualification and operational qualification, including environmental monitoring and control described previously, should be performed and followed by performance qualification.

Figure 16.7. *Major construction features for an aseptic processing area.*

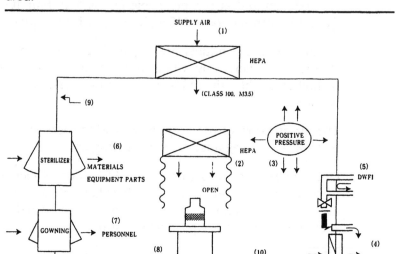

SCALE–UP TECHNOLOGY

As product proceeds through development, it is inevitable that production output will increase from laboratory small scale to a commercial scale. The final scale must be in place before Phase III clinical trials begin. Production scale should be increased gradually. A stepwise increase in production (i.e., two times each the previous scale) to full-scale production is accomplished by evaluating at each step every quality parameter of the microspheres. In each step, the quality of the microspheres should meet all product specifications. Examination of the microspheres by electron microscopy has also proven to be beneficial. By gradually increasing steps during scale-up, all of the technical challenges can be overcome.

SUMMARY

The scale-up and technology transfer of sustained-release injectable products is a very complex process. The following outlines the most important principles in the scale-up, validation, and manufacturing of sustained-release polymer microspheres.

- Specifications should be established throughout the manufacturing processes. The microsphere manufacturing process specifications should include the raw materials and intermediate processes.

- Continuous monitoring of the processes and the environment are critical in the manufacturing of a quality product. By using a computer-controlled monitoring system, continuous data can be generated for statistical review.

- Construction of the aseptic processing facility is very important.

- Scale-up of production capacity should be performed in small increments with careful stepwise evaluation of production quality, particularly the in vitro release characteristics.

- Information obtained from laboratory-scale preparations should be fully utilized during the scale-up and technology transfer process. A thorough understanding of the laboratory-scale process will result in a more successful scale up.

- Regulatory guidelines have been published by many agencies. See Table 16.4 for a listing of these documents. Following these guidelines and thorough understanding of regulatory requirements will expedite the product approval process.

Table 16.4. *Major Guidelines from Regulatory Agencies and Related Information*

Aseptic Processing of Healthcare Products: ISO/TC 198 N218

USP: <1116> Microbial Evaluation of Clean Rooms and Other Controlled Environments

Guide to Inspections of Sterile Drug Substance Manufacturers (FDA) (July 1994)

Fundamentals of a Microbiological Environmental Monitoring Program, Technical Report No. 13, PDA Environmental Task Force, *Journal of Science and Technology*, Vol. 44, Supplement (1990)

Validation of Aseptic Drug Powder Filling/Processes, PDA Technical Report No. 6 (1984)

Guildeine on Sterile Drug Products Produced by Aseptic Processing (FDA) (1987)

Sterile Drug Process Inspection, FDA *Compliance Program Guidance Manual*, 7356.002A (1993)

The revised EU GMP Annex on the Manufacture of Sterile Products (June 1996)

Process Simulation Testing for Sterile Bulk Pharmaceutical Chemicals, PDA Technical Report No. 28, Vol. 52, No. S3 (1998)

REFERENCES

Chiu, Y., and S. Sobel. 1989. Protein and peptide drug delivery systems: A regulatory viewpoint: Therapeutic peptides and proteins, formulation, delivery, and targeting. *Cur. Comm. Molecul. Biol.* 187–193.

Heya, T., H. Okada, Y. Tanigawara, Y. Ogawa, and H. Toguchi. 1991a. Effects of counteranion of TRH and loading amount on control of TRH release from copoly(DL-lactic/glycolic acid) microspheres prepared by an in-water drying method. *Int. J. Pharm.* 69:69–75.

Heya, T., H. Okada, Y. Ogawa, and H. Toguchi. 1991b. Factors influencing the profiles of TRH release from copoly(d,l-lactic/glycolic acid) microspheres. *Int. J. Pharm.* 72:199–205.

Heya, T., H. Okada, Y. Ogawa, and H. Toguchi. 1994. In vitro and in vivo evaluation of thyrotrophin releasing hormone release from copoly(dl-lactic/glycolic acid) microspheres. *J. Pharm. Sci.* 83:636.

Ingber, D., T. Fujita, S. Kishimoto, K. Sudo, T. Kanamaru, H. Brem, and J. Folkman. 1990. Synthetic analogues of fumagillin that inhibit angiogenesis and suppress tumor growth. *Nature* 348:555–557.

Kamei, S., H. Okada, Y. Inoue, T. Yoshioka, Y. Ogawa, and H. Toguchi. 1993. Antitumor effects of angiogenesis inhibitor TNP-470 in rabbits bearing VX-2 carcinoma by arterial administration of microspheres and oil solution. *J. Pharmacol. Exp. Ther.* 264:469–474.

Kent, J. S., D. H. Lewis, L. M. Sanders, and T. R. Tice. 1981. Microencapsulation of water-soluble active polypeptides. U.S. Patent 4,675,189 (1987); U.S. Patent 184,342 (1981).

Nonomura, M., N. Takechi, K. Morikawa, Y. Akagi, and H. Shimizu. 1994. Application of a spray drying technique in the production of TRH-containing injectable sustained release microparticles of biodegradable polymers II. *Proceedings of the PDA Asian Symposium,* 14–16 November Tokyo, p. 255.

Ogawa, Y., M. Yamamoto, H. Okada, T. Yashiki, and T. Shimamoto. 1988a. A new technique to efficiently entrap leuprolide acetate into microcapsules of polylactic acid or copoly(lactic/glycolic) acid. *Chem. Pharm. Bull.* 36:1095–1103.

Ogawa, Y., M. Yamamoto, S. Takada, H. Okada, and T. Shimamoto. 1988b. Controlled-release of leuprolide acetate from polylactic acid or copoly (lactic/glycolic) acid microcapsules: Influence of molecular weight and copolymer ratio of polymer. *Chem. Pharm. Bull.* 36:1502–1507.

Ogawa, Y., H. Okada, M. Yamamoto, and T. Shimamoto. 1988c. In vivo release profiles of leuprolide acetate from microcapsules prepared with polylactic acids or co-poly (lactic/glycolic) acids and in vivo degradation of these polymers. *Chem. Pharm. Bull.* 36:2576–2581.

Okada, H. 1989. One-month release injectable microspheres of leuprolide acetate, a superactive agonist of LHRH. *Proc. Int. Symp. Control. Rel. Bioact. Mater.* 16:12.

Okada, H., Y. Ogawa, and T. Yashiki. 1983. Prolonged release microcapsule and its production, U.S. Patent 4,652,441 (1987); Japanese Patent 207,760 (1983).

Okada, H., T. Heya, Y. Ogawa, and T. Shimamoto. 1988. One-month release injectable microcapsules of a luteinizing hormone-releasing hormone agonist (leuprolide acetate) for treating experimental endometriosis in rats. *J Pharmacol. Exp. Ther.* 244:744.

Okada, H., Y. Inoue, Y. Ogawa, and H. Toguchi. 1992. Three-month release injectable microspheres of leuprorelin acetate. *Proc. Int. Symp. Control. Rel. Bioact. Mater.* 19:52–53.

Okada, H., Y. Doken, Y. Ogawa, and H. Toguchi. 1994a. Preparation of three-month depot injectable microspheres of leuprorelin acetate using biodegradable polymers. *Pharm. Res.* 11,:1143.

Pavanetto, F., I. Genta, P. Giunchedi, and B. Conti. 1993. Evaluation of spray drying as a method for polylactide and polylactide-co-glycolide microsphere preparation. *Microencap. J.* 10:487–497.

Redding, T. W., A. V. Schally, T. R. Tice, and W. E. Meyers. 1984. Long-acting delivery systems for peptides: Inhibition of rat prostate tumors by controlled release of [D-TRP6] luteinizing hormone-releasing hormone from injectable microcapsules. *Proc. Natl. Acad. Sci. USA* 81:5845–5848.

Ruiz, J. M. and J. P. Benoit. 1991. In-vivo peptide release from poly(DL-lactic acid-co-glycolic acid) copolymer 50/50 microspheres. *J. Controll. Rel.* 16:177–186.

Sanders, L. M., J. S. Kent, G. I. McRae, B. H. Vickery, T. R. Tice, and D. H. Lewis. 1984. Controlled release of a luteinizing hormone-releasing hormone analogue from poly(D,L-lactide-co-glycolide) microspheres. *J. Pharm. Sci.* 73:1294–1297.

Schwope, A. D., D. L. Wise, and J. F. Howes. 1975. Lactic/glycolic acid polymers as narcotic antagonist delivery systems. *Life Sci.* 17:1877–1886.

Takada, S., Y. Uda, H. Toguchi, and Y. Ogawa. 1994. Application of a spray drying technique in the production of TRH-containing injectable sustained release microparticles of biodegradable polymers. *Proceedings of the PDA Asian Symposium,* 14–16 November, Tokyo, p. 249.

Toguchi, H. 1996. Practical examples of DDS studies: Dosage-form design for leuprorelin acetate and other drugs. *Bull. Pharm. Res. Technol. Inst.* 5:1–13.

Wise, D. L., H. Rosenkrantz, J. B. Gregory, and H. J. Esber. 1980. Long-term controlled delivery of levonorgestrel in rats by means of small biodegradable cylinders. *J. Pharm. Pharmacol.* 32:399–403.

Yamamoto, M., S. Takada, and Y. Ogawa. 1985. Method for producing microcapsules. Japanese Patent 22,978.

17

Quality Control Methods and Specifications

Johanna K. Lang
Lang Consulting
Fremont, California

The concept of a sustained-release formulation was first developed and applied to oral dosage forms. So called "extended-release" or "delayed-release" dosage forms are now widely available for a variety of oral drugs. The development of these dosage forms has progressed to a point where drug monographs and detailed assay methods have appeared in pharmacopeias, the widely used and accepted reference texts governing pharmaceutical practice in most countries of the world. The U.S. Pharmacopeia (USP) not only describes the properties of a variety of extended-release dosage forms in the form of product monographs (e.g., aspirin, indomethacin, propranolol) but also includes "modified-release" dosage forms in a general chapter on the characterization of dosage forms (USP 23 <1088>). As demonstrated by many chapters in this book, sustained-release injectable formulations are a heavily researched field, from which several commercial products and a host of product candidates have emerged. The technologies, however, are new and diverse, and no official standard practices for the characterization of such products have been defined.

This chapter intends to provide a conceptual and practical guide to the in vitro characterization of sustained-release injectables as it relates to the selection, development, and validation of assays for product release and stability testing (quality control [QC] methods). As a separate but closely related topic, the setting of product specifications, is discussed.

Innovative pharmaceutical products emerge most often from academic or industrial research, which by design or serendipity result in promising technological concepts. While pharmaceutical research is traditionally accompanied by numerous scientific publications, the ensuing product development is not. As a rule, formulas, manufacturing procedures, assay methods, and specifications for new product candidates are proprietary information that often do not become public, even after the product has reached the market. Thus, the literature references in this chapter are general in nature and focus not on specific products or product candidates but on applicable development concepts and regulatory texts.

GENERAL STRATEGIES FOR THE SELECTION OF QUALITY CONTROL METHODS

Product Characteristics of a Sustained-Release Injectable Formulation

Sustained-release injectable products typically consist of the drug (the active ingredient), a carrier, and a suspending medium. Since the formulation is applied via an injection, the product is either a liquid or a gel with sufficient fluidity to be filled and applied from a syringe, i.e., "syringeable." Sustained-release injectable dosage forms are defined by physical, chemical, and biological characteristics, although biological parameters are not always tested, especially in the early stages of product development. Physical parameters are often determined by the carrier or, in the cases of salt forms and complexes, by the physical state of the drug or drug complex. Typical chemical parameters are the concentration of the active ingredient, pH, or the osmolality of the solution. Biological characteristics relate to typical effects of the product and often mirror the therapeutic application, e.g., enzyme induction or cytostasis. Biological parameters are determined in vitro, in cell culture, or in test animals. Table 17.1 lists some examples.

Release Assays

The term *release assay* is destined to lead to confusion in a book devoted to "sustained-release" products, because the term *release* has quite a different meaning in each case. In the context of this chapter, "release" refers to the cGMP laws of drug product manufacturing in compliance with the U.S. Code of Federal Regulations (CFR) Title 21, Part 211.165: Testing and Release for Distribution. "Release assays" are test methods used to decide whether a manufactured pharmaceutical product is of sufficient quality for the intended use, and "product release" is the process by which a manufactured batch of product is made available to the users (clinic, pharmacy suppliers). These assays therefore measure product characteristics that are relevant to the manufacturing process and the use of the product. The USP (USP 23 <1086>), an FDA guideline (FDA 1987), and an ICH

Table 17.1. *Examples of Product Characteristics of Sustained-Release Injectable Products*

Physical Parameters	Chemical Parameters	Biological Parameters
viscosity	total drug concentration	enzyme inhibition
opalescence	carrier concentration	enzyme induction
density	drug-carrier association	cell growth inhibition
mean particle size	pH	cell growth promotion
particle size distribution	osmolality	transfection efficacy
	residual process solvents	vasoconstriction

(International Conference on Harmonisation) draft guideline (FR 1997) describe general expectations for the release testing of drug products. These are geared toward traditional, intravenous injectable and oral dosage forms and emphasize quantitative assays for the content and the purity of the active ingredient. In evaluating new pharmaceutical dosage forms, these traditional standards remain valid, since the amount and the purity of the active ingredient determines (among others) the quality of the product, even if it is applied and released in a new way. In contrast to traditional formulations, however, sustained-release injectables contain drug carriers (e.g., microparticulates) or other drug forms (e.g., complexes) that bring about a sustained release of the active ingredient. Hence, the properties of the drug carrier, the type and extent of drug-carrier association, and drug liberation are also a matter of product quality. Since sustained-release products come in many forms, no set of release assays can fit every product. The appropriate release parameters have to be selected on a case-by-case basis from the characteristic features of the product.

Figure 17.1 illustrates this selection process. The following is a guide to selecting those tests that constitute useful, necessary, or essential release assays for the variety of physical, chemical, and biological characteristics of a given sustained-release dosage form. Table 17.2 provides an overview.

Drug Substance–Related Assays

Drug substance–related assays measure the identity, the concentration, and the impurity profile of the active ingredient. Often, the drug substance is a well-known, commercially available pharmaceutical. Occasionally, a drug substance is not readily available, mostly because it is a proprietary drug, manufactured by a single company for its own products. Assays that are used (by the manufacturer) to release the bulk drug are generally a good starting point for developing QC assays for the active ingredient of a new dosage form. The Certificate of Analysis from the supplier will show the identity, content, and impurity profile of the bulk drug, and most suppliers are willing to share their assay methods with a customer. If a compendial monograph (USP, British or European Pharmacopoeia) exists,

Figure 17.1. *Selecting quality control and stability parameters from product characteristics.*

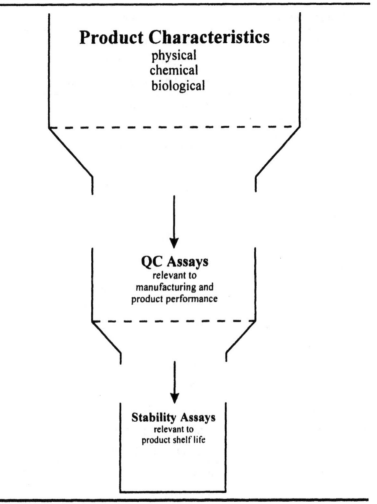

Table 17.2. *Selection Scheme of Product Release Assays for Sustained-Release Injectable Dosage Forms*

Release assays	Test
Drug-related assays active ingredient	identity test for drug
	quantitative assay for drug
	drug-related impurities (related substances)
Carrier-related assays (carrier components or complexing agents)	identity of carrier
	quantitative assay for carrier
	known carrier-related impurities
	mean particle size
	particle size distribution
Manufacturing-related assays (process contaminants)	residual solvents
	heavy metals
Drug-carrier association	drug binding to carrier
Drug liberation	time dependent release of bound drug from the carrier into a relevant medium
Other common attributes	description of visual appearance
	pH
	osmolarity
	viscosity
Uniformity of dosage units	weight variation or content uniformity
Preservatives	antimicrobial preservative
	antioxidant preservative
Microbiology	sterility
	apyrogenicity

compendial assays should be considered first, although there may be good reasons why they will not work with the new dosage form. A major reason for deviating from the monograph or established drug assays is the interference of the carrier. Another one is the obsolescence of the pharmacopeial method.

Carrier-Related Assays

The carrier(s) or complexing agents are usually new and patented technologies, for which little, if any, literature exists. Sometimes rudimentary assays methods have emerged from product-oriented research, but mostly this is a case for developing of a new assay from scratch. A typical release profile will contain a qualitative assay for the presence of the carrier(s) and an assay determining the amount of carrier and any impurities that are known to be present in significant quantities. For suspensions of particulates or vesicles, the particle size is often a critical parameter and therefore included in release testing. Usually, a mean particle size is determined. In rare cases, the mean particle size is not sufficient to control manufacturing or efficacy, thus necessitating a complete particle size distribution profile.

Manufacturing-Related Assays

The drug product manufacturing process may use substances such as buffer salts, metal catalysts, or organic solvents that are removed in subsequent processing steps but may be carried into the final product in small amounts. Special attention is paid to substances such as heavy metals and residual organic solvents, where even trace amounts may be toxic. For these substances, the release assay will either be a quantitative assay or a limit test. Contrary to quantitative assays, limit tests will only determine if a substance is present above or below a certain limit level. Organic solvent residues are best determined by direct injection or headspace gas chromatography. The USP offers several gas chromatographic procedures for "organic volatile impurities" (USP 23<467>). Heavy metals can be determined individually or as a group. Atomic spectroscopy (atomic absorption, flame photometry, ICP [inductively coupled plasma spectrophotometry] or

ICP-MS [inductively coupled plasma mass spectroscopy]) are best suited for quantitative or limit tests of individual heavy metals. Limit tests for lead, iron, mercury, or selenium as well as general heavy metal assays can be found in USP (USP 23 <231>).

Drug-Carrier Association

An essential parameter for every sustained-release product, since it determines the amount of unbound (free) and therefore immediately active drug, is drug-carrier association. If the product is a prodrug or a stoichiometric complex between the drug and the carrier, little or no free drug may exist. If a drug-carrier association assay is not feasible, it is worth pursuing alternative means of demonstrating that only the drug-carrier complex is present.

Drug Liberation

The drug liberation assay shows the ability of the sustained-release formulation to liberate the active ingredient in a controlled, time-dependent way. For sustained-release oral dosage forms, the assays are prescribed in great detail in the USP and involve dissolution of the coated tablet or capsule in a buffered medium at controlled temperature and agitation. For sustained-release injectable dosage forms, well-defined and accepted procedures do not exist, and the appropriate "liberation medium" is not obvious. Designing a relevant and technically feasible liberation assay requires a thorough understanding of the product technology and the injection site.

Other Common Product Attributes

There are a variety of straightforward tests. Visual appearance, pH, osmolality, and viscosity describe useful parameters that can be measured with good reproducibility and relatively little effort. Nevertheless, it is worth carefully considering whether they are relevant to the product. For waterlike, highly fluid suspensions, viscosity is not a relevant parameter, while it is essential for a viscous gel that must be manufactured not to exceed a limit viscosity, lest it cannot be drawn and injected from a syringe. Including irrelevant parameters in the product release should always be avoided, since the extra effort does not improve the assessment and control of product quality.

Uniformity of Dosage Units

The uniformity of dosage units (USP 23 <905>) is a straightforward compendial assay that is required for all products, where the individual container provides a defined quantity (unit dose) of the drug. A set number of containers is randomly pulled from the filling line. In the assay, the fill quantity, as determined either from the fill weight or the quantitative drug assay, must not exceed predetermined compendial limits (see below). This release assay applies to ready-made liquid formulations as well as concentrates and dry formulations (e.g., lyophilized products) that are reconstituted to a defined drug concentration and dose.

Preservatives—Antimicrobial

Under certain circumstances, antimicrobial preservatives are included in injectable dosage forms. Since injectables are manufactured to be sterile, antimicrobial preservatives are not required if the product is withdrawn directly from the product container into a sterile syringe and the remaining product is destroyed. If this is not the case, preservatives are part of the product formula, and their presence at minimum effective levels must be confirmed by a release assay.

Preservatives—Antioxidants

Antioxidants are added to extend the shelf life of products that are prone to air oxidation. It is expected that products be assayed to confirm antioxidant presence at the intended level.

Microbiology—Sterility

By definition, all injected products have to be sterile. Manufacturing methods are therefore designed to exclude microbes at all stages of production. Nevertheless, the absence of viable microbes (bacteria and fungi) from the finished product must be confirmed by a sterility test. The USP describes a standardized sterility testing procedure. It involves applying the product or a filtration residue (an appropriate membrane will filter bacteria and fungi from the product and concentrate them on the membrane surface) on a growth medium in which viable organisms can grow to form colonies, which are then counted. While the assay is well standardized, new dosage forms can cause problems. A well-recognized source of error is the possibility that the carrier or its suspending medium inhibits the growth of the test organisms and creates false negative results. This can be eliminated by the membrane filtration method, which works if the formulation is not viscous, and the carrier will pass through the filter. However, the presence of bacterial/fungal growth inhibitors in the product can be tested in a bacteriostatis/fungistasis (B/F validation) test. Sterility and the B/F validation test are best performed in a specialized testing facility. There are a number of qualified contract laboratories from which to choose.

Microbiology—Apyrogenicity

Common pyrogens are by-products of bacterial growth and cause fever. Their absence from injectable products must be shown by a validated assay. The USP offers two tests, a qualitative pyrogen test and a quantitative bacterial endotoxins assay. The pyrogen test (USP 23 <151>) uses rabbits as test animals and measures the increase in body temperature (fever) after administration of an excess of the typical drug dose. The bacterial endotoxins test (USP 23 <85>) is an in vitro assay for bacterial endotoxins, a certain class of lipopolysaccharides and a pyrogenic by-product of bacterial growth. Both assays are equally acceptable, but the choice should be considered carefully. In general, the bacterial endotoxins assay is preferable, since it is a standardized in vitro test that is quantitative or a least semiquantitative. However, it is prone to interference from drug product components and has to be validated for every new dosage form. Often, the interference can be overcome by dilution. The validation procedures for bacterial endotoxin assays are well described in the USP (USP 23 <85>), and contract labs will offer those services along with routine testing. If the interferences are not manageable, the rabbit pyrogen test is an alternative.

Stability Assays

Stability assays are those test procedures used to determine product shelf life. They are typically a subset of release assays and encompass those assays that are relevant to the stability of the product (see Figure 17.1). Thus the results of the release assays may also serve as the starting point (time 0) for stability testing. Stability assays will detect changes in the product that can occur over time. A good understanding of the science behind the sustained-release technology employed in the product will help to make educated guesses about relevant stability parameters. Yet new technologies usually have surprises in store. Nevertheless, a general, rational approach to stability testing of new formulations exists. Table 17.3 provides an overview.

Quantitative Assays for Total Drug Content

Total drug content is a key stability parameter. The "stability-indicating" assay must be capable of determining the amount of active drug in fresh and partially degraded samples. Typically, a drug content of less than 90 percent of the labeled amount is not acceptable in a marketed product (see "Regulatory Expectations and Changing Industry Standards"). For degradation kinetics, drug degradation to as low as 50 percent of the initial amount may be followed. Often, this is a challenging assay to develop, since it must be specific for the intact drug, not only in the presence of the formulation excipients but also when the product becomes chemically more complex through degradation of the active ingredient and possibly even the carrier.

Quantitative Assays of Drug-Related Impurities and Degradation Products

Total drug and related impurities assays are often combined in a single HPLC (high performance liquid chromatography)

Table 17.3. *Scheme for the Selection of Stability Assays*

	Release Assays	Stability Assays
Drug-related assays (active ingredient)	identity test for drug	no
	quantitative assay for drug	yes
	drug-related impurities	yes
Carrier-related assays (carrier components or complexing agents)	identity of carrier	no
	quantitative assay for carrier	PS[1]
	carrier-related impurities	PS
	mean particle size	PS
	particle size distribution	PS
Manufacturing-related assays (process contaminants)	residual solvents	no
	heavy metals	no
Drug-carrier association	drug binding to carrier	yes
Drug liberation (release)	time-dependent liberation of bound drug from carrier into relevant medium	yes
Other common attributes	description of visual appearance	yes
	pH	yes
	osmolality	yes
	viscosity	yes
Uniformity of dosage units	weight variation or content uniformity	no
Preservatives	antimicrobial preservative	yes
	antioxidant preservative	yes
Microbiology	sterility	no/yes[2]
	apyrogenicity	no

[1]PS = product specific, determined by specific product technology
[2]see "Microbiology"

assay. This is not essential and sometimes not feasible, especially with more complex degradation processes that generate a variety of degradation products with different chemical properties. Assays for impurities/degradation products are usually quantitative HPLC assays, but limit tests are equally acceptable.

Carrier-Related Assays

The need to include carrier-related assays in stability testing depends entirely on the nature of the carrier and its propensity to degrade during storage. For instance, polymer microspheres can be manufactured with excellent physical and chemical stability, while liposome formulations are sensitive to both chemical degradation of the carrier components and changes in vesicle size. Lyophilization sometimes offers improved stability. More often than not, it is necessary to determine the extent of chemical degradation of carrier components or changes in critical physical parameters such as particle size or viscosity. Impurities that arise from the chemical degradation of carrier components are often assayed to gain a better understanding of the Achilles' heel of the carrier technology. It is usual to include them in registration-oriented stability testing, if toxic effects are a concern. Identity tests are not stability indicating and, therefore, never included in stability testing.

Manufacturing-Related Assays

Trace residues of heavy metals or solvents do not increase or disappear during storage and are, therefore, not relevant to product stability. They are not tested outside initial (release) testing.

Drug-Carrier Association and Drug Liberation

Since both drug-carrier association and drug liberation are key parameters of sustained-release formulations that are sensitive to either drug and/or carrier degradation, they are usually also key stability parameters. Even in cases where no drug or carrier degradation is observed, demonstration of unchanged drug-carrier association and drug liberation may be necessary.

Other Common Attributes

Most common attributes have some relevance to stability. If their relevance is not certain, they are best checked out in exploratory stability studies. In the absence of historical data, it makes sense to include these attributes in stability studies.

Uniformity of Dosage Units

The uniformity of dosage units is not relevant to stability. Designed to test the uniformity of the filling of a bulk product into individual containers, it is a typical release assay. Once the uniformity of the dosage units is confirmed, no further uniformity testing is required.

Preservatives

Federal regulatory guidelines for submitting documentation for the manufacture of and controls for drug products clearly state the expectation of regulatory agencies that the content of preservatives be monitored over the entire shelf life of the product (FR 1997).

Microbiology

In the author's experience, the inclusion of microbiological tests in stability testing has been a matter of debate. Usually, sterile products are packaged in airtight container/closure systems, in which bacterial contamination during storage cannot occur. Sterility and apyrogenicity could be compromised during storage if container/closure systems are defective or sealed improperly. It is preferable to test the intactness of the container/closure system immediately after manufacturing by more convenient means.

METHOD DEVELOPMENT AND VALIDATION

Method development is usually a resource-intensive, time-consuming endeavor. It therefore pays to comb pharmacopeias, company product bulletins, and the scientific literature for methods that could be applied to the new dosage form with little or no modification. Even if no published assays appear directly applicable to the new dosage form, a thorough literature search often provides good background information in deciding how to develop a new assay. Whether adopted from existing methods, modified, or developed from scratch, every assay must be confirmed to perform properly. Even the very general cGMP regulations state explicitly that pharmaceutical manufacturers must have "data that establish that the methods used in the testing of the (product) sample meet proper standards of accuracy and reliability as applied to the product tested" (CFR 1993). The procedure by which these accuracy and reliability data are determined is called assay validation. The term *validation* has led to much misunderstanding, since it is used freely within many disciplines of pharmaceutical product development. The validation concept itself is generally applicable to any form of testing. In the context of this chapter, the term *validation* strictly applies to a set of procedures that demonstrate that a QC assay is "suitable for its intended purpose" as described in the ICH Harmonized Tripartite Guideline (ICH 1995). The most extensive body of validation literature exists for quantitative assays, especially HPLC assays. Individual authors, pharmacopeias, and regulatory documents have described in varying detail similar approaches but without consistent terminology. Most recently, ICH has issued two validation guidelines that define uniform terminology and standards of European, U.S., and Japanese regulatory agencies: ICH Topic Q2A (1995) and ICH Topic Q2B (1996). The following discussion on the validation of QC assays is based on these guidelines.

Assays Covered Under the ICH Guidelines

ICH has provided a uniform set of terminology and well-defined expectations for assay validation. However, only the more common types of assays are covered: quantitative assays for the active ingredient in a drug product, quantitative assays for impurities, limit tests for impurities, and identification tests. For these assays, the validation requirements as stated in the ICH guideline are summarized in Table 17.4 in ICH terminology, with other commonly used terms in parentheses. Table 17.4 may be used as a checklist to ensure suitable validation assays are completed, since it contains the individual validation parameters as clearly and concisely defined in ICH Guideline Q2A. For convenience, these definitions are provided in an appendix to this chapter.

Guideline Q2B describes in detail how the validation parameters shown in Table 17.4 are determined. Although the guidelines were developed with traditional pharmaceutical preparations in mind, most of the parameters can be readily applied to sustained-release formulations. Some notable exceptions and frequently encountered method development issues with sustained-release formulations are discussed below.

Sample Homogeneity

The vast majority of quantitative assays and limit tests require a homogeneous liquid sample, from which a well-defined portion is analyzed for the concentration of drug and/or impurity. If the sample is not macroscopically homogeneous (e.g., precipitated or phase separated), a homogenization step (e.g., sonication) must precede the sampling.

Sample Preparation

In typical sustained-release formulations, only a small portion of the drug is present in solution, while the majority is sequestered by the carrier. In total drug assays, the binding of drug to the carrier will result in the presence of two or more forms of the drug. This may cause assay errors if the drug forms are separated by the sample preparation or the chromatographic assay or if the two drug forms show different assay responses. This is common for ultraviolet (UV) and HPLC-UV assays, since the binding of drug to a particle surface or vesicle membrane often leads to changes in the spectrum of the chromophore and the molar absorptivity. This problem is best dealt with in a sample

Table 17.4. *Validation for Common Types of Assays as Defined in ICH Guideline Q2A*

Validation Parameters	Identification Test	Testing for Impurities		Quantitative Assays	
		Quantitative Assay	Limit Test	Content (potency)	Content assay for dissolution tests
Accuracy	–	+	–	+	+
Precision					
Repeatability (intra-assay)	–	+	–	+	+
Intermediate precision	–	+	–	+	+
Specificity	+	+	+	+	+
Detection limit	–	–	+	–	–
Quantitation limit	–	+	–	–	–
Linearity	–	+	–	+	+
Range	–	+	–	+	+

(–) = the validation parameter is not determined

(+) = the validation parameter is determined

preparation step that removes the bound drug from the carrier (i.e., extraction) or destroys the carrier. Polymeric or lipidic carriers can often be conveniently destroyed by dilution with organic solvents or detergents. This approach is fast and convenient but requires that the drug remains soluble in the resulting matrix and that no matrix components interfere with the drug assay.

Assays Not Covered Under the ICH Guidelines

Among the typical QC assays for sustained-release injectables (see Table 17.2), some assays that require validation are not covered by the ICH guidelines. For these formulation-specific assays, the author suggests the validation parameters listed in Table 17.5.

Particle Size

Apart from microscopic assays, typical particle size assays (e.g., static or dynamic light scattering) do not measure absolute particle size; they determine particle size through mathematical algorithms that estimate the mean particle size. The accuracy of such assays cannot be verified directly, since well-defined standard preparations of the individual carrier formulations usually

Table 17.5. *Recommended Validation for Assays Not Covered by the ICH Guidelines*

Validation Parameters	Particle Size	Drug-Carrier Association	Drug Liberation (Release)
Accuracy	–	+	–
Precision			
Repeatability (intra-assay)	+	+	+
Intermediate precision	+	+	+
Range	+	+	+

(–) = the validation parameter is not determined

(+) = the validation parameter is determined

do not exist. Commercially available particle size standards (latex particles of defined size), however, are useful to ensure proper calibration and performance (system suitability) of the particle sizer. The precision of the particle size measurement is readily determined by repeat analyses of a typical sample. The reliable measuring range of the instrument should be defined. For very large, charged particles (> 3 μm), hemocytometry can be used as an independent reference method.

Drug-Carrier Association

Drug-carrier association assays typically consist of a separation step, followed by a quantitative assay. In the separation step, the free drug is separated from the carrier-associated drug. Subsequently, either free drug, carrier-associated drug, or both are measured (e.g., by HPLC). For the HPLC portion of the assay, all of the validation parameters for quantitative drug assays apply. The accuracy of the association assay can be determined by preparing samples of a known degree of drug association (e.g., by spiking free drug into the sustained-release formulation). In cases where the free drug is in rapid equilibrium with the bound drug, the accuracy of the association assay should be shown by alternative means or theoretical considerations. The evaluation of precision and range is accomplished according to Guideline Q2B.

Drug Liberation

In drug liberation, the parameter of interest is the increase of drug concentration in the release medium over time. For the drug concentration assay, the validation parameters for quantitative assays from Guideline Q2A apply. For the time/concentration curve, range and precision can be readily determined from repeat assays of a sample under identical conditions.

APPROACHES TO SETTING PRODUCT SPECIFICATIONS

Terminology and the concept of product specifications originate from the cGMP regulations (21 CFR 211.160) "Laboratory controls shall include the establishment of scientifically sound and appropriate specifications . . . ." Specifications are acceptance limits that are intimately tied to QC (release) assays. The specifications for a drug product therefore state the range of QC assay results that are acceptable for any batch of manufactured drug product. Drug product batches that assay within the limits are released for distribution; those that do not are rejected and cannot be distributed. Some acceptable approaches to the setting of specifications are discussed below.

Manufacturing History

Specification settings can be based on the QC assay results of several manufacturing batches. The range of actual results plus some safety margin define the limits of what is acceptable. This is a good approach when several representative batches have been prepared with a well-defined manufacturing procedure.

Fitness for Use

Early pilot batches of a drug product that passed safety and maybe efficacy tests in animals or humans can be used to define acceptable limits. This is particularly useful in the setting of safety-related specifications, such as limits for impurities. In this context, it is not advisable to conduct animal safety studies with unusually pure or a specially purified drug and/or carrier.

Analytical Method Limits

Every specification must accommodate normal assay variability, i.e., the specified range should be much wider than the normal variability of the assay. For this purpose, the "worst" of the precision parameters from the assay validation is the guide. If for other reasons specification ranges need to be tighter, a more precise assay must be used. A statistical model for the relationship of method precision, system precision, and specifications has been described for quantitative HPLC assays of the active ingredient (Debesis et al. 1982).

Compendial Limits

Some specification limits are stated in the pharmacopeias. The USP states acceptable limits for sterility, apyrogenicity, the uniformity of dosage units, and some residual solvents. These limits are also applied to noncompendial drugs and dosage forms. Any specification limit that is wider than the compendial limits should be supported by convincing evidence that the compendial limit cannot be attained with reasonable effort and that a wider limit does not pose a safety risk.

Regulatory Expectations and Changing Industry Standards

A glance through compendial monographs shows a wide range of acceptable specifications for drug content (potency). Typically expressed as percent of label claim (the drug content stated on the label), they range from values as low as 75 percent up to 125 percent. The vast majority of drug products, however, and specifically recently included or updated monographs, list specifications of 90–110 percent or tighter. This can be explained by progress in analytical technology as well as the increased focus of regulatory agencies on product characterization. The general trend is toward higher drug product quality and tighter limits on drug content.

QC ASSAYS, VALIDATION, AND
SPECIFICATIONS IN REAL–LIFE
PRODUCT DEVELOPMENT

In conclusion, it should be made clear that the discussion about QC assays, validation, and specifications applies to marketed product. Yet the cGMP is also applied to all product manufactured for clinical use, which includes product manufactured early in the product development process. It is understood, however, and various regulatory guidelines contain wording to this effect, that final assay methods and specifications will not be in place at the time early clinical batches are manufactured. Early in the development process, the choice of assay methods is determined by the need to gauge critical formulation parameters. Typically, these include purity and stability-indicating methods for the active ingredient and one or more key parameters of the carrier technology. In the same vein, assay validation is performed only to the level that is adequate for this stage of development. As a rule, it is not necessary to worry about the robustness of a method until a transfer of the analytical method to an off-site QC laboratory is contemplated. Also, investigations of intermediate precision should be limited to those parameters that are relevant to laboratory practice at this time. Rules of thumb suggest that by the time Phase III clinical material is manufactured, release and stability testing should use a set of assays that is deemed adequate for the release of commercial product. Similar considerations apply to specifications. While specifications must be set and met or all batches intended for clinical trials, it is, however, understood that the specification ranges may be wider in the beginning and tightened as more manufacturing experience is gained.

REFERENCES

CFR. 1993. *Code of Federal Regulations,* Title 21, Part 211: Current Good Manufacturing Practice for Finished Pharmaceuticals. Washington, D.C.: U.S. Government Printing Office.

Debesis, E., J. P. Boehlert, T. E. Givand, and J. C. Sheridan. 1982. Submitting HPLC methods to the compendia and regulatory agencies. *Pharm. Tech.* (September): 120–138.

FDA. 1987. *Guideline for submitting documentation for the manufacture of and controls for drug products.* Rockville, Md., USA: Food and Drug Administration, Center for Drugs and Biologics.

FR. 1997. International Conference on Harmonisation; Draft guidance on specifications: Test procedures and acceptance criteria for new drug substances and new drug products: Chemical substances; Notice. *Federal Register* 62 (227):62894–62896 (November 25).

ICH. 1995. *ICH Harmonized Tripartite Guideline: Text on Validation of Analytical Procedures:* ICH Topic Q2A. Geneva, Switzerland: International Conference on Harmonisation.

ICH. 1996. *ICH Harmonized Tripartite Guideline: Validation of Analytical Procedures: Methodology,* ICH Topic Q2B. Geneva, Switzerland: International Conference on Harmonisation.

USP 23. 1995. *The United States Pharmacopeia,* <85> Bacterial endotoxins test. Taunton, Mass., USA: Rand McNally, pp. 1696–1697.

USP 23. 1995. *The United States Pharmacopeia,* <151> Pyrogen test. Taunton, Mass., USA:, Rand McNally, p. 1718.

USP 23. 1995. *The United States Pharmacopeia,* <231> Heavy metals. Taunton, Mass., USA: Rand McNally, pp. 1727–1732.

USP 23. 1995. *The United States Pharmacopeia,* <467> Organic volatile impurities. Taunton, Mass., USA: Rand McNally, pp. 1746–1748.

USP 23. 1995. *The United States Pharmacopeia,* <905> Uniformity of dosage units. Taunton, Mass., USA: Rand McNally, pp. 1838–1839.

USP 23. 1995. *The United States Pharmacopeia,* <1086> Impurities in official articles. Taunton, Mass., USA: Rand McNally, pp. 1922–1924.

USP 23. 1995. *The United States Pharmacopeia,* <1088> In vitro and in vivo evaluation of dosage forms. Taunton, Mass., USA: Rand McNally, pp. 1924–1929.

APPENDIX: DEFINITIONS OF VALIDATION PARAMETERS FROM ICH HARMONIZED TRIPARTITE GUIDELINE: TEXT ON VALIDATION OF ANALYTICAL PROCEDURES, ICH TOPIC Q2A

1. Analytical Procedure

The analytical procedure refers to the way of performing the analysis. It should describe in detail the steps necessary to perform each analytical test. This may include but is not limited to: the sample, the reference standard and the reagents preparations, use of the apparatus, generation of the calibration curve, use of the formulae for the calculation, etc.

2. Specificity

Specificity is the ability to assess unequivocally the analyte in the presence of components which may be expected to be present. Typically these might include impurities, degradants, matrix, etc. Lack of specificity of an individual analytical procedure may be compensated by other supporting analytical procedure(s).

This definition has the following implications:

Identification:	to ensure the identity of an analyte
Purity Tests:	to ensure that all the analytical procedures performed allow an accurate statement of the content of impurities of an analyte, i.e. related substances test, heavy metals, residual solvents content, etc.
Assay (content or potency):	to provide an exact result which allows an accurate statement on the content or potency of the analyte in a sample.

3. Accuracy

The accuracy of an analytical procedure expresses the closeness of agreement between the value which is accepted either as a conventional true value or an accepted reference value and the value found. This is sometimes termed trueness.

4. Precision

The precision of an analytical procedure expresses the closeness of agreement (degree of scatter) between a series of measurements obtained from multiple sampling of the same homogeneous sample under the prescribed conditions. Precision may be considered at three levels: repeatability, intermediate precision and reproducibility.

Precision should be investigated using homogeneous, authentic samples. However, if it is not possible to obtain a homogeneous sample it may be investigated using artificially prepared samples or a sample solution.

The precision of an analytical procedure is usually expressed as the variance, standard deviation or coefficient of variation of a series of measurements.

4.1   Repeatability

Repeatability expresses the precision under the same operating conditions over a short interval of time. Repeatability is also termed intra-assay precision.

4.2.   Intermediate precision

Intermediate precision expresses within laboratories variations: different days, different analysts, different equipment, etc.

4.3.   Reproducibility

Reproducibility expresses the precision between laboratories (collaborative studies, usually applied to standardization of methodology).

5. Detection Limit

The detection limit of an individual analytical procedure is the lowest amount of analyte in a sample which can be detected but not necessarily quantitated as an exact value.

6. Quantitation Limit

The quantitation limit of an individual analytical procedure is the lowest amount of analyte in a sample which can be quantitatively determined with suitable precision and accuracy. The quantitation limit is a parameter of quantitative assays for low levels of compounds in sample matrices, and is used particularly for the determination of impurities and/or degradation products.

7. Linearity

The linearity of an analytical procedure is its ability (within a given range) to obtain test results which are directly proportional to the concentration (amount) of analyte in the sample.

8. Range

The range of an analytical procedure is the interval between the upper and lower concentration (amounts) of analyte in the sample (including these concentrations) for which it has been demonstrated that the analytical procedure has a suitable level of precision, accuracy and linearity.

9. Robustness

The robustness of an analytical procedure is a measure of its capacity to remain unaffected by small, but deliberate variations in method parameters and provides an indication of its reliability during normal usage.

18

Regulatory Perspectives and Product Approval Processes

Natalie L. McClure
IntraBiotics Pharmaceuticals, Inc.
Mountain View, California

In the preceding chapters, the different types of sustained-release injectable products and approaches for their characterization and control have been discussed. In this chapter, the regulatory aspects of dosage form development and strategies for market registration, the ultimate goal of every drug development program, are addressed.

GENERAL PRINCIPLES OF DRUG DEVELOPMENT

Drug development is divided into six phases: preclinical testing; Phase I, II, and III clinical testing; approval process; and postmarketing activities. In the preclinical testing phase, the emphasis is on animal and in vitro studies designed to characterize the drug product and assess its potential for use in human clinical studies. The studies include development of a robust dosage form with sufficient characterization to allow the consistent manufacture of multiple batches. Pharmacology studies are generally conducted to demonstrate the mechanism of action as well as to determine the absorption, metabolism, and distribution of the drug. Toxicology studies are also required prior to the initiation of human clinical studies. As a general rule of thumb, the duration of dosing used in the toxicology studies needs to exceed twice the anticipated human clinical study treatment duration.

Phase I clinical studies are typically conducted in healthy, normal human volunteers and are designed to assess the safety of the proposed dosage form. These studies are generally single-dose studies and are often designed as dose escalation studies. In addition to monitoring side effect profiles, Phase I clinical tests can provide data on the pharmacokinetics of the dosage form.

Phase II clinical testing occurs in patients who suffer from the condition that the drug is intended to treat. The emphasis of Phase II clinical studies is on the safety of the dosage form, although preliminary effectiveness data can also be obtained.

From the Phase II clinical data, Phase III clinical studies can be developed and designed. Phase III studies are generally large clinical studies intended to demonstrate the safety and efficacy of the drug in a more varied patient population and at many clinical sites.

Since most, if not all, sustained-release dosage forms are line extensions for drugs previously approved in simpler dosage forms, several of the drug development phases can be shortened or eliminated. Figure 18.1 provides a schematic of a typical drug development timeline for a new chemical entity and for a sustained-release dosage form. The preclinical testing phase is often compressed because the primary requirement is to characterize the release profile and compare and contrast it to the known properties of the drug. Similarly, early clinical testing can be shortened. Again, the side effect profile of the drug will be well known from earlier, nonsustained-release dosage form studies. The early clinical testing of a sustained-release product will often focus on pharmacokinetics and assess the presumably improved safety profile. This compression of the drug development

Figure 18.1. *Comparison of development timelines.*

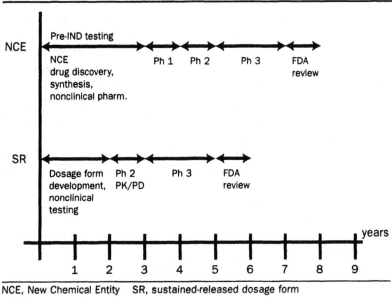

NCE, New Chemical Entity   SR, sustained-released dosage form
PK/PD, phamacokinetics/pharmacodynamics

timeline can save 2–4 years in overall development, which should translate into significant savings for the sponsoring corporation.

Case Study: Development of the Various Formulations of Leuprolide Acetate

The development of gonadotropin releasing hormone (GnRH) agonist injectable formulations, such as Lupron Depot® (leuprolide acetate), illustrates a regulatory approach to the development of a sustained-release injectable drug product (FDA 1997c). GnRH is a naturally occurring decapeptide hormone. GnRH derivatives have been developed for a wide variety of indications, including palliation of advanced prostate cancer, treatment of central precocious puberty, palliative treatment of advanced breast cancer, and management of endometriosis. As synthetic analogs were developed, the optimal mechanism of drug delivery was one of the major hurdles in the development programs.

The peptide is readily delivered by subcutaneous (SC) injections, but this route of administration leaves much to be desired for products that are used on a chronic basis. Since the peptides do not survive the acidic conditions of conventional oral administration, a variety of alternative routes of administration were examined for this class of compounds, including nasal sprays and poly(lactic acid-co-glycolic acid) (PLGA) microspheres.

The development history of Lupron Depot® is summarized in Table 18.1. As can be seen from this table, the first approval was for an SC injectable formulation.

The first New Drug Application (NDA) for leuprolide was submitted to the U.S. Food and Drug Administration (FDA) in December 1983. This NDA included data from 2 pivotal clinical studies, which are summarized in Table 18.2, and 9 additional studies that provided additional safety information. Six Phase I studies in a total of 48 patients were conducted to characterize the tolerance and safety profile of leuprolide. Phase II studies in 100 pre- and postmenopausal breast cancer patients were conducted. These studies showed a possible significant improvement but were not pursued for registration. The initial NDA also included multiple pharmacology and toxicology studies. For the first indication of a New Chemical Entity (NCE), such as leuprolide acetate, both acute and long-term (2-year) toxicology studies were required.

Table 18.1. *Approval History for Prostate Indications of Leuprolide Acetate[1]*

Product	FDA Approval Date	Indication
Lupron Injection, 1 mg/0.2 mL	4/85	Advanced prostate cancer
Lupron Depot®, 7.5 mg/vial	1/89	Palliative treatment of advanced prostate cancer
Lupron Depot®, 3-Month, 22.5 mg/vial	12/95	Palliative treatment of advanced prostate cancer
Lupron Depot®, 4-Month, 30 mg/vial	5/97	Palliative treatment of advanced prostate cancer

[1]Abstracted from the Approval Package for Lupron Depot®, application 020517/S002, dated 5/30/97.

Table 18.2. *Clinical Studies Required for the Registration of Lupron Injection and Lupron Depot® for the Treatment of Prostate Cancer*

Formulation	Approval Date	Study Number	Sample Size	Study Design
Daily Injection	4/85	M80-036	118	Open label, dose titration study, retrospective control
Daily Injection		M81-017	199	Comparative trial, endpoints: hormonal control, objective tumor response
1-Month Depot	1/89	M85-097	56	Open label, historical control
1-Month Depot		M88-124	14	Pharmacodynamic
3-Month Depot	12/95	M91-583	60	Open label, historical control
3-Month Depot		M91-653	33	Therapeutic equivalence of clinical formulation to marketed formulation
4-Month Depot	5/97	M93-013	49	Open label, historical control
4-Month Depot		M93-012	24	Pharmacokinetic analysis of plasma levels of leuprolide following a single intramuscular (IM) injection of the depot

In contrast to the development program for the first leuprolide indication, the development product for the 1-, 3-, and 4-month depot products emphasized the pharmacokinetics and pharmacodynamics of the depot formulations rather than the overall safety and efficacy of leuprolide. The development program for the 4-month depot formulation consisted of only 2 clinical studies: one open label, multicenter study in 49 patients and a pharmacokinetic study of 24 patients. The approval was based on the results of these two studies and on a determination that the safety and efficacy of this formulation is similar to that of other leuprolide depot formulations.

The development of sustained-release formulations of GnRH agonist is simplified by the pharmacology of this class of compound. There is a clearly accepted surrogate marker for product performance—the suppression of serum testosterone. Exploitation of this surrogate marker simplifies the clinical and preclinical program.

AREAS OF EMPHASIS FOR SUSTAINED–RELEASE DRUG PRODUCTS

Drug Release Profile Studies

Any regulatory application must include a characterization of the drug release profile for the drug product. For immediate-release oral systems, this characterization is a relatively simple dissolution or disintegration profile designed to demonstrate comparability of a variety of clinical formulations and different manufacturing lots. For sustained-release products, drug release as a function of time is a critical parameter of the drug product's performance. An analytical method to assess drug release is an important development tool and can be used to screen formulations for release characteristics and demonstrate comparability of the product following minor changes to the quantitative composition or manufacturing process during product development. The U.S. Pharmacopeia (USP) suggests that the drug release profile be controlled at several time points: an early time point to demonstrate the absence of "dose dumping" and at two time points during the anticipated release time (USP 1995). The drug release profile usually becomes a specification parameter and ideally can be correlated with in vivo performance.

In Vitro/In Vivo Correlation Studies

The ability to predict anticipated bioavailability characteristics for a sustained-release product from release profile characteristics is a long-sought goal that, if achieved, leads to an in vitro/in vivo correlation (FDA 1997b). If the in vitro/in vivo correlation can be established, regulatory agencies and the sponsor have more flexibility in approving future manufacturing changes in both process and scale. Three levels of correlation have been established by the FDA:

1. *Level A:* Comparison of the fraction of drug absorbed to the fraction of drug dissolved. A correlation of this type is generally linear and represents a point-to-point relationship between in vitro dissolution and the in vivo input rate (e.g., the in vivo dissolution of the drug from the dosage form).

2. *Level B:* The mean in vitro dissolution time is compared either to the mean residence time or to the mean in vivo dissolution time. A Level B correlation, like a Level A correlation, uses all of the in vitro and in vivo data but is not considered to be a point-to-point correlation. A Level B correlation does not uniquely reflect the actual in vivo plasma level curve because a number of different in vivo curves will produce similar mean residence time values.

3. *Level C:* A Level C correlation establishes a single point relationship between a dissolution parameter, for example, the percent dissolved in 4 hours and a pharmacokinetic parameter (e.g., AUC, C_{max}, T_{max}). A Level C correlation does not reflect the complete shape of the plasma concentration time curve, which is the critical factor that defines the performance of extended-release products.

To date, very few, if any, extended-release drug products have established an in vitro/in vivo correlation to the satisfaction of the FDA. However, once established, the existence of the correlation would allow manufacturing and formulation changes without the need for additional bioequivalence studies in humans. The FDA has stated that a Level A correlation is the most informative and is recommended, if possible. Level C correlations can be useful in the early stages of formulation development when pilot formulations are being selected. Level B correlations are the least useful for regulatory purposes.

The most commonly used approach to developing a Level A correlation is to develop formulations with different release rates and obtain in vitro dissolution profiles and in vivo plasma concentration profiles for these formulations. Mathematical deconvolution of the data should establish the correlation, if it exists.

Excipient Characterizations

Many sustained-release formulations include novel excipients that are critical to the performance of the formulation. These excipients may not be compendial items and may require custom manufacture or purification to be suitable for use in human drug products. It will often be necessary to provide toxicological assessments of the excipient alone as well as the formulated drug product. In some cases, it may be necessary to provide evidence, either in clinical or preclinical studies, that the excipient is truly necessary to the formulation. This is generally demonstrated in preclinical studies by comparing the pharmacodynamics and pharmacokinetics of the sustained-release drug product to an immediate-release formulation.

In addition, the regulatory application should contain full characterization (specifications, source or manufacturing process, and stability considerations) of the excipient and the formulated drug product. This information can be either included directly within the sponsor's application or included by reference to a Drug Master File.

U.S. REGULATORY CONSIDERATIONS

Drug, Device, or Biologic Product?

One of the first considerations for the development of a new sustained-release product is an assessment of whether the product will be regulated as a drug, a device, or a biologic product. Since the requirements and submission formats differ markedly for these three areas, the proper regulatory status should be clarified very early in the development process. The FDA has issued a series of intercenter agreements that help clarify which center will have review responsibility (FDA 1991a, b, c). The primary review jurisdiction is assigned to a single center based on the primary mode of action of the combination product. Consultation between centers is allowed for combination products.

Drugs are primarily regulated by the Center for Drug Evaluation and Research (CDER). Drugs are loosely defined as products intended for human use that act by pharmacological, immunological, or metabolic means. CDER is primarily responsible for small molecule drugs, all antibiotics, and all hormone products. Biological products are regulated by the Center for Biologics Evaluation and Research (CBER). CBER is responsible for all vaccines, blood products, and most recombinant products. Devices are regulated by the Center for Devices and Radiological Health (CDRH). Devices are primarily defined as products that have physical attributes (i.e., instruments or apparatus) and do not achieve their primary intended purpose through chemical action. Obviously, there are several areas of potential overlap between these three centers. Some examples of the jurisdictional assignment of drugs, biologics, and devices are given in Table 18.3 to assist in the decision process.

The New Drug Application

A New Drug Application (NDA) is prepared from the data gained from preclinical and clinical studies. One popular analogy for an NDA submission is shown in Figure 18.2, the NDA pyramid. An NDA is a very large document that includes raw clinical data, a

---

Table 18.3. *Examples of the FDA Classification of Drugs, Biologics, and Devices*

CDER

- Products that act by pharmacological, immunological, or metabolic means
- Combinations that consist of a biological material used as a mode of localization or to affect the distribution of a drug
- Combination products that consist of a biological component and a drug where the biological component enhances the efficacy or ameliorates the toxicity of the drug
- Transdermal delivery patches

CBER

- Combinations that consist of a biological and a radioactive component
- Combinations that consist of a drug and biologic, where the drug enhances the efficacy or ameliorates the toxicity of the biological component
- In vitro reagents such as blood grouping reagents, antibodies to hepatitis B, antibodies to HIV

CDRH

- Devices incorporating drugs with the combination fulfilling a device function (e.g., wound dressings with antimicrobial agents)
- Dyes for tissues used in conjunction with laser surgery
- Tissue processing equipment and solutions used to transport organs

---

variety of data summaries and line listings, study reports, and the proposed package insert text. The FDA review process is very data driven. The review team, which consists of chemists, pharmacologists, pharmacokineticists, and medical officers, will generally start with the raw data and perform extensive validation of the conclusions presented in the sponsor's summaries and final reports. The entire NDA should be considered a large extensive document designed to support the proposed package insert, which will form the basis for the drug's ultimate marketing and usage. A BLA (Biologics License Application) should be viewed in the same fashion, although some of the elements and the format for the submission differ from that of the NDA. An idealized timeline for the FDA review process is shown in Table 18.4.

EUROPEAN REGULATORY CONSIDERATIONS

If the drug development program is successfully designed, the same studies that support the U.S. application should also suffice for European registrations. Unlike the FDA review process, which is a data-driven, bottom-up approach, the European approval process relies much more heavily on three expert reports. In the application for a marketing authorization, an expert report is

Figure 18.2. *The NDA pyramid.*

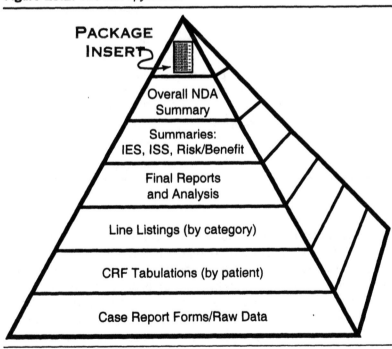

IES, integrated efficacy summary

ISS, integrated safety summary

CRF, case report form

Table 18.4. *An Idealized Timeline for NDA Reviews*

Day	Action	Comments
0	Sponsor submits NDA	
60	FDA agrees to file	If the application is incomplete, the FDA will refuse to file.
90	Sponsor can request a "90-day" meeting	This meeting is allowed per regulations. It is not frequently used.
Before day 180	Preapproval inspection of manufacturing facilities	May be waived by Field office if the facility has been recently inspected and found to be in good compliance with cGMPs.
120–180	Advisory committee review	This step is optional and at the discretion of the FDA reviewing division. Division policies vary widely on the use of the advisory committees.
180	User fee clock expires for priority applications	FDA comments are due. Clock is stopped until all responses are submitted.
270–360	FDA action deadline	The length of the FDA review cycle is determined by the scope and magnitude of the responses.

submitted describing the clinical, toxicological, and chemical aspects of the drug. This expert report is intended to be a critical, objective assessment of the product. Its authors are often not employees of the sponsor but rather academic experts in the area. The European review process entails a careful assessment of the product and the expert report, with the sponsor's study reports and data used to validate the opinions expressed in the expert reports.

The regulatory approval process in Europe is a dynamic environment, with new approaches to drug approval introduced recently as the European Community develops and the regulatory processes gel. A new pan-European regulatory agency has been established, the European Agency for the Evaluation of Medicinal Products (EMEA). Based in London, this group is responsible for administering of the review processes and arbitration hearings. Guidance documents and review decisions are issued by the Committee for Proprietary Medicinal Products (CPMP), which includes members from each EU (Europeon Union) country's health authority. The actual review and product approvals are conducted by the CPMP or each country's Ministry of Health, depending on the procedure used.

At the time of writing, there are two approaches to drug approvals: the centralized procedure and the mutual recognition procedure. Either process can be used for sustained-release products if the manufacturer uses an innovative delivery system. Since these are relatively new approaches to drug regulation, the jury is still out on which approach is optimal, and a careful case-by-case assessment is needed to make the best selection for registration. Luckily, the information required for approval is identical for either approach, so this decision can be made relatively late in the drug development process in order to take advantage of current trends. After January 1, 1998, it is no longer possible to submit national applications for a drug product, unless the sponsor is only intending to sell the product in a single EU country.

Centralized Application Process

Under the centralized procedure, the dossier is submitted to the EMEA and is reviewed by representatives from all EU countries via the CPMP (CPMP 1996b). The dossier is assigned to two countries who will serve as the primary reviewers (the rapporteurs) for the application. The rapporteur assessors prepare their review of the application. Upon completion of their review (which will likely include at least one round of questions to the sponsors), their assessment report and recommendation is presented to the CPMP, who issues their opinion. Once all issues have been satisfactorily resolved, a European Public Assessment Report (EPAR) is issued by the EMEA. This public document outlines the reasons in favor of granting the marketing authorization.

To be eligible for the centralized procedure, the drug product has to fit into one of two categories: List A, medicinal products derived from biotechnology, or List B, innovative medicinal products. Table 18.5 lists the types of products eligible for centralized submissions. List A products must be approved using the centralized process. List B products include all New Chemical Entities and innovative medicinal products with novel characteristics. List B products can be registered by either the centralized process or the mutual recognition procedure (see below). It is strongly recommended that the sponsor solicit advice from either the EMEA or a member state regulatory authority regarding the eligibility of its product for consideration as a List B product. This advice can be obtained through submission of a draft SPC (Summary of Product Characteristics) and a brief two- to three-

Table 18.5. *Medicinal Products Eligible for the Centralized Procedure*

List Type	Product description
A	Prepared using recombinant DNA technology
A	Prepared using controlled expression of gene coding for biologically active proteins
A	Prepared using hybridoma and monoclonal antibody methods
B	New Chemical Entities
B	Developed using other biotechnological processes that constitute a significant innovation
B	Products administered by means of new delivery systems
B	Products for an entirely new indication that is of significant therapeutic interest
B	Products based on radioisotopes that are of significant therapeutic interest
B	Products derived from human blood or human plasma
B	Products whose manufacture employs processes that demonstrate a significant technical advance, such as two-dimensional electrophoresis under microgravity

page product profile, including a justification of why the product should qualify for List B status.

Four to six months prior to the market application submission, the sponsor should notify the EMEA of their intent to submit a centralized submission. The notification letter should include a brief product profile, a justification for eligibility for the centralized process, a summary of any previous advice provided by the EMEA, and the sponsor's preferences for the rapporteur. The rapporteur and corapporteur (the primary reviewers of the application) and an EMEA project manager will be identified approximately three months prior to the submission. All countries within the EU are expected to have equal opportunity to serve as rapporteur. Although the guidelines state that the sponsor's preferences will be considered, the EMEA is under no obligation to follow the stated preferences. For the centralized applications considered to date, the EMEA seems to be teaming experienced rapporteurs with corapporteurs from less experienced countries on a rotating basis. A schematic and idealized timeline of the centralized review process is presented in Table 18.6.

The centralized process offers the sponsor several major advantages over the mutual recognition approval process. First, once completed, approval is achieved in all EU member states simultaneously. Second, a product approved through the centralized process is granted 10 years of market exclusivity.

The inability to select the rapporteur is one of the weaknesses seen by the pharmaceutical industry of the centralized process. However, the other advantages it offers, including the potential for pan-European approval and longer market exclusivity, generally outweigh the problems associated with the choice of the rapporteur. Another disadvantage of the centralized process is that a single trade name is required for all EU countries. This is a challenge to obtain and requires that the sponsor either identify the trade name very early in the development process so that all issues can be resolved prior to regulatory approvals, or that the sponsor risk filing the marketing application concurrently with the trade name applications and face potential delays if the trade name application runs into problems.

Table 18.6. *Timeline for Evaluation of a Centralized Application*

Day	Activity	Comment
Day -11	Submission of MAA to EMEA.	One full copy to EMEA; 2 copies of labeling in the 11 official languages of the EU; copies to rapporteur and corapporteur.
Day -1	Validation complete.	Clock stops if validation process identifies any issues.
Day -1	Facility inspection(s) requested.	Report due in 6 months.
0	Scientific evaluation begins.	
70	Assessment reports distributed from rapporteurs to CPMP.	Copy also sent to sponsor. Does not yet reflect all CPMP comments.
100	CPMP members send their comments back to rapporteurs and EMEA.	
120	List of questions sent to sponsors.	Clock stops until responses are submitted. Normally responses are due back within 6 months.
121	Restart of clock after submission of responses.	Can only happen on 11 official dates per year. Requires submission of proposed mock-up of package in official languages.
150	Response assessment report sent by rapporteurs to EMEA and CPMP.	
170	Final comments from CPMP member states due back to EMEA.	

Table 18.6 continued on next page.

Table 18.6 continued from previous page.

Day	Activity	Comment
180	CPMP discussion, decision on need for oral explanations by sponsor.	If oral explanations requested from sponsor, clock stopped for preparation time.
181	Clock restart, oral explanation presented.	
210	CPMP opinion and draft assessment report.	
Day 5 post-opinion	Translations of SPC, labeling, and package inserts due from sponsor.	
Day 15 post-opinion	Deadline for sponsor appeal of decision.	Appeals are to be reconsidered within 60 days of EMEA receipt of grounds for appeal.
240	Final CPMP assessment report sent to applicant.	
300	Finalization and publication of EPAR.	Product can now be sold.

Mutual Recognition Application Procedure

Under the mutual recognition procedure, the dossier is initially submitted to one EU member state (CPMP 1996a). After approval by the first country, an assessment report is prepared by the approving health authority, called the reference member state. This assessment report and a copy of the dossier, updated to reflect any changes requested by the reference member state, is then submitted to all EU countries where approval is sought. These countries have 14 days to accept the application and conduct an additional review of the data. Any additional questions or comments are sent to the reference member state, who serves as the advocate for the product. If needed, at day 90 of the review process, the EMEA holds a mutual recognition review meeting. The sponsoring company does not attend this meeting, although they are advised to be readily available to respond to any questions. The meeting is chaired by the reference member state, and the objective is to negotiate a Summary of Product Characteristics (the European "package insert") incorporating the comments of all countries. If agreement on approval and labeling of the product cannot be reached at this meeting, there is an arbitration procedure to resolve disputes. Few products have entered arbitration at the time of writing, so this process is relatively untested. Assuming no arbitration is required, individual product approvals should be issued by each country within 90 days of the agreement to mutually recognize the reference member state's assessment. A schematic and idealized timeline of the mutual recognition approval process is presented in Table 18.7.

One major challenge for line extension products in the mutual recognition process is the need for a harmonized SPC. If the sustained-release drug product is a line extension of a product already marketed, the chances are high that the SPC already approved for the drug will be different in the different European markets. The differences could be in the approved indication, in the warnings section, or simply in wording. Before a mutual recognition application for the line extension can even be filed, all countries need to agree on the SPCs. This is also true for second indications for an approved product.

The mutual recognition process has not been as successful to date as the centralized process. Fewer submissions have been

Table 18.7. *Evaluation of a Mutual Recognition Application*

Day	Activity	Comment
0	Submission of application to reference member state (RMS).	Reference member state is selected by applicant.
210	First authorization.	This clock may be stopped if the RMS has questions. Product can be sold in the first national market at this point.
210–300	Applicant requests mutual recognition. The RMS will prepare an assessment report. The applicant prepares an updated dossier and expert reports, including any changes requested by the RMS during their review.	Identical submissions are sent to all concerned member states by the applicant. The RMS forwards their assessment report.
300	The RMS submits the dossier for mutual recognition.	
314	The dossiers are validated and accepted by the member states. The 90-day review clock cannot start until all concerned member states have confirmed receipt of a valid dossier.	All questions and correspondence is sent to the RMS, who forwards any issues to the sponsor.
315–405	Clarification and dialogue phase. All questions are to be submitted to the RMS within 60 days of the start date.	The dossier will be discussed at a mutual recognition facilitation group meeting scheduled to coincide with the CPMP meetings.

Table 18.7 continued on next page.

Table 18.7 continued from previous page.

Day	Activity	Comment
405	National applications approved.	Each member state must issue an individual approval before product can be sold in that market.
405	Alternatively, arbitration begins.	Arbitration is required if issues of potential risk to human health cannot be resolved during the 90-day review period.
495	Arbitration to be discussed at CPMP meeting.	Clock stops are possible for input from the sponsor.
555	A CPMP opinion is due within 60 days.	If this opinion is unfavorable, the applicant has an opportunity to appeal, which starts another 60-day clock.

made, and the arbitration process has not yet been fully tested. One strategy that has been used to avoid arbitration is to withdraw the application at the last opportunity from the country(s) with objections, effectively defeating the spirit of mutual recognition. The entire mutual recognition process is under review, and changes will undoubtedly be proposed to improve the process.

Combination Products

The EU has also recently issued guidelines to help clarify whether a product is a drug (and thus reviewed as described above), a device (subject to CE marking), or both. In general, products are regulated either as devices or as drugs (*Europe Drug & Device Report*). The product is classified based on an analysis of the product's intended purpose and mechanism of action. The manufacturer is expected to provide a scientific justification of the rationale behind its classification decision. Some examples of drug and device classifications are presented in Table 18.8 which may help the manufacturer to prepare the classification decision.

It is generally simpler and less time-consuming to obtain CE marks for devices. If it is possible to classify a borderline product as a device, this strategy should be carefully considered.

CONCLUSION

This chapter has summarized some of the particular points to be addressed during the development of a sustained-release injectable product. As with any other drug development program, a carefully planned series of studies is critical to success. Close communication with the relevant regulatory agencies is essential to success. Discussions should be initiated:

- prior to submission of the IND (Investigational New Drug) application,

- prior to the initiation of Phase III registration studies (e.g., "end of phase II meeting"),

- at the "pre-NDA" meeting, primarily intended to determine data presentation format, and

---

Table 18.8. *Examples of the EU Classification of Drugs and Devices*

Drugs: products that act by pharmacological, immunological, or metabolic means

- Hemostatic agents where the primary mode is not physical, like collagens with a molecular structure capable of pharmacological interaction with receptors
- Topical disinfectants
- Antacids
- Solutions administered in vivo to local circulation to cool organs during surgery
- Agents for transport and storage of organs intended for transplantation

Devices

- Materials for adhesion of tissues
- Bone void fillers where primary action is physical (e.g., collagen matrix incorporating proteins)
- Wound dressings, which may be liquids, gels, and pastes
- Hemostatic products where the hemostatic effect results from physical characteristics or is due to the surface properties of the material (e.g., collagen, calcium alginate, or oxidized cellulose)

Devices Incorporating Medicinal Substances: classified as a device if the substance's action is secondary

- Catheters coated with heparin or antibiotic
- Soft tissue fillers incorporating local anesthetics
- Intrauterine contraceptives containing copper or silver
- Hemostatic devices containing collagen, which may pharmacologically react with platelet receptors (see also drugs and devices)

Drug Delivery Systems: products where drug and device components are separately regulated

- Drug delivery pumps
- Nebulizers

Drug Delivery Systems: regulated as drugs with device and must comply with device directives

- Prefilled syringes
- Aerosols containing drugs
- Transdermal delivery patches
- Implants containing drugs in polymer matrix to release drugs

---

- To discuss study design or the regulatory approach to all unusual aspects of the drug development program.

With careful planning, and some good luck, successful worldwide registrations should be readily achievable.

REFERENCES

CPMP. 1996a. Notice to Applicants, Volume IIA, Chapter 2, Mutual Recognition. European Commission, III/5371/96 and updates published in draft. London: Committee for Proprietary Medicinal Products.

CPMP. 1996b. Notice to Applicants, Volume IIA, Chapter 4, Centralised Process. European Commission, III/5371/96 and updates published in draft. London: Committee for Proprietary Medicinal Products.

Europe Drug & Device Report. Washington Business Information, Inc., March 2, 1998.

FDA. 1997a. Center for Drug Evaluation and Research, Approval Package for Lupron Depot®, 4 month, 30 mg, 5/30/97.

FDA. 1997b. *Guidance for Industry: Extended Release Oral Dosage Forms: Development, Evaluation, and Application of In Vitro/In Vivo Correlations.* Rockville, MD, USA: Food and Drug Administration.

FDA. 1991a. Intercenter Agreement between the Center for Biologics Evaluation and Research and the Center for Devices and Radiological Health. Effective date: 10/31/1991.

FDA. 1991b. Intercenter Agreement between the Center for Drug Evaluation and Research and the Center for Devices and Radiological Health. Effective date: 10/31/1991.

FDA. 1991c. Intercenter Agreement between the Center for Drug Evaluation and Research and the Center for Biologics Evaluation and Research. Effective date: 10/31/1991.

FDA. 1997c. Summary basis for approval for Lupron Depot®. Application 020517/S002, dated 05/30/97. Rockville, MD, USA: Food and Drug Administration, Center for Drug Evaluation and Research.

USP. 1995. *The U.S. Pharmacopeia,* 23rd ed., <1088> In Vitro and In Vivo Evaluation of Dosage Forms. Taunton, Mass., USA: Rand McNally, pp. 1924–1929.

Index